Lecture Notes of the Institute for Computer Sciences, Social-Informatics and Telecommunications Engineering 36

Alexander Sergienko   Saverio Pascazio
Paolo Villoresi (Eds.)

Quantum Communication and Quantum Networking

First International Conference, QuantumComm 2009
Naples, Italy, October 26-30, 2009
Revised Selected Papers

 Springer

Volume Editors

Alexander Sergienko
Boston University
Department of Electrical & Computer Engineering
and Department of Physics
8 Saint Mary' s Street
Boston, MA 02215-2421, USA
E-mail: alexserg@bu.edu

Saverio Pascazio
Unversity of Bari
Dipartimento di Fisica
70126 Bari, Italy
E-mail: saverio.pascazio@ba.infn.it

Paolo Villoresi
University of Padova
Department of Information Engineering
CNR-INFM LUXOR Laboratory for Ultravilolet
and X-Ray Optical Research
via Gradenigo 6, 35131 Padova, Italy
E-mail: paolo.villoresi@dei.unipd.it

Library of Congress Control Number: 2009943506

CR Subject Classification (1998): J.2, C.2, E.3, K.6.5, B.4.3, C.2.5, C.3

ISSN   1867-8211
ISBN   3-642-11730-9 Springer Berlin Heidelberg New York
ISBN   978-3-642-11730-5 Springer Berlin Heidelberg New York

springer.com

© ICST Institute for Computer Sciences, Social-Informatics and Telecommunications Engineering 2010

Typesetting: Camera-ready by author, data conversion by Scientific Publishing Services, Chennai, India
Printed on acid-free paper   SPIN: 12841332   06/3180   5 4 3 2 1 0

Preface

QUANTUMCOMM 2009—the International Conference on Quantum Communication and Quantum Networking (from satellite to nanoscale)—took place in Vico Equense near Naples, Italy, during October 26–30, 2009.

The conference made a significant step toward stimulating direct dialogue between the communities of quantum physics and quantum information researchers who work with photons, atoms, and electrons in pursuit of the common goal of investigating and utilizing the transfer of physical information between quantum systems.

This meeting brought together experts in quantum communication, quantum information processing, quantum nanoscale physics, quantum photonics, and networking. In the light of traditional approaches to quantum information processing, quantum communication mainly deals with encoding and securely distributing quantum states of light in optical fiber or in free space in order to provide the technical means for quantum cryptography applications. Exciting advances in the area of quantum communication over the last decade have made the metropolitan quantum network a reality. Several papers presented at this meeting have demonstrated that quantum cryptography is approaching the point of becoming a high-tech application rather than a research subject. The natural distance limitation of quantum cryptography has been significantly augmented using ideas of global quantum communication with stable-orbit satellites. The results presented at this conference demonstrated that practical secure satellite communication is clearly within reach.

The rapid advancement of modern nanoscale technology makes it clear that quantum photonic devices naturally communicate with each other either electromagnetically or by virtue of quantum electron transport. Basic laws of quantum physics and quantum electromagnetics govern this type of communication at the nanoscale. To a macroscopic observer such nanoscale systems could appear in the form of a structured or unstructured network. The cross-disciplinary merger of quantum physics and nanophotonics with principles of networking opens new perspectives for developing modern quantum communication applications in the form of practical devices. Several technical sections at this conference confirmed that the interaction of quantum systems could provide not only an exciting research opportunity but will also create a base for interesting and novel applications, ranging from quantum repeaters to near-field nanoscale optical detectors.

This conference was devoted to the discussion of new challenges in quantum communication and quantum networking that extends from nanoscale devices to global satellite communication networks. It placed particular emphasis on basic quantum science effects and on emerging technological solutions leading to practical applications in the communication industry, culminating with a special section on hybrid information processing.

Alexander Sergienko
Saverio Pascazio
Paolo Villoresi

Organization

Steering Committee Chair

Imrich Chlamtac   Create-Net, Trento, Italy

Steering Committee Co-chair

Andrea Simoni   Fondazione Bruno Kessler, Trento, Italy
Alexander Sergienko   Boston University, USA
Steve Bush   GE Global Research, USA
Fabrizio Illuminati   University of Salerno, Italy

General Chair

Alexander Sergienko   Boston University, USA

Publicity Chair

Saverio Pascazio   University of Bari, Italy

Conference Coordinator

Gergely Nagy   ICST

Website Chair

Shawn Doria   Boston University, USA

Technical Program Committee

Free-Space Quantum Communication on Earth and Space

Paolo Villoresi   University of Padova, Italy

Long-Distance Quantum Communication in Telecom Fiber Networks

Harald Weinfurter   University of Munich, Germany

Quantum Electron Transport in Nanoscale Semiconductor Structures

Axel Lorke   University of Duisburg-Essen, Germany

Nanophotonics Devices for Quantum Communication

Lorenzo Pavesi   University of Trento, Italy

Hybrid Quantum Systems: Connecting Photons, Atoms, Solid-State Devices

Jörg Schmiedmayer   Technical University of Vienna, Austria

Superconducting Nanoscale Devices in Quantum Communication

Andrea Fiore   Technical University Eindhoven, The Netherlands

Nanomechanical Quantum Systems and Their Interaction with Nonclassical Light

Markus Aspelmeyer   University of Vienna, Austria

Entanglement as a Resource for Quantum Communication

Saverio Pascazio   University of Bari, Italy

Practical Applications of Quantum Communication and Networking

Andreas Poppe   Austrian Institute of Technology – AIT

Workshop on Quantum and Classical Information Security

Quantum and Classical Information Security

Romain Alléaume   Institut télécom / Télécom ParisTech, France

Table of Contents

QuantumComm 2009

Session 1

Session 2

Session 3

Session 4

Session 5

Session 6

Session 7

Session 8

Special Session on Hybrid Information Processing (HIP)

Poster Session

Workshop on Quantum and Classical Information Security

Tailoring the Spatio-temporal Bandwidth of Biphotons via the Non-factorable Structure of Entanglement

L. Caspani, E. Brambilla, L.A. Lugiato, and A. Gatti

CNR-INFM and CNISM, Dipartimento di Fisica e Matematica,
Università dell'Insubria,
Via Valleggio 11, Como, Italy
alessandra.gatti@mi.infn.it

Abstract. We investigate the spatio-temporal structure of the biphoton entanglement in Parametric Down Conversion (PDC). In particular we study the biphoton amplitude at the output face of the nonlinear crystal (near-field) and we demonstrate its X-shaped geometry in the space-time dimensions, i.e. the non-factorability of the state with respect to spatial and temporal variables. Our analysis provides a precise and quantitative characterization of this structure in various regimes and types of phase matching of PDC. The key elements of novelty emerging from our analysis are the non-factorability of the state with respect to spatial and temporal variables, and the extreme relative localization of the entangled photons, both in space (few microns) and time (few femtoseconds). This extreme localization is connected to our ability to resolve the photon positions in the source near-field. The non factorability opens the possibility of tailoring the temporal entanglement by acting on the spatial degrees of freedom of twin photons.

Keywords: Quantum entanglement, Parametric Down-Conversion, Biphoton correlation.

1 Introduction

Parametric down-conversion (PDC) is probably the most efficient and widely used source of entangled photon pairs, which have been employed in several successful implementations of quantum communication and information schemes. At the very heart of such technologies lies the quantum interference between photonic wave functions, which depends crucially on the spatio-temporal mode structure of the photons. In dependence of the application under consideration, different properties of the entangled photons (e.g the spectral bandwidth) are enhanced. For example, recently proposed protocols for long-distance quantum communication based on photon-atom interaction require narrowband biphotons [1, 2], with linewidth less than the atomic linewidth ($\sim$ MHz [3, 4]), whereas ultra-broadband biphotons ($\sim$ GHz) are required for high-axial-resolution quantum optical coherence tomography [5, 6], or for clock synchronization protocols [7, 8].

A. Sergienko, S. Pascazio, and P. Villoresi (Eds.): QuantumCom 2009, LNICST 36, pp. 1–16, 2010.

Different methods were proposed in last years to generate broadband biphotons, and more in general to taylor their spectral properties. Some of them
exploits the properties of the nonlinear medium, by choosing a very thin crystal
[9], or using chirped quasi-phase-matched nonlinear gratings [10, 11, 12, 13]. In
other works the spectral properties of biphotons were controlled thanks to pump
engineering, by choosing appropriately the pump wavelength [14, 15], acting on
its spatial properties and exploiting the non-collinear phase matching configuration [16, 17, 18] or, more recently, by mean of a tilted pump pulse [19].

In this work, the issue of controlling and tailoring the biphoton spatio-temporal
structure is addressed from a peculiar and novel point of view, that is, the non factorability in space and time of the PDC biphoton entanglement. The idea comes
from the context of nonlinear optics, where recent studies [20, 21] outlined how in
nonlinear media the angular dispersion relations impose a hyperbolic geometry involving both temporal and spatial degrees of freedom in a non-factorable way. The
wave object that captures such a geometry is the so-called X-wave (the X being
formed by the asymptotes of the hyperbola), which is a localized and propagation-
invariant wave-packet, non separable in space and time. The statistical counterpart
of the X-wave was recently shown [22, 23] to emerge in the X-shaped structure of
the coherence function, describing the classical phase coherence of the individual
signal (idler) field.

In this paper, we turn our attention to the genuine quantum level, investigating the spatio-temporal structure of the biphoton cross-correlation, the quantity
that is at the heart of the photon-pair PDC entanglement. We shall demonstrate
that a X-geometry emerges in this microscopic context as well. With few excep-
tions [24, 25], the PDC entanglement has been to date investigated mostly either
in a purely temporal [26, 27, 28] or spatial [29, 30, 31, 32] framework. Our approach, based on the non-factorability in space and time of the state, will point
out a key element of novelty, i.e. the possibility of tailoring the temporal bandwith of the biphotons by manipulating their spatial degrees of freedom. In particular, by resolving their near-field positions, we will show that the X-structure
opens the access to an ultra-broad bandwidth entangled photonic source, with a
temporal localization in the femtosecond range. Our results compare with recent
findings [12], where a $\sim$ 7fs Hong-Ou-Mandel dip was observed through the use
of a quasi-phase-matched nonlinear grating.

2   Type I Parametric Down Conversion

We first focus on type I parametric down conversion, i.e. we consider the signal
and idler fields having ordinary polarization. We consider basically the same
model adopted by Gatti *et al.*[25, 32, 31]. A coherent and quasi-monochromatic
pump field propagates along the z axis in a $\chi^{(2)}$ nonlinear crystal of lentgh l_c.
We denote by $\hat{A}_p(\vec{x}, t, z)$ and $\hat{A}_s(\vec{x}, t, z)$ the envelope operator for the pump and
signal field, of central frequencies ω_p and $\omega_s = \omega_p/2$, respectively:

$$\hat{A}_j(\vec{x}, t, z) = \hat{E}_j^{(+)}(\vec{x}, t, z)\mathrm{e}^{i\omega_j t} \quad (j = s, p),$$

(1)

where $\hat{E}_j^{(+)}$ is the positive frequency part of the field operator, and $\vec{x} = (x, y)$ represents the transverse coordinates, while t is the time. We next pass to the Fourier domain:

$$\hat{A}_j(\vec{q}, \Omega, z) = \int \frac{d^2\vec{x}}{2\pi} \int \frac{dt}{\sqrt{2\pi}} \, e^{-i\vec{q}\cdot\vec{x}+i\Omega t} \hat{A}_j(\vec{x}, t, z) \,, \tag{2}$$

where $\vec{q}$ represents the transverse component of the wave vector and Ω is the frequency offset from the carriers ω_j. In addition we extract the fast variation along z due to the linear propagation inside the crystal:

$$\hat{A}_j(\vec{q}, \Omega, z) = e^{ik_{jz}(\vec{q}, \Omega) z} \hat{a}_j(\vec{q}, \Omega, z) \,, \tag{3}$$

where $k_{jz}(\vec{q}, \Omega) = \sqrt{k_j^2(\vec{q}, \omega_j + \Omega) - q^2}$ is the projection of the wave vector of the j-th field along the z axis and $k_j(\vec{q}, \omega_j + \Omega)$ is the corresponding wave number. The fields $\hat{a}_j$ defined in Eq. (3) have thus a slow variation along the crystal, due only to the nonlinear interaction. In the following we assume that the pump beam is undepleted by the nonlinear interaction, i.e. $\hat{a}_p(\vec{q}, \Omega, z) = \hat{a}_p(\vec{q}, \Omega, 0)$. Moreover, since we consider as pump an intense coherent beam, the field operator can be replaced by its classical mean value: $\hat{a}_p(\vec{q}, \Omega, 0) \to \alpha_p(\vec{q}, \Omega)$. In this way the pump evolution along the crystal is simply detrmined by the linear propagation, $\mathcal{A}_p(\vec{q}, \Omega, z) = e^{ik_{pz}(\vec{q}, \Omega) z} \alpha_p(\vec{q}, \Omega)$.

The evolution of the signal field $\hat{a}_s$ due to the non linear interaction is described by the following equation [25]:

$$\frac{\partial \hat{a}_s(\vec{q}, \Omega, z)}{\partial z} = \frac{g}{l_c} \int \frac{d^2\vec{q}'}{2\pi} \int \frac{d\Omega'}{\sqrt{2\pi}} \bar{\alpha}_p(\vec{q}+\vec{q}', \Omega + \Omega') \hat{a}_s^\dagger(\vec{q}', \Omega', z) e^{-i\Delta(\vec{q}, \Omega; \vec{q}', \Omega') z} \,, \tag{4}$$

where $\bar{\alpha}_p$ represents the pump field (at the input face of the crystal) normalized to its peak value α_p, and $g = \chi^{(2)} \alpha_p l_c$ is the dimensionless parametric gain, with $\chi^{(2)}$ being a parameter proportional to the second-order susceptibility of the medium. The phase mismatch function, defined as

$$\Delta(\vec{q}, \Omega; \vec{q}', \Omega') = k_{sz}(\vec{q}, \Omega) + k_{sz}(\vec{q}', \Omega') - k_{pz}(\vec{q}+\vec{q}', \Omega + \Omega') \,, \tag{5}$$

determines the efficiency of the down-conversion process, in which a pump photon of frequency $\omega_p + \Omega + \Omega'$ with transverse wave vector $\vec{q}+\vec{q}'$, splits into two signal photons of frequencies $\omega_s + \Omega$ and $\omega_s + \Omega'$, with wave vectors $\vec{q}$ and $\vec{q}'$, respectively.

We can find an analytic solution of Eq. (4), if we consider a plane and monochromatic pump, i.e. $\bar{\alpha}_p(\vec{q}, \Omega) = (2\pi)^{3/2}\delta(\vec{q})\delta(\Omega)$ (an analysis for a nearly monochromatic pump that leads to similar results can be found in [33, 34]). In this case, indeed, the solution can be expressed by mean of the usual input-output relations [25]:

$$\hat{A}_s(\vec{q}, \Omega) = U(\vec{q}, \Omega)\hat{A}_s(\vec{q}, \Omega, 0) + V(\vec{q}, \Omega)\hat{A}_s^\dagger(-\vec{q}, -\Omega, 0) \,, \tag{6}$$

where, if not differently stated, the fields are evaluated at $z = l_c$ (near-field), and

$$U(\vec{q}, \Omega) = e^{i\frac{k_p l_c}{2}} e^{i\varphi(\vec{q}, \Omega)} \left\{ \cosh\left[\Gamma(\vec{q}, \Omega)\right] + i\frac{\Delta_{\mathrm{pw}}(\vec{q}, \Omega) l_c}{2\Gamma(\vec{q}, \Omega)} \sinh\left[\Gamma(\vec{q}, \Omega)\right] \right\}, \quad (7a)$$

$$V(\vec{q}, \Omega) = e^{i\frac{k_p l_c}{2}} e^{i\varphi(\vec{q}, \Omega)} \frac{g}{\Gamma(\vec{q}, \Omega)} \sinh\left[\Gamma(\vec{q}, \Omega)\right], \quad (7b)$$

$$\varphi(\vec{q}, \Omega) = \frac{k_s(\vec{q}, \Omega) - k_s(-\vec{q}, -\Omega)}{2} l_c, \quad (7c)$$

$$\Gamma(\vec{q}, \Omega) = \sqrt{g^2 - \frac{\Delta_{\mathrm{pw}}^2(\vec{q}, \Omega) l_c^2}{4}}, \quad (7d)$$

$$\Delta_{\mathrm{pw}}(\vec{q}, \Omega) = k_{sz}(\vec{q}, \Omega) + k_{sz}(-\vec{q}, -\Omega) - k_p, \quad (7e)$$

where $k_p = k_{pz}(0, 0)$.

We note that the functions $U(\vec{q}, \Omega)$ and $V(\vec{q}, \Omega)$ satisfy the conditions:

$$\begin{aligned} |U(\vec{q}, \Omega)|^2 - |V(\vec{q}, \Omega)|^2 &= 1, \\ U(\vec{q}, \Omega)V(-\vec{q}, -\Omega) &= U(-\vec{q}, -\Omega)V(\vec{q}, \Omega), \end{aligned} \quad (8)$$

necessary to preserve the commutation relations from the input to the output face of the crystal.

Once we have found the solution to the propagation equation, i.e. we have an explicit expression for the field at the output face of the crystal, we can evaluate the quantity of primary interest when dealing with the PDC entangled state, the so called *biphoton amplitude*:

$$\psi(\vec{x}, t, \vec{x}', t') = \langle \hat{A}_s(\vec{x}, t)\hat{A}_s(\vec{x}', t') \rangle. \quad (9)$$

The biphoton amplitude, i.e. the biphoton cross-correlation, is the quantity at the heart of the photon-pair PDC entanglement: the "anomalous" propagator into Eq. (9) is, indeed, characteristic of processes where particles are created in pairs.

Thanks to Eq. (6) we can easily evaluate the plane wave pump (PWP) result for the biphoton amplitude:

$$\begin{aligned} \psi(\vec{x}, t, \vec{x}', t') &= \langle \hat{A}_s(\vec{x}, t)\hat{A}_s(\vec{x}', t') \rangle \\ &= \int \frac{d^2\vec{q} d\Omega}{2\pi^{3/2}} \int \frac{d^2\vec{q}' d\Omega'}{2\pi^{3/2}} \langle \hat{A}_s(\vec{q}, \Omega)\hat{A}_s(\vec{q}', \Omega') \rangle e^{i(\vec{q}\cdot\vec{x} - \Omega t)} e^{i(\vec{q}'\cdot\vec{x}' - \Omega't')} \\ &= \int \frac{d^2\vec{q} d\Omega}{(2\pi)^3} e^{i\vec{q}\cdot(\vec{x} - \vec{x}') - i\Omega(t - t')} U(\vec{q}, \Omega)V(-\vec{q}, -\Omega), \end{aligned} \quad (10)$$

where we exploit the well known commutation relation

$$[\hat{A}_s(\vec{q}, \Omega, 0), \hat{A}_s^\dagger(\vec{q}', \Omega', 0)] = \delta(\vec{q} - \vec{q}')\delta(\Omega - \Omega'), \quad (11)$$

and the identities $\langle \hat{A}_s\hat{A}_s \rangle = \langle \hat{A}_s^\dagger\hat{A}_s \rangle = \langle \hat{A}_s^\dagger\hat{A}_s^\dagger \rangle = 0$, deriving from the fact that the signal field at input face of the crystal is in the vacuum state. We notice that the biphoton amplitude is proportional to g. We note that the biphoton amplitude depends only on the relative coordinates $\vec{\xi} = \vec{x} - \vec{x}'$ and $\tau = t - t'$, as it is natural to expect for an homogeneous and stationary pump.

We need therefore to calculate the Fourier transform of the function $F(\vec{q}, \Omega) = U(\vec{q}, \Omega)V(-\vec{q}, -\Omega)$:

$$\psi(\vec{\xi}, \tau) = \int \frac{d^2\vec{q}\,d\Omega}{(2\pi)^3} e^{i\vec{q}\cdot\vec{\xi} - i\Omega\tau} F(\vec{q}, \Omega), \tag{12a}$$

$$F(\vec{q}, \Omega) = \frac{g e^{ik_p l_c}}{\Gamma(\vec{q}, \Omega)} \sinh\left[\Gamma(\vec{q}, \Omega)\right] \left\{ \cosh\left[\Gamma(\vec{q}, \Omega)\right] + i\frac{\Delta_{pw}(\vec{q}, \Omega)l_c}{2\Gamma(\vec{q}, \Omega)} \sinh\left[\Gamma(\vec{q}, \Omega)\right] \right\} \tag{12b}$$

where in Eq. (12b) we used the identity $\varphi(-\vec{q}, -\Omega) = -\varphi(\vec{q}, \Omega)$.

We note that $F(\vec{q}, \Omega)$ depends on variables $(\vec{q}, \Omega)$ only through the function $\Delta_{pw}(\vec{q}, \Omega)$, that for the case of e-oo type I parametric down-conversion considered in this section, has radial symmetry with respect to $\vec{q}$: $\Delta_{pw}(\vec{q}, \Omega) = \Delta_{pw}(q, \Omega)$, where $q = |\vec{q}|$. We can therefore evaluate Eq. (12) by mean of a Fourier-Hankel transform:

$$\psi(\xi, \tau) = \int_{-\infty}^{\infty} \frac{d\Omega}{2\pi} e^{-i\Omega\tau} \int_0^{\infty} \frac{dq}{2\pi} q F(q, \Omega) J_0(q \cdot \xi), \tag{13}$$

where

$$J_0(x) = \frac{1}{2\pi} \int_0^{2\pi} e^{i\sin(\theta)x} d\theta \tag{14}$$

is the fisrt order Bessel function of the first kind. It is evident from Eq. (13) that the biphoton amplitude depends only on the radial coordinate ξ, i.e. on the relative distance of the down-converted photons.

We numerically evaluate Eq. (13) for a type I 4mm long BBO crystal, pumped at $\lambda_p = 352$nm, cut at $33.426°$ for collinear phase matching at degeneracy, with $g = 10^{-3}$. In order to select an appropriate range of wavelengths, consistent for example with the finite bandwidth of detection, we introduce a symmetrical frequency filter centered at degeneracy with a super-gaussian profile with a full width at half maximum (FWHM) of 1370 THz, corresponding to a wavelength band from 550 nm to 950 nm.

The results are shown in Fig.1: in panel (a) we report the two dimensional cut of the biphoton amplitude at $\xi_y = 0$, clearly displaying its X-structure, non separable in space and time. The whole 3-dimensional plot has a rotational symmetry in space, as shown in Fig.1(b), where we plot the isosurface obtained for the particular value $|\psi(\vec{\xi}, \tau)|/|\psi(0, 0, 0)| = 0.15$. A clear X-shaped structure emerges: this non-trivial shape of the spatio-temporal two-photon correlation, that we shall call *X-entanglement* can be considered the counterpart, at the quantum level, of the nonlinear X-waves [20].

A remarkable characteristic of the X-entanglement is the unusually small width of the spatio-temporal correlation peak, which corresponds to a strong relative localization of twin photons both in time and space. The two lower frames of Fig.1 plot cuts of the two-photon coincidence rate $|\psi|^2$ along the temporal

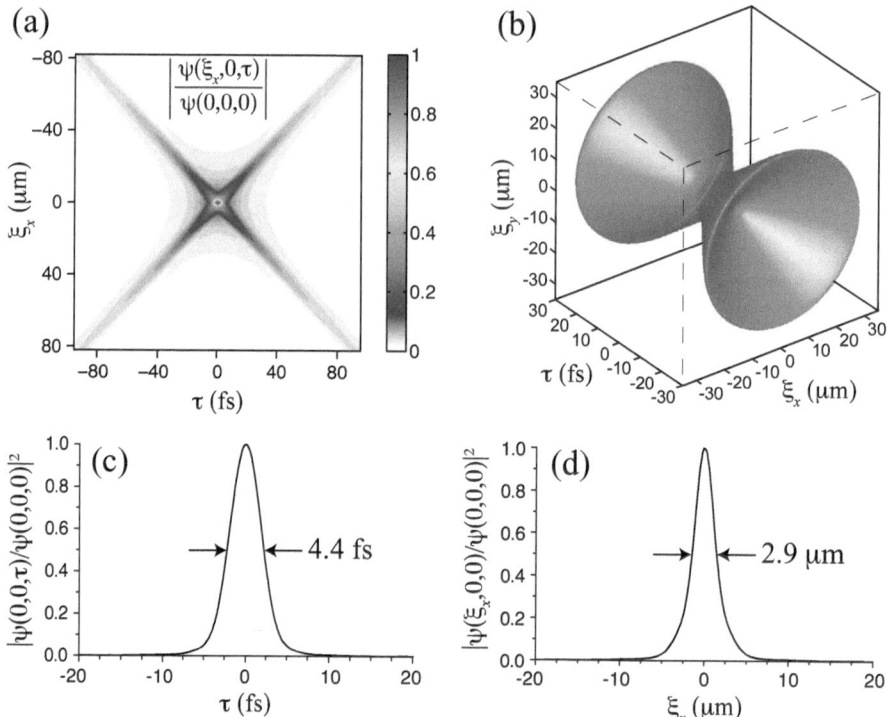

Fig. 1. (color online) (a) Two dimensional plot of the section of the biphoton amplitude for $\xi_y = 0$, normalized to its peak value $\psi(0,0,0)$, clearly showing its X-shaped geometry. (b) Isosurface representing the full 3-dimensional structure of the biphoton amplitude (in modulus), for the particular value $|\psi(\vec{\xi},\tau)|/|\psi(0,0,0)| = 0.15$. In the lower panels we report the one dimensional section of the coincidence rate $|\psi|^2$ along the temporal (c) and spatial (d) coordinate axis. The widths of the peaks shows the relative temporal and spatial localization of biphotons. We consider a 4mm-long type I BBO crystal cut at 33.436° for collinear phase matching at degeneracy, $g = 10^{-3}$, $\lambda_p = 352$nm.

and spatial axis, respectively. The relative spatial localization is remarkable but not impressive, as displayed in Fig.1(d) by the $|\psi(\xi_x,0,0)|^2$ spatial profile, obtained for $\tau = 0$, which has a FWHM of $\sim 2.9\,\mu$m. More impressive, and in a sense, unexpected, is the relative localization in time of twin photons, which can be appreciated from the temporal profile $|\psi(0,0,\tau)|^2$ in Fig.1(c), which is as narrow as 4.4 fs. Such an ultra-short two-photon localization emerges spontaneously from a monochromatic pump, as a consequence of the ultra-broad bandwidth of PDC phase-matching, which in principle extends over the optical frequency $\omega_p \sim 5 \cdot 10^{15}$ Hz.

It is interesting to compare our results with the typical ~ 100 fs temporal localization of the coincidence rate measured in the far-field zone, by collecting twin photons that propagate at symmetric directions $\vec{q}$ and $-\vec{q}$, within a

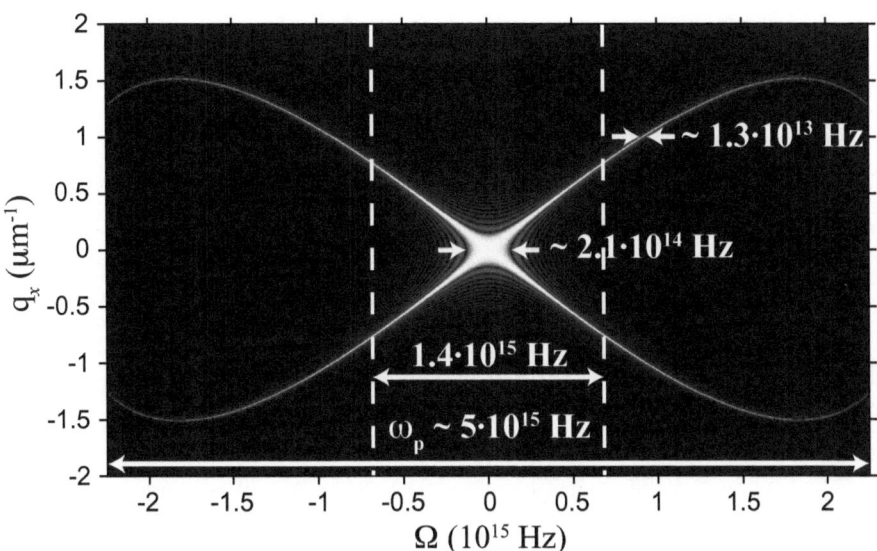

Fig. 2. (Color online) Plot of $|F|$, showing the phase matching curve in the (q_x, Ω) plane. The dashed lines show the bandwidth selected by the filter and the arrows indicate the different bandwidths involved (same parameters as in Fig.1).

small angular bandwidth. In that case, the measured quantity is proportional to $|F(\vec{q}, \tau)|^2 = |\int \frac{d\Omega}{2\pi} e^{-i\Omega\tau} F(\vec{q}, \Omega)|^2$. Its temporal width is determined by the inverse of the bandwidth of $F(\vec{q}, \Omega)$ at fixed q, i.e. the narrow $(10^{13} - 10^{14}$Hz) thickness of the curve in Fig.2. This bandwidth can be roughly evaluated as Ω_0 for $q/q_0 \ll 1$, and as $\Omega_0 q_0/q$ for $q/q_0 > 1$, where $\Omega_0 = \sqrt{1/k_s'' l_c}$, $q_0 = \sqrt{k_s/l_c}$, and we used the short-hand notation $k_s = k_s(0,0)$, $k_s'' = d^2 k_s/d\Omega^2|_{0,0}$. Clearly, since Ω_0 scales as $l_c^{-1/2}$, the shorter the crystal, the stronger the temporal localization in the far-field; however, a far-field localization in the femtosecond range would require a crystal as short as $\sim 50\,\mu$m, with a strongly reduced down-conversion efficiency. Conversely, in our case, the detection of coincidences in the near-field gives in principle access to the full $(\sim 10^{15}$Hz) bandwidth of phase matching even for a long crystal.

It is however important to stress that such an extreme temporal localization of twin photons relies on the ability to resolve their relative position in the near-field plane. Indeed, a measurement collecting all the photons over the beam cross-section, without discriminating their positions, is characterized by the integrated coincidence rate $\int d^2\vec{\xi} |\psi(\vec{\xi}, \tau)|^2$, which for the case considered here, has a width of $\sim$100fs.

In Fig.3 we compare the standard deviation of the coincidence rate when re-solving the near-field position of twin photons $|\psi(\vec{\xi} = 0, \tau)|^2$ (red solid line), with the integrated coincidence rate $\int d^2\vec{\xi} |\psi(\vec{\xi}, \tau)|^2$ (black dashed line), for different crystal lengths. For a thin crystal with $l_c = 0.1$mm (Fig.3(a)), the standard deviations of the two distribution are almost the same, while it is clearly evident the

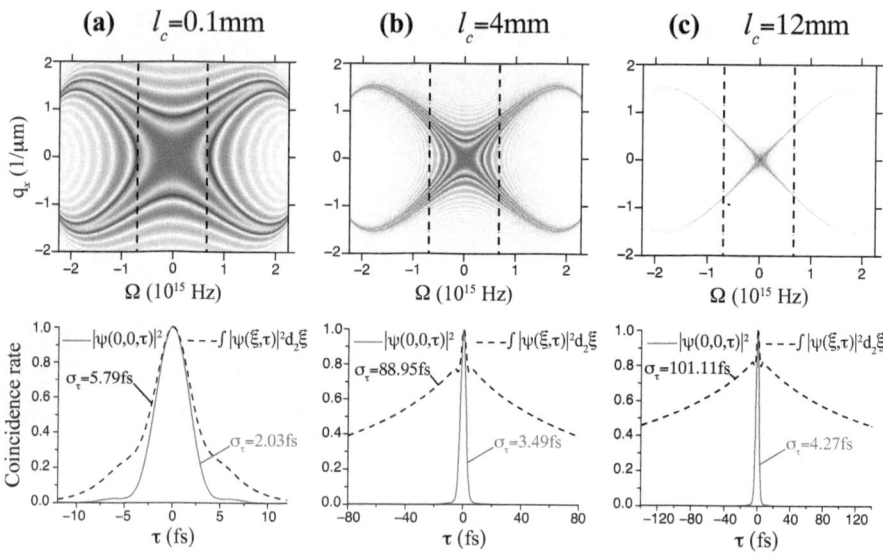

Fig. 3. (color online) Section of $|F(\vec{q}, \Omega)|$ in the (q_x, Ω) plane (upper panels), for different crystal lengths: (a) $l_c = 0.1$mm, (b) $l_c = 4$mm and (c) $l_c = 12$mm. Lower panels: comparison between the coincidence rate when twin photons are collected at the same near-field position ($|\psi(\xi = 0, \tau)|^2$, solid line) and the coincidence rate measured without resolving photon positions ($\int d^2\xi |\psi(\xi, \tau)|^2$, dashed line), for the different crystal lengths considered.

difference as the crystal length increases: even for a long crystal, that means an high efficiency for the down conversion process, the temporal localization of twin photons is on the order of few fs if we resolve their near-field position, while it is two order of magnitude greater if we consider the integrated coincidence rate (Fig.3(c)).

It may appear surprising that when the photons are collected without discriminating their near-field position, their temporal localization is determined by the inverse of the bandwidth of $F(q, \Omega)$ at fixed q rather than by the inverse of the full phase-matching bandwidth, because in this measurement all the photons at the different frequencies within the phase-matching are collected. However, the identity $\int d^2\xi |\psi(\vec{\xi}, \tau)|^2 = \int d^2\vec{q}|F(\vec{q}, \tau)|^2$ shows that in this case the coincidence rate takes the form of an *incoherent superposition* of the probabilities of detecting a pair of photons at a given q and has therefore the same ~ 100fs temporal localization as the far-field coincidence rate at fixed q. Conversely, by resolving the near-field positions of twin photons, the measured quantity is $|\psi(0, 0, \tau)|^2 = |\int \frac{d^2\vec{q}}{(2\pi)^2} F(\vec{q}, \tau)|^2$, which corresponds to a *coherent superposition* of the probability amplitudes at a given q (i.e. at a given frequency due to the the angle-frequency relation imposed by phase matching), and therefore allows a stronger temporal localization.

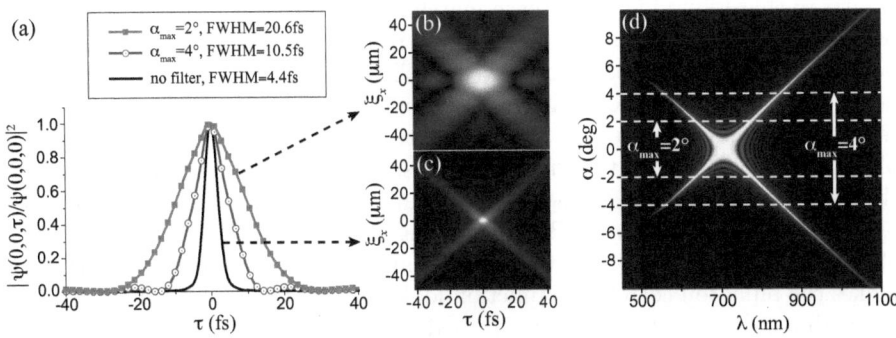

Fig. 4. (Color online) Effect of spatial filtering on the X-entanglement: (a) Temporal correlation peak $|\psi(0,0,\tau)|^2$ in the presence of a spatial filter, that cuts the angular spectrum at an angle $\alpha_{\max}$. The two insets show the full X-correlation for $\alpha_{\max} = 2^0$ (b), and in the absence of the spatial filter (c). As the angular cut-off α_{max} is reduced, the temporal peak broadens, as a consequence of the non-factorability in space-time, as illustrated by panel (d) that shows how an angular cut-off also cuts the temporal spectrum, due to the angular dispersion relation of PDC.

The non-factorability in space and time of the X-entanglement thus opens the relevant possibility of tailoring the temporal bandwidth of the biphotons by acting on their spatial degrees of freedom. As a specific example, let us consider the effects of spatial filtering on the temporal correlation. Let us assume that a $4f$ lens system is employed to image the near-field of the PDC fluorescence, and that a circular aperture of radius r_a is located in the far-field $2f$ plane, acting as a filter that cuts all the angular spectrum at $\alpha > \alpha_{\max} = \arcsin(r_a/f)$. Fig. 4 shows the effect of such a spatial filter on the temporal correlation peak. While in the absence of any spatial filter the correlation shows a strong temporal localization, as the angular bandwidth is reduced by spatial filtering, the two-photon correlation broadens in time.

This is a clear effect of the non-factorability of the correlation in space and time, as can be understood by inspection of Fig. 4(d), that shows how a spatial filter that cuts the angular spectrum at an angle α_{max} has also the effect of cutting the temporal bandwith. In turns, this implies a broadening of the temporal correlation peak, so that the relative temporal localization of twin photons deteriorates.

3   Type II Parametric Down Conversion

In this section we shall investigate how the spatio-temporal structure of the biphoton amplitude behaves in a type II phase-matching configuration, outlining the similarity and the differences with respect to the type I phase-matching case. We will show in particular that the nonfactorable character with respect to the space-time coordinates persists and the strong temporal localization is

similar to that found in type I phase-matching. However, at low gains the bipho-
tonic structure is highly asymmetric with respect to the temporal dimension,
because of the group velocity mismatch (GVM) between the signal and the idler
fields. It is known, indeed, that the GVM affecting type II PDC produces this
temporal asymmetry (see e.g. [35]), because of the temporal lag of the "slow"
photons with respect to the "fast" photons. This asymmetry is not present in
the type I configuration, as the GVM is vanishingly small in the neighborhood
of the degenerate frequency. By considering the full spatio-temporal structure
of entanglement, we shall see that the asymmetry characterizing type II PDC
makes the biphoton correlation function looks like a truncated X which is somewhat similar to a Y (the truncation regarding the temporal region corresponding
to anticipitated arrival times of the "slow" photons). It turns out, however, that
this asymmetry tends to disappear as the parametric gain is raised: a transition from an Y-shaped to an X-shaped structure is indeed observed, recovering
thereby the symmetrical structure encountered in type I phase-matching.

Following also for the type II PDC the model described by Gatti
et al.[25, 32, 31], we can find an equation similar to Eq. (4) for both the signal
and idler field:

$$\frac{\partial \hat{a}_j(\vec{q}, \Omega, z)}{\partial z} = \frac{g}{l_c} \int \frac{d^2\vec{q}'}{2\pi} \int \frac{d\Omega'}{\sqrt{2\pi}} \bar{\alpha}_p(\vec{q} + \vec{q}', \Omega + \Omega') \hat{a}_l^\dagger(\vec{q}', \Omega', z) e^{-i\Delta_{jl}(\vec{q}, \Omega; \vec{q}', \Omega')z} ,$$

$$j \neq l = s, i ,$$
$$(15)$$

where the subscripts s and i represent the signal and idler fields, respectively,
and

$$\Delta_{jl}(\vec{q}, \Omega; \vec{q}', \Omega') = k_{jz}(\vec{q}, \Omega) + k_{lz}(\vec{q}', \Omega') - k_{pz}(\vec{q} + \vec{q}', \Omega + \Omega'). \qquad (16)$$

For a monochromatic and plane wave pump, as shown in the previous section,
the propagation equation (15) has an analytic solution, that takes the form:

$$\hat{A}_s(\vec{q}, \Omega) = U_s(\vec{q}, \Omega)\hat{A}_s(\vec{q}, \Omega, 0) + V_i(\vec{q}, \Omega)\hat{A}_i^\dagger(-\vec{q}, -\Omega, 0) ,$$
$$\hat{A}_i(\vec{q}, \Omega) = U_i(\vec{q}, \Omega)\hat{A}_i(\vec{q}, \Omega, 0) + V_s(\vec{q}, \Omega)\hat{A}_s^\dagger(-\vec{q}, -\Omega, 0) ,$$
$$(17)$$

where

$$U_s(\vec{q}, \Omega) = U(\vec{q}, \Omega)e^{i\varphi(\vec{q}, \Omega)}$$
$$V_s(\vec{q}, \Omega) = V(\vec{q}, \Omega)e^{i\varphi(\vec{q}, \Omega)}$$
$$U_i(\vec{q}, \Omega) = U(-\vec{q}, -\Omega)e^{-i\varphi(-\vec{q}, -\Omega)}$$
$$V_i(\vec{q}, \Omega) = V(-\vec{q}, -\Omega)e^{-i\varphi(-\vec{q}, -\Omega)}$$
$$(18)$$

with the same definitions as in Eq. (7), with

$$\Delta_{\mathrm{pw}}(\vec{q}, \Omega) = k_{sz}(\vec{q}, \Omega) + k_{iz}(-\vec{q}, -\Omega) - k_p , \qquad (19)$$

and

$$\varphi(\vec{q}, \Omega) = \left[k_{sz}(\vec{q}, \Omega) - k_{iz}(-\vec{q}, -\Omega) \right] \frac{l_c}{2}. \tag{20}$$

It can be shown [36] that in this case the biphoton amplitude takes the form:

$$\psi(\vec{\xi}, \tau) = \int \frac{d^2\vec{q}}{(2\pi)^2} \int \frac{d\Omega}{2\pi} e^{i\vec{q}\cdot\vec{\xi} - i\Omega\tau} U_s(\vec{q}, \Omega) V_i(-\vec{q}, -\Omega). \tag{21}$$

As in the previous section we need to numerically evaluate Eq. (21), but in this case it is not straightforward the reduction to a 2-dimensional integral, since we can not exploit the radial symmetry, because of the different polarization of signal and idler fields. Nevertheless, considering the paraxial approximation and an appropriate change of variables, we have:

$$\psi(\vec{\xi}, \tau) = \frac{1}{\beta_{xx}\beta_{yy}} \int \frac{d\Omega}{2\pi} e^{-i\Omega t} e^{iq_c(\Omega)\xi} \int_0^\infty \frac{dq}{2\pi} q J_0(q \cdot \xi) F_{si}(q, \Omega), \tag{22}$$

where

$$F_{si}(\vec{q}, \Omega) \equiv U_s(\vec{q}, \Omega) V_i(-\vec{q}, -\Omega), \tag{23}$$

and an explicit expression for the coefficients β_{xx}, β_{yy} and for the function $q_C(\Omega)$ can be found in [36].

For the numerical evaluation of ψ given by Eq. (22), we first evaluated the Fourier-Hankel transform by using the quadrature method illustrated in [37] which offers an excellent precision level. The Fourier transform with respect to the temporal frequency was then evaluated using the FFT based on Bluestein algorithm which allows to select an appropriate time window within the Nyquist interval without resorting to zero-padding. Especially at low gain, the slow decaying and strongly oscillating behaviour of F_{si} requires a large number of quadrature points, so that the direct evaluation of the (2+1)D transform given in Eq. (21) would have been much more demanding, if not prohibitive, from the point of view of both computation times and memory use.

In order to evaluate numerically the biphoton amplitude $\psi(\xi, \tau)$ we focus our attention on the specific case of a 4mm long BBO crystal cut for type II phase-matching with an injected plane-wave pump field at $\lambda_p = 352$nm. We first consider a regime of low parametric gain, by taking $g = 0.001$. Assuming the pump axis z is taken at an angle $\theta_p = 49.05°$ with the crystal axis, collinear phase-matching occurs at the degenerate frequency $\lambda_s = \lambda_i = 704$nm.

Figure 5(a) maps the large scale phase-matching curves $\Delta_{pw}(q_x, q_y = 0, \Omega) = 0$ of both the signal and idler waves in the plane corresponding to the walk-off direction. For this particular phase-matching configuration the two curves are almost tangent close to degeneracy, in the region where nearly collinear emission takes place (i.e. for $\lambda \sim 2\lambda_p$ and $\vec{q} \sim 0$). In Fig.5(b) the modulus of the complex function we have to Fourier transform, namely $F_{si}(\vec{q}, \Omega)$, is mapped with a grayscale image in the in the (q_x, Ω) plane. Its maximum values (in white) clearly reproduces the signal phase-matching curve shown in the left panel. We included also in this case the multiplication by the frequency filter described into the previous section.

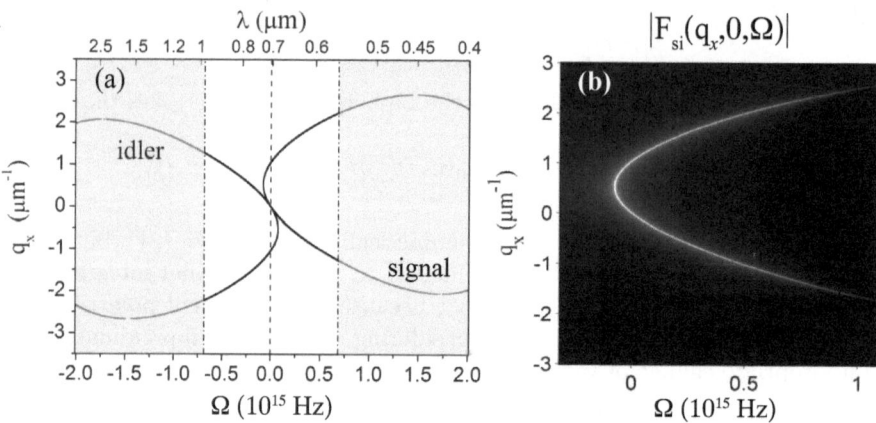

Fig. 5. (a) Large scale plot of the phase-matching curves $\Delta_{\mathrm{pw}}(q_x, q_y = 0, \Omega) = 0$ of the signal and idler waves in the walk-off plane. The unshaded region centered at degeneracy corresponds to the chosen frequency range selected by the super-Gaussian filter. (b) Grayscale plot of $|F_{si}(q_x, q_y = 0, \Omega)|$ in the same plane, modulated by the super-Gaussian frequency filter.

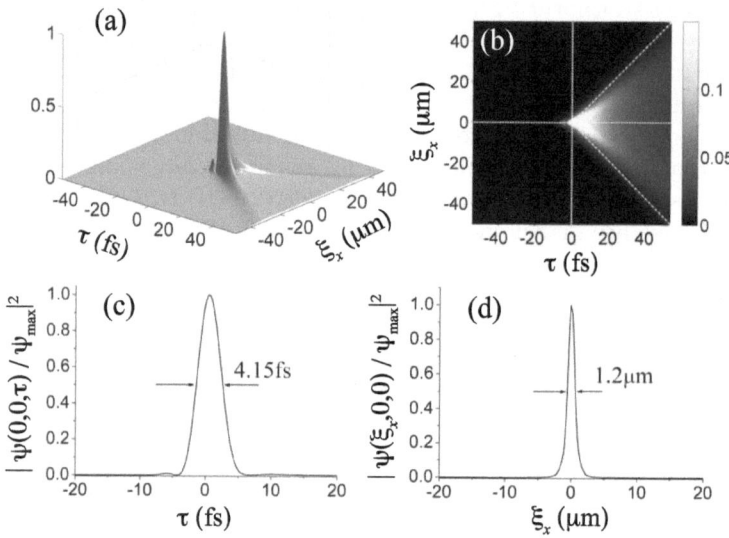

Fig. 6. (Color online) (a) and (b) Structure of the biphoton amplitude on a fs/μm scale. The plots in panels (c) and (d) show the profile of the coincidence rate $|\psi(\xi, \tau)|^2$ along the temporal and the transverse ξ_x-axis, respectively [the cuts are indicated by the two dashed lines in frame (b)]. The FWHM of the correlation peaks (indicated in the figures) gives a measure the relative localization of biphotons in space and in time.

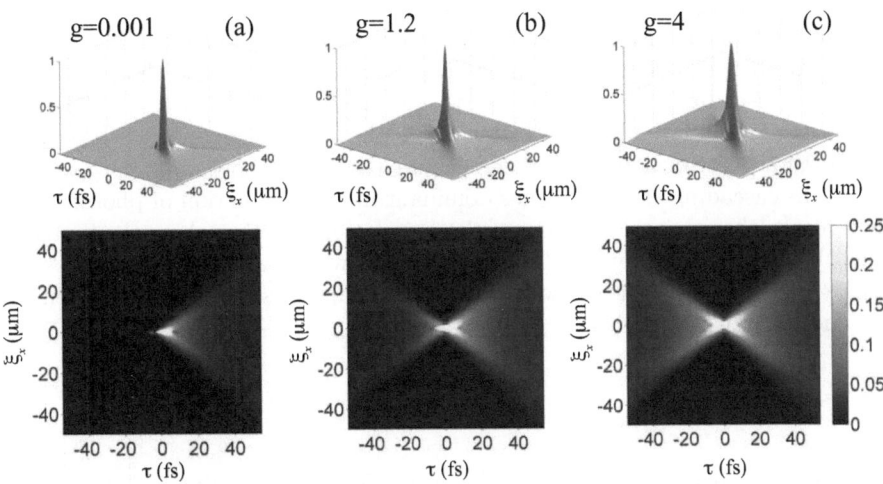

Fig. 7. (Color online) Behavior of the biphoton correlation in the (ξ_x, τ) plane for increasing value of the parametric gain. The color map of the 2D plot below is truncated to $1/4$ of the peak value of $|\psi|$ to enhance the contrast of the tails.

Figure 6 displays the modulus of biphoton amplitude in the (ξ_x, τ) plane on a femtosecond/micrometer scale in order to put in evidence the strong relative localization of the twin photons in space and time. We have a kind of Y-shaped structure (the Y being clock-wise rotated to 90° in the plot of Fig.6(b)) that clearly does not factorize with respect to the space-time coordinates and that extends mainly in the positive τ half-plane. Regarding this last feature, the structure is qualitatively different from that found in the previous section for type I phase-matching configuration, which is perfectly symmetric both with respect to the temporal and spatial relative position and appears as X-shaped in the (ξ_x, τ) plane. The tails direction in the (ξ_x, τ) plane is related to the asymptotic behaviour of the phase-mismatch function $\Delta_{\mathrm{pw}}(\vec{q}, \Omega)$ for large Ω. If we consider the paraxial and quadratic dispersion approximation for $\Delta_{\mathrm{pw}}(\vec{q}, \Omega)$ (for details see Brambilla *et al.*[36]), we obtain the relation $\tau = \pm\sqrt{\frac{2k_s k_i (k_s'' + k_i'')}{k_s + k_i}}\xi$ corresponding to the two diagonals indicated with the dashed lines in Fig.6(b).

As can be seen from temporal profile shown in Fig.6(c) the biphoton correlation is trongly localized with a FWHM as narrow as 4.15fs, which is quite close to that found for the type I configuration.

To go deep into the understanding of the biphoton amplitude spatiotemporal structure, we evaluated numerically the biphoton amplitude by increasing the parametric gain. As can be seen from Fig.7, when g is raised, diagonal tails that mirror those in the $\tau > 0$ half-plane develops in the $\tau < 0$ half-plane. Thus, the strong asymmetry along the temporal axis that characterizes the low gain regime (see Fig.7(a)) progressively disappears and a transition from an Y-shaped to an X-shaped structure is observed. We can explain qualitatively this behaviour noticing that the vacuum fluctuations that trigger spontaneous PDC are exponentially amplified in a regime of high parametric gain, as the produced photon pairs stimulate the generation of many other photon pairs in a kind of cascading process. The "slow" signal photons can thus be generated in processes stimulated by the "faster" idler photons and vice versa, so that the arrival time distributions of the signal and the idler photons tend to overlap when this cascading effect becomes dominant in the production of photon pairs. As a result, the strong assymetry of the correlation function along the temporal axis, that reveals the temporal delay of the "slow" signal photons with respect to the idler photons in the low gain regime, disappears pregressively when the parametric gain is raised.

4   Conclusions

In conclusion, this work demonstrates that the X-geometry is intrinsic to PDC at the microscopic quantum level of photon-pair entanglement. As for the macroscopic X-waves and the classical phase coherence of PDC [22, 23], the non-factorability is imposed by the phase-matching mechanism governing the wave-mixing processes. Following this analogy, we coined the name of X-entanglement. The key element of novelty that emerges in the microscopic context is the extreme relative localization of twin photons, with correlation times and correlation lengths in the femtosecond and micrometer range, respectively. The strong temporal localization is determined by the full extent of the PDC bandwidth, rather than by the bandwidth $\sim \Omega_0$ characterizing the PDC far-field. For this reason, a near-field measurement scheme able to resolve spatially the coincidences would provide a powerful tool for high-precision measurements, capable of improving substantially the resolution power in the temporal domain with respect to standard schemes.

Moreover this extreme temporal localization emerges spontaneously from a monochromatic pump, and persists even for long crystals, where on the contrary usual measurement schemes give correlation times on the order of hundreds of fs, for both type I and type II parametric down conversion. This is very striking, as the GVM characterizing the type II PDC leads to signal-idler temporal walk-off typically in the picosecond range. We verified that in the coincidence regime the GVM generates an aymmetry in the biphoton amplitude along the temporal dimension which progressively disappears as the parametric gain is increased, and a transition from a Y to an X-shaped structure is observed. On the other hand the temporal and the spatial localizations are almost preserved as the parametric gain increases.

Acknowledgments

Work supported by the FET programme of the EC, under the GA HIDEAS FP7-ICT-221906.

References

[1] Duan, L.-M., Lukin, M.D., Cirac, J.I., Zoller, P.: Long-distance quantum communication with atomic ensembles and linear optics. Nature 414, 413 (2001)

[2] Lloyd, S., Shahriar, M.S., Shapiro, J.H., Hemmer, P.R.: Long distance, unconditional teleportation of atomic states via complete bell state measurements. Phys. Rev. Lett. 87(16), 167903 (2001)

[3] Du, S., Kolchin, P., Belthangady, C., Yin, G.Y., Harris, S.E.: Subnatural linewidth biphotons with controllable temporal length. Phys. Rev. Lett. 100(18), 183603 (2008)

[4] Neergaard-Nielsen, J.S., Nielsen, B.M., Takahashi, H., Vistnes, A.I., Polzik, E.S.: High purity bright single photon source. Opt. Express 15(13), 7940–7949 (2007)

[5] Abouraddy, A.F., Nasr, M.B., Saleh, B.E.A., Sergienko, A.V., Teich, M.C.: Quantum-optical coherence tomography with dispersion cancellation. Phys. Rev. A 65(5), 053817 (2002)

[6] Nasr, M.B., Saleh, B., Sergienko, A., Teich, M.: Dispersion-cancelled and dispersion-sensitive quantum optical coherence tomography. Opt. Express 12(7), 1353–1362 (2004)

[7] Giovannetti, V., Lloyd, S., Maccone, L.: Quantum-enhanced positioning and clock synchronization. Nature 412, 417–419 (2001)

[8] Giovannetti, V., Lloyd, S., Maccone, L., Wong, F.N.C.: Clock synchronization with dispersion cancellation. Phys. Rev. Lett. 87(11), 117902 (2001)

[9] Dauler, E., Jaeger, G., Muller, A., Migdall, A., Sergienko, A.: Tests of a two-photon technique for measuring polarization mode dispersion with subfemtosecond precision. J. Res. Natl. Inst. Stand. Technol. 104(1), 1 (1999)

[10] Carrasco, S., Torres, J.P., Torner, L., Sergienko, A., Saleh, B.E.A., Teich, M.C.: Enhancing the axial resolution of quantum optical coherence tomographyby chirped quasi-phase matching. Opt. Lett. 29(20), 2429–2431 (2004)

[11] Harris, S.E.: Chirp and compress: Toward single-cycle biphotons. Phys. Rev. Lett. 98(6), 063602 (2007)

[12] Nasr, M.B., Carrasco, S., Saleh, B.E.A., Sergienko, A.V., Teich, M.C., Torres, J.P., Torner, L., Hum, D.S., Fejer, M.M.: Ultrabroadband biphotons generated via chirped quasi-phase-matched optical parametric down-conversion. Phys. Rev. Lett. 100(18), 183601 (2008)

[13] Nasr, M.B., Minaeva, O., Goltsman, G.N., Sergienko, A.V., Saleh, B.E., Teich, M.C.: Submicron axial resolution in an ultrabroadband two-photon interferometer using superconducting single-photon detectors. Opt. Express 16(19), 15104–15108 (2008)

[14] Nasr, M.B., Giuseppe, G.D., Saleh, B.E., Sergienko, A.V., Teich, M.C.: Generation of high-flux ultra-broadband light by bandwidth amplification in spontaneous parametric down conversion. Opt. Comm. 246(4-6), 521–528 (2005)

[15] O'Donnell, K.A., U'Ren, A.B.: Observation of ultrabroadband, beamlike parametric downconversion. Opt. Lett. 32(7), 817–819 (2007)

[16] Carrasco, S., Torres, J.P., Torner, L., Sergienko, A., Saleh, B.E.A., Teich, M.C.: Spatial-to-spectral mapping in spontaneous parametric down-conversion. Phys. Rev. A 70, 043817 (2004)

[17] Carrasco, S., Sergienko, A.V., Saleh, B.E.A., Teich, M.C., Torres, J.P., Torner, L.: Spectral engineering of entangled two-photon states. Phys. Rev. A 73(6), 063802 (2006)

[18] Carrasco, S., Nasr, M.B., Sergienko, A.V., Saleh, B.E., Teich, M.C., Torres, J.P., Torner, L.: Broadband light generation by noncollinear parametric downconversion. Opt. Lett. 31(2), 253–255 (2006)

[19] Hendrych, M., Shi, X., Valencia, A., Torres, J.P.: Broadening the bandwidth of entangled photons: A step towards the generation of extremely short biphotons. Phys. Rev. A 79(2), 023817 (2009)

[20] Conti, C., Di Trapani, P., Trillo, S.: X-waves in self-focusing of ultra-short pulses. In: Boyd, R.W., Lukishova, S.G., Shen, Y.R. (eds.) Self-focusing: Past and Present. Topics in Applied physics, ch. 18, vol. 114, pp. 439–456. Springer, Heidelberg (2009)

[21] Gaižauskas, E., Dubietis, A., Kudriašov, V., Sirutkaitis, V., Couairon, A., Faccio, D., Di Trapani, P.: On the role of conical waves in self-focusing and filamentation of femtosecond pulses with nonlinear losses. In: Boyd, R.W., Lukishova, S.G., Shen, Y.R. (eds.) Self-focusing: Past and Present, ch. 19, vol. 114, pp. 457–479. Springer, Heidelberg (2009)

[22] Jedrkiewicz, O., Picozzi, A., Clerici, M., Faccio, D., Trapani, P.D.: Emergence of x-shaped spatiotemporal coherence in optical waves. Phys. Rev. Lett. 97(24), 243903 (2006)

[23] Jedrkiewicz, O., Clerici, M., Picozzi, A., Faccio, D., Trapani, P.D.: X-shaped space-time coherence in optical parametric generation. Phys. Rev. A 76(3), 033823 (2007)

[24] Atatüre, M., Di Giuseppe, G., Shaw, M.D., Sergienko, A.V., Saleh, B.E.A., Teich, M.C.: Multiparameter entanglement in femtosecond parametric down-conversion. Phys. Rev. A 65(2), 023808 (2002)

[25] Gatti, A., Zambrini, R., Miguel, M.S., Lugiato, L.A.: Multiphoton multimode polarization entanglement in parametric down-conversion. Phys. Rev. A 68, 053807 (2003)

[26] Hong, C.K., Ou, Z.Y., Mandel, L.: Measurement of subpicosecond time intervals between two photons by interference. Phys. Rev. Lett. 59(18), 2044 (1987)

[27] Law, C.K., Walmsley, I.A., Eberly, J.H.: Continuous frequency entanglement: Effective finite hilbert space and entropy control. Phys. Rev. Lett. 84(23), 5304 (2000)

[28] Grice, W.P., U'Ren, A.B., Walmsley, I.A.: Eliminating frequency and space-time correlations in multiphoton states. Phys. Rev. A 64(6), 063815 (2001)

[29] Rubin, M.H.: Transverse correlation in optical spontaneous parametric down-conversion. Phys. Rev. A 54(6), 5349 (1996)

[30] Law, C.K., Eberly, J.H.: Analysis and interpretation of high transverse entanglement in optical parametric down conversion. Phys. Rev. Lett. 92(12), 127903 (2004)

[31] Gatti, A., Brambilla, E., Lugiato, L.A.: Quantum Imaging. In: Wolf, E. (ed.) Progress in Optics, vol. 51, ch. 5, pp. 251–348. Elsevier B.V., Amsterdam (2008)

[32] Brambilla, E., Gatti, A., Bache, M., Lugiato, L.: Simultaneous near-field and far field spatial quantum correlations in the high-gain regime of parametric down-conversion. Phys. Rev. A 69, 023802 (2004)

[33] Gatti, A., Brambilla, E., Caspani, L., Jedrkiewicz, O., Lugiato, L.A.: X entanglement: The nonfactorable spatiotemporal structure of biphoton correlation. Physical Review Letters 102(22), 223601 (2009)

[34] Caspani, L., Brambilla, E., Lugiato, L.A., Gatti, A.: Tailoring the spatio-temporal structure of biphoton entanglement in type I parametric down conversion (submitted, 2009)

[35] Rubin, M.H., Klyshko, D.N., Shih, Y.H., Sergienko, A.V.: Theory of two-photon entanglement in type-II optical parametric down-conversion. Phys. Rev. A 50(6), 5122 (1994)

[36] Brambilla, E., Caspani, L., Lugiato, L.A., Gatti, A.: Spatio-temporal structure of biphoton entanglement in type II PDC (in preparation, 2009)

[37] Yu, L., Huang, M., Chen, M., Chen, W., Huang, W., Zhu, Z.: Quasi-discrete hankel transform. Opt. Lett. 23(6), 409 (1998)

Entanglement Generation by a Three-Dimensional Qubit Scattering: Concurrence vs. Path (In)Distinguishability

Yasser Omar[1], Yuichiro Hida[2], Hiromichi Nakazato[2], and Kazuya Yuasa[3]

[1] CEMAPRE, ISEG, Universidade Técnica de Lisboa, P-1200-781 Lisbon, Portugal,
and SQIG, Instituto de Telecomunicações, P-1049-001 Lisbon, Portugal
[2] Department of Physics, Waseda University, Tokyo 169-8555, Japan
[3] Waseda Institute for Advanced Study, Waseda University, Tokyo 169-8050, Japan

Abstract. A scheme for generating an entangled state in a two spin-1/2 system by means of a spin-dependent potential scattering of another qubit is presented and analyzed in three dimensions. The entanglement is evaluated in terms of the concurrence both at the lowest and in full order in perturbation with an appropriate renormalization for the latter, and its characteristics are discussed in the context of (in)distinguishability of alternative paths for a quantum particle.

Keywords: Entanglement, scattering, quantum indistinguishability, renormalization.

1 Introduction

Entanglement is one of the most peculiar features of quantum theory with no classical analog and plays an essential role in quantum information and technology [1], though its acquisition or controlled generation is by no means a trivial matter. There are proposals for its generation and one often makes use of their mutual interaction to let the two quantum systems entangled [2]. On the other hand, when they are far apart and/or their mutual interaction is (in theory for all practical purposes) absent, one may resort to another quantum system ("mediator") to make the two parties entangled, through the individual and successive local interactions of the former with the latters [3,4,5,6]. This kind of scheme has been investigated for simple systems of qubits (quantum two-level systems), usually under the assumption that the strength of the interaction between the mediator and each qubit is completely under our control [3,4].

Even though the assumption is considered to be legitimate, for example, when the interaction is well controlled by switching on/off the external parameters [4], there are still cases in which it is untenable or its applicability is questionable. In particular, when the interaction time is not well defined or its definition necessarily requires a resolution in some conceptual issues, like the moments of the beginning and the end of the interaction for a particle described by a wave packet with a finite width and scattered by a static potential, we would be forced to

A. Sergienko, S. Pascazio, and P. Villoresi (Eds.): QuantumCom 2009, LNICST 36, pp. 17–25, 2010.

treat the process as a quantum mechanical scattering process of a mediator system off the target, where additional (internal, e.g., spin) degrees of freedom are duly taken into account. In the scattering problem, the interaction strength is in a sense automatically and implicitly given and we have no choice of defining or controlling the interaction duration once the initial conditions have been fixed. It is therefore an interesting and nontrivial matter of physical relevance to examine whether the schemes for entanglement generation or extraction based on the interaction between the mediator and the subsystems could remain valid and function properly even when one has little controllability on such parameters as time.

In this paper, a three-dimensional scattering process, in which a mediator spin-1/2 qubit is scattered off a fixed target composed of two other spin-1/2 qubits by spin-flipping δ-shaped potentials, is considered to examine the ability of obtaining an entangled state in the target system. The same setup has already been considered, but essentially only in one (spatial) dimension, to generate a maximally entangled state in the two-qubit system by tuning the interaction strengths [3] or, as a one-dimensional scattering process, by adjusting the incident momentum of the mediator (or the distance between two target qubits) [6].

Notice that the treatments of the scattering processes so far are not considered to be completely satisfactory because in one dimension, there would be no way to incorporate such important physical parameters as the incident and scattering angles, the size of the wave packet and the detector resolution. The purpose here is to take these elements into account in the scattering process and to show that they actually rule the resulting entanglement. We shall also see an interesting connection between the concurrence and the path (in)distinguishability of the particles.

2   Setup, Scattering, and Concurrence

We want to generate entanglement between two spin-1/2 qubits A and B, initially in the separable state $|\downarrow\downarrow\rangle_{AB}$. Assume that these qubits are fixed at positions $-d/2$ and $d/2$, and that there is no interaction between them. In order to make them entangled, a third spin-1/2 qubit X is prepared in the up state $|\uparrow\rangle_X$ and sent toward the system A+B. X is then scattered by spin-flipping δ-shaped potentials produced by A and B and is finally detected by a spin-sensitive detector with a finite opening, as described in Fig. 1. This physical process can be described by the total Hamiltonian

$$H = \frac{p^2}{2m} + g\delta^3(r + d/2)(\sigma^{(X)} \cdot \sigma^{(A)}) + g\delta^3(r - d/2)(\sigma^{(X)} \cdot \sigma^{(B)}), \quad (1)$$

where p and r are the momentum and the position of qubit X of mass m, $\sigma^{(J)}$ $(J = X, A, B)$ the Pauli matrices acting on qubit J, and g the coupling constant. We shall treat two qubits A and B symmetrically, for simplicity.

Since the Hamiltonian H preserves the total number of ups ($\uparrow$) among the three qubits X, A, and B, if we find X in the spin-flipped (i.e., down) state at the detector after scattering, we are sure that one of the qubits, A *or* B,

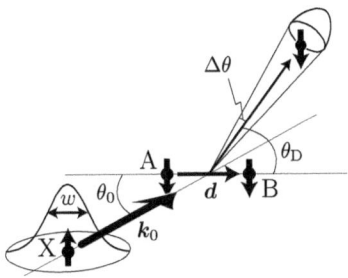

Fig. 1. Qubit X, with wave-packet width w and central momentum $\boldsymbol{k}_0$, is sent toward qubits A and B (fixed at positions $-\boldsymbol{d}/2$ and $\boldsymbol{d}/2$, respectively) with an incident angle θ_0. X is then scattered by A and B with an angle θ_D (measured from $\boldsymbol{d}$), and detected by a detector with a finite resolution characterized by an opening angle $\Delta\theta$.

must be in a spin-flipped state, that is, system A+B is either in $|\uparrow\downarrow\rangle_{AB}$ *or* in $|\downarrow\uparrow\rangle_{AB}$. If there is no way to judge which spin has been flipped in the scattering process with X, the qubits A and B are certainly in their superposed state, i.e., an entangled state. Notice, however, that the state finally to be extracted can no longer be a pure state, because it shall be given by a reduced density matrix after being traced over possible spatial or momentum degrees of freedom of X within the detector resolution and therefore it becomes mixed in general. We assume adiabatic switchings of interaction and no (abrupt) changes of the parameters in the Hamiltonian are considered to occur. As it is well known, the quantity of physical relevance in the scattering problem is the element of the scattering matrix (S) which describes the transition from the remote past $t \to -\infty$ to the remote future $t \to +\infty$ under the Hamiltonian H.

Let the incident qubit X be described by a Gaussian wave packet, with a central momentum $\boldsymbol{k}_0$ and a (spatial) width w, so that the initial state of the total system reads as

$$|\psi_0\rangle = \int d^3\boldsymbol{k}\, \psi_0(\boldsymbol{k})|\boldsymbol{k}\uparrow\downarrow\downarrow\rangle_{XAB}, \quad \psi_0(\boldsymbol{k}) = \left(\frac{2w^2}{\pi}\right)^{3/4} e^{-w^2(\boldsymbol{k}-\boldsymbol{k}_0)^2}. \tag{2}$$

The spin-sensitive detector detects X scattered in the direction $\hat{\boldsymbol{n}}_D$ seen from the origin (scattering center) with an opening angle $\Delta\theta$. When X has been found in $|\downarrow\rangle_X$ in the detector that covers the solid angle

$$\Delta\Omega = 4\pi \sin^2(\Delta\theta/2) \tag{3}$$

around direction $\hat{\boldsymbol{n}}_D$, the state of the target system A+B is given by the reduced density matrix of the form

$$\begin{aligned}
\rho = P^{-1}\int_{\Delta\Omega} d^3\boldsymbol{k}\, {}_X\langle\boldsymbol{k}\downarrow|S|\psi_0\rangle\langle\psi_0|S^\dagger|\boldsymbol{k}\downarrow\rangle_X \\
= P^{-1}\Big(a_{11}|\uparrow\downarrow\rangle_{AB}\langle\uparrow\downarrow| + a_{22}|\downarrow\uparrow\rangle_{AB}\langle\downarrow\uparrow| \\
+ a_{12}|\uparrow\downarrow\rangle_{AB}\langle\downarrow\uparrow| + a_{12}^*|\downarrow\uparrow\rangle_{AB}\langle\uparrow\downarrow|\Big),
\end{aligned} \tag{4}$$

where S denotes the S matrix. The normalization constant

$$P = a_{11} + a_{22} \tag{5}$$

is nothing but the probability that this particular event occurs, i.e., the yield. Given the state ρ for two qubits, the degree of entanglement can be measured in terms of its concurrence $C(\rho)$ [7], which reads, for the above ρ in (4), as

$$C(\rho) = \frac{2|a_{12}|}{a_{11} + a_{22}}. \tag{6}$$

The S-matrix element is given by

$$\langle \mathbf{k}'\zeta'|S|\mathbf{k}\zeta\rangle = \delta^3(\mathbf{k}' - \mathbf{k})\delta_{\zeta'\zeta} - 2\pi i\delta(E_{k'} - E_k)\langle \mathbf{k}'\zeta'|V|\Psi_{\mathbf{k}}\zeta\rangle, \tag{7}$$

where $|\mathbf{k}\zeta\rangle$ is the eigenstate of the free Hamiltonian $H_0 = p^2/2m$ and $|\Psi_{\mathbf{k}}\zeta\rangle$ that of the total Hamiltonian $H = H_0 + V$, both belonging to the same eigenvalue $E_k = \hbar^2 k^2/2m$, with ζ denoting the spin degrees of freedom. The coordinate representation of the latter reads as

$$\langle \mathbf{r}|\Psi_{\mathbf{k}}\zeta\rangle = \langle \mathbf{r}|\mathbf{k}\zeta\rangle - \int d^3 r' \, G_k(\mathbf{r} - \mathbf{r}')\frac{2m}{\hbar^2}V(\mathbf{r}')\langle \mathbf{r}'|\Psi_{\mathbf{k}}\zeta\rangle, \tag{8}$$

where $G_k(\mathbf{r})$ is the retarded Green function.

Up to the first order in g (Born approximation), the relevant S-matrix elements are given by (subscripts x,A,B for spin states shall be suppressed)

$$\langle \mathbf{k}\downarrow\uparrow\downarrow|S|\psi_0\rangle = -\langle \mathbf{k}\downarrow\downarrow\uparrow|S|\psi_0\rangle^* \equiv A(\mathbf{k})$$

$$= -\frac{ig}{2\pi^2} \int d^3 k' \, \psi_0(\mathbf{k}')\delta(E_k - E_{k'})e^{i(\mathbf{k}-\mathbf{k}')\cdot \mathbf{d}/2}. \tag{9}$$

The matrix elements a_{ij} are simply expressed as

$$a_{11} = a_{22} = \int_{\Delta\Omega} d^3\mathbf{k} \, |A(\mathbf{k})|^2, \qquad a_{12} = -\int_{\Delta\Omega} d^3\mathbf{k} \, A^2(\mathbf{k}). \tag{10}$$

When the incident wave packet $\psi_0(\mathbf{k})$ is well monochromatized, $wk_0 \gg 1$, $A(\mathbf{k})$ is approximately evaluated analytically and we end up with the following expressions for the concurrence and the yield at the lowest order [9]

$$C(\rho) \simeq \frac{1}{\Delta\Omega}\left|\int_{\Delta\Omega} d^2\hat{\mathbf{k}} \, e^{ik_0(\hat{\mathbf{k}}\cdot\mathbf{d}) - [(\hat{\mathbf{k}}-\hat{\mathbf{k}}_0)\cdot\mathbf{d}]^2/8w^2}\right|, \tag{11}$$

$$P \simeq \frac{m^2 g^2}{\pi^3 \hbar^4 w^2} \, \Delta\Omega \, e^{-[d^2 - (\hat{\mathbf{k}}_0\cdot\mathbf{d})^2]/8w^2}. \tag{12}$$

3   Entanglement and Path (In)Distinguishability

The angle integrations over $\hat{\mathbf{k}}$ in the concurrence $C(\rho)$ in (11) are numerically performed and we find the following characteristics of it [Figs. 2(a)–(c)]. When the incident qubit X is accompanied with a large wave packet $w \gg d$ [Figs. 2(a)–(b)], (i) the concurrence $C(\rho)$ depends on in which direction the scattered qubit X is detected and (ii) takes the maximal value ~ 1 when X is captured in the direction $\pm d$, i.e., on the line connecting target qubits A and B, while (iii) no apparent dependence is seen on the incident angle θ_0. (iv) It is reduced considerably as $\Delta\theta$ is increased, while it keeps the maximal value in the $\pm d$ directions [Fig. 2(a)]. Furthermore, (v) it strongly depends on k_0 and reduces considerably for a large k_0 or a short wavelength compared with the distance d between the two qubits A and B in the target [Fig. 2(b)]. If the incident wave packet is small compared with the distance between two qubits A and B, that is, $w \lesssim d$, the concurrence $C(\rho)$ becomes deteriorated in general and the incident-angle dependence, which is almost absent in the opposite cases with $w \gg d$, appears [Fig. 2(c)]. We observe that (vi) there are directions where the concurrence still takes the same value as that in the case of $w \gg d$, i.e., when X is detected in the same direction as the incident direction $\hat{k}_0$ and in its symmetric direction with respect to d. As for the probability P in (12), it does not depend on scattering (detection) angle θ_D and Young-like interference is not observed.

The characteristics of the concurrence $C(\rho)$ can be understood on the basis of its mathematical expression in (11) or its approximate expression [9]

$$C(\rho) \sim 1 - \frac{1}{8}(k_0 d)^2 (\Delta\theta)^2 \sin^2\theta_D \quad \text{for} \quad w \gg d \text{ and } \Delta\theta \ll 1. \tag{13}$$

We understand that this expression indeed well describes the characteristics (i)–(v) mentioned above. The condition for obtaining a higher concurrence is also attainable by evaluating the range of the variation of the phase of the integrand in (11) over the opening of the detector mouth, which should be smaller than 2π,

$$2 k_0 d \sin\Delta\theta \sin\theta_D \lesssim 2\pi \quad \text{for} \quad w \gg d. \tag{14}$$

On the other hand, if the incident wave packet is small $w \lesssim d$, we have to keep the second term $-[(\hat{k} - \hat{k}_0) \cdot d]^2/8w^2$ in the exponent in (11). Since this term would entail an exponential reduction of $C(\rho)$, the condition for keeping a higher concurrence becomes

$$|(\hat{k} - \hat{k}_0) \cdot d| \ll w \quad \text{for} \quad w \lesssim d, \tag{15}$$

which explains well the characteristics (vi).

It would be interesting to interpret the above conditions (14) and (15) for higher entanglement, in the context of the (in)distinguishability of the paths taken by particle X. As a general rule in quantum theory, the (in)distinguishability of alternative paths of a particle results in the (non)vanishing of quantum interference. Since the concurrence $C(\rho)$ is proportional to the absolute value of the off-diagonal matrix element a_{12}, its value is rather dependent on the information about which qubit A or B has changed its spin state in the scattering process. Notice also that in the lowest-order perturbation, since qubit X with its spin flipped is scattered only and surely once, either by qubit A or B, and the interaction certainly changes their spin states, the information about which qubit has scattered X is the same as that about which spin has flipped. That is why the probability P does not exhibit Young-like interference: the path taken by X can *in principle* be disclosed by checking the spin state of A and B after the scattering. This distinguishability erases the interference. The same, applies to the concurrence. If one could distinguish the two alternative paths of X, originating from A or B, one is able to know which spin has been flipped. This knowledge results in a reduction of the off-diagonal matrix elements and therefore of the concurrence $C(\rho)$.

We understand that the conditions for higher concurrence (entanglement) (14) and (15) would be interpreted as those for the indistinguishability of the two alternative paths from A or B. Indeed, if the incident wave packet is long $w \gg d$ and therefore is approximately considered as a plane wave of wavelength $\lambda_0 = 2\pi/k_0$, the condition for higher concurrence (14) may be understood in comparison with the resolving power of an optical device. It is known in classical optics that the optical device that has an aperture $\Delta\theta$ seen from an object composed of two optical sources with mutual distance d is unable to distinguish the two sources if the condition

$$\Delta\theta \lesssim \frac{\lambda_0}{d \sin \theta_{\mathrm{D}}} \tag{16}$$

is satisfied (for $\Delta\theta \ll 1$). This is essentially the same as the condition (14). For the opposite case with $w \lesssim d$, since the quantity $|(\hat{\boldsymbol{k}} - \hat{\boldsymbol{k}}_0) \cdot \boldsymbol{d}|$ is nothing but the difference in length between the two paths via A or B, if the condition (15) is not satisfied, one could determine the path the particle X has passed through on its way to the detector, for the path length difference is certainly larger than the size of the particle w. In both cases, the conditions for higher entanglement are interpreted as those for indistinguishability of particle paths.

4 Full-Order Contributions and Renormalization

Notice that this interpretation and the connection with the indistinguishability are limited to the lowest-order results, since in higher order in perturbation, multiple scatterings that invalidate the one to one correspondence between the knowledge of the particle paths and that of the spin flips come into play. One also realizes that higher-order terms require a proper treatment, for otherwise the result would become trivial, that is, no scattering could occur by the δ-shaped potentials in (spatial) dimensions greater than one. This is a famous issue in quantum theory, and Jackiw has proposed a prescription how to deal with such systems. According to Jackiw [8], we have to renormalize the strength of the δ-shaped potential so that the source term becomes nonvanishing. The coupling constant in the Hamiltonian, g in our case, has to be regarded as a bare one and absorb possible divergences arising from the Green function at the origin $G_k(\mathbf{0})$.

Furthermore, in the present case, another element that was absent at that time, i.e., the spin degrees of freedom, turns out to require another care in dealing with higher-order terms. In this respect, it is important to notice that the interaction of the form $g\boldsymbol{\sigma}^{(X)} \cdot \boldsymbol{\sigma}^{(A)}$ inevitably causes another type of interaction in its higher-order terms. For example, in its second order, a term proportional to unit operator in spin space, that is not proportional to the original form, appears,

$$(g\boldsymbol{\sigma}^{(X)} \cdot \boldsymbol{\sigma}^{(A)})^2 = 3g^2 - 2g^2\boldsymbol{\sigma}^{(X)} \cdot \boldsymbol{\sigma}^{(A)}. \tag{17}$$

The situation is more clearly understood from the fact that the interaction can be written as

$$\boldsymbol{\sigma}^{(X)} \cdot \boldsymbol{\sigma}^{(A)} = \mathcal{P}_{3XA} - 3\mathcal{P}_{1XA} \tag{18}$$

in terms of projection operators on the spin-triplet and singlet spaces, $\mathcal{P}_{3XA}$ and $\mathcal{P}_{1XA}$. This means that in any higher-order terms in perturbation, there are only two types of interactions proportional to $\mathcal{P}_{3XA}$ or $\mathcal{P}_{1XA}$ present and we need to renormalize these two terms simultaneously. That is, we need two counter terms to obtain sensible results. We therefore introduce another bare coupling constant g', start with a bare interaction Hamiltonian

$$V(\boldsymbol{r}) = \delta^3(\boldsymbol{r} + \boldsymbol{d}/2)(g\boldsymbol{\sigma}^{(X)} \cdot \boldsymbol{\sigma}^{(A)} + g') + \delta^3(\boldsymbol{r} - \boldsymbol{d}/2)(g\boldsymbol{\sigma}^{(X)} \cdot \boldsymbol{\sigma}^{(B)} + g'), \tag{19}$$

and renormalize the bare coupling constants

$$\frac{1}{g_r + g_r'} = \frac{1}{g + g'} + \frac{2m}{\hbar^2} \lim_{\Lambda \to \infty} \frac{1}{2\pi^2}\Lambda, \tag{20a}$$

$$\frac{1}{-3g_r + g_r'} = \frac{1}{-3g + g'} + \frac{2m}{\hbar^2} \lim_{\Lambda \to \infty} \frac{1}{2\pi^2}\Lambda, \tag{20b}$$

where the linear divergence is due to the divergent part of $G_k(\boldsymbol{0})$ in three dimensions. After the renormalization, everything is finite and nonvanishing and we obtain the relevant S-matrix elements in full order, which can be evaluated numerically and even analytically for the case of large wave packet $w \gg d$, and the concurrence [9].

Figure 2(d) shows the concurrence as a function of the scattering (detection) angle θ_D, evaluated both at the lowest order (Born approximation) and in full order in perturbation. We observe here that the characteristics of the concurrence at the lowest order mentioned above are largely maintained in full order even though the effect of multiple scatterings is manifest as an appearance of wrinkles in the central part. This implies that when qubit X is detected in some particular directions, one can expect a higher concurrence or entanglement produced in the target qubits A and B owing to the multiple scatterings, though the optimal one is still obtainable in the direction parallel to $\boldsymbol{d}$ (i.e., $\theta_D = 0, \pi$).

5 Summary

We have studied a scheme for the entanglement generation between two fixed spins by scattering a mediator spin in three dimension. The close connection

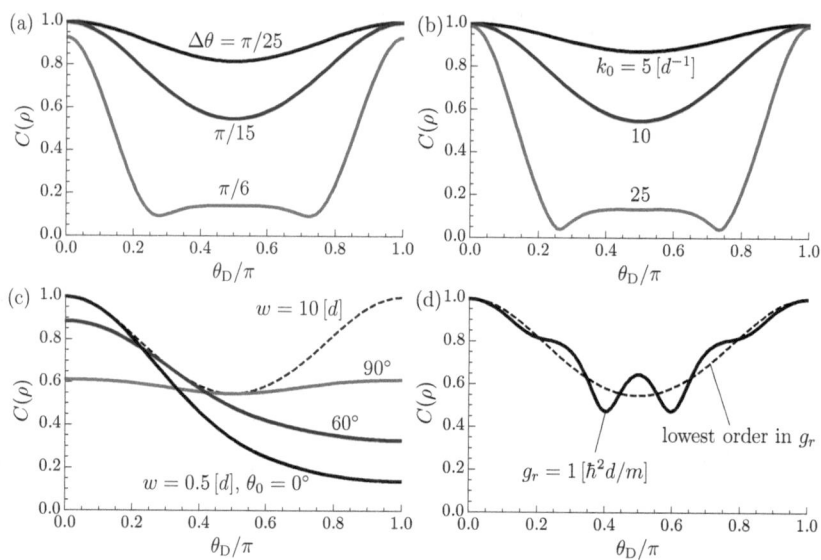

Fig. 2. (Color online) Concurrence $C(\rho)$ as a function of the scattering (or detecting) angle θ_D, (a)–(c) evaluated at the lowest and (d) in full order (solid line) in perturbation. (a),(b),(d) are for $w \gg d$ and (c) is for $w = 0.5d$. The other parameters are (if not specified in the figures): $\theta_0 = \pi/2$, $\Delta\theta = \pi/15$, $k_0 d = 10$, $g_r' = 0$.

between the concurrence, a measure of entanglement, and the path (in) distinguishability, is disclosed, which seems to be valid in a global sense even in higher orders in perturbation. It is remarkable that we would not need to control the interaction time to obtain the higher entanglement and this possibility may strengthen the way of entanglement generation by means of quantum scatterings.

Acknowledgments

This work is partly supported by a Grant-in-Aid for Scientific Research (C) from the Japan Society for the Promotion of Science, by the Grant-in-Aid for Young Scientists (B) (No. 21740294) and a Special Coordination Fund for Promoting Science and Technology both from the Ministry of Education, Culture, Sports, Science and Technology, Japan, by the the bilateral Italian-Japanese Projects II04C1AF4E on "Quantum Information, Computation and Communication" of the Italian Ministry of Education, University and Research, and by the Joint Italian-Japanese Laboratory on "Quantum Information and Computation" of the Italian Ministry for Foreign Affairs. Y. Omar thanks the support from project IT QuantTel and the MMQIRT initiative, and from Fundação para a Ciência e a Tecnologia (Portugal), namely through project QuantPrivTel (PTDC/EEA-TEL/103402/2008) and programs POCTI/POCI/PTDC, partially funded by FEDER (EU).

References

1. Nielsen, M.A., Chuang, I.L.: Quantum Computation and Quantum Information. Cambridge University Press, Cambridge (2002); In: Bouwmeester, D., Zeilinger, A., Ekert, A.(eds.) The Physics of Quantum Information: Quantum Cryptography. Quantum Teleportation, Quantum Computation. Springer, Berlin (2000); Galindo, A., Martín-Delgado, M.A.: Rev. Mod. Phys. 74, 347 (2002); Bennett, C.H., DiVincenzo, D.P.: Nature (London) 404, 247 (2000)
2. Loss, D., DiVincenzo, D.P.: Phys. Rev. A. 57, 120 (1998); Kane, B.E.: Nature (London) 393, 133 (1998)
3. Bergou, J.A., Hillery, M.: Phys. Rev. A. 55, 4585 (1997); Messina, A.: Eur. Phys. J. D. 18, 379 (2002); Browne, D.E., Plenio, M.B.: Phys. Rev. A 67, 012325 (2003); Compagno, G., Messina, A., Nakazato, H., Napoli, A., Unoki, M., Yuasa, K.: ibid 70, 052316 (2004); Yuasa, K., Nakazato, H.: Prog. Theor. Phys. 114, 523 (2005); Migliore, R., Yuasa, K., Nakazato, H., Messina, A.: Phys. Rev. B. 74, 104503 (2006)
4. Hagley, E., Maître, X., Nogues, G., Wunderlich, C., Brune, M., Raimond, J.M., Haroche, S.: Phys. Rev. Lett. Rev 79, 1 (1997); Raimond, J.M., Brune, M., Haroche, S.: Rev. Mod. Phys. 73, 565 (2001)
5. Chanelière, T., Matsukevich, D.N., Jenkins, S.D., Lan, S.-Y., Kennedy, T.A.B., Kuzmich, A.: Nature (London). 438, 833 (2005); Matsukevich, D.N., Chanelière, T., Jenkins, S.D., Lan, S.-Y., Kennedy, T. A. B., Kuzmich, A.: Phys. Rev. Lett. 96, 030405 (2006)
6. Costa Jr., A.T., Bose, S., Omar, Y.: Phys. Rev. Lett. 96, 230501 (2006); Giorgi, G. L., de Pasquale, F.: Phys. Rev. B 74, 153308 (2006); Ciccarello, F., Palma, G.M., Zarcone, M., Omar, Y., Vieira, V.R.: New J. Phys. 8, 214 (2006); Yuasa, K., Nakazato, H.: J. Phys. A 40, 297 (2007); Ciccarello, F., Palma, G.M., Zarcone, M., Omar, Y., Vieira, V.R.: ibid 40, 7993 (2007); Laser Phys. 17, 889 (2007); Ciccarello, F., Paternostro, M., Kim, M.S., Palma, G.M.: Phys. Rev. Lett. 100, 150501 (2008); Habgood, M., Jefferson, J.H., Briggs, G.A.D.: Phys. Rev. B 77, 195308 (2008); J. Phys.: Condens. Matter 21, 075503 (2009)
7. Wootters, W.K.: Phys. Rev. Lett. 80, 2245 (1998)
8. Jackiw, R.: M. A. B. Beg Memorial Volume. In: Ali, A., Hoodbhoy, P. (eds.) Diverse Topics in Theoretical and Mathematical Physics, p. 35. World Scientific, Singapore (1991/1995)
9. Hida, Y., Nakazato, H., Yuasa, K., Omar, Y.: Phys. Rev. A. 80, 012310 (2009)

Multipartite Entangled Codewords for Gaussian Channels with Additive Noise and Memory

C. Lupo*, L. Memarzadeh, and S. Mancini

University of Camerino, via Madonna delle Carceri 9, I-62032 Camerino (MC), Italy
cosmo.lupo@unicam.it

Abstract. We study a bosonic Gaussian channel with classical additive noise and memory effects. In particular we consider correlations in the added noise which are asymmetric in the phase space quadratures and show in such a case the usefulness of entangled codewords for reliable communication. We explicitly demonstrate that optimal rates can be achieved with multiparty entangled codewords.

Keywords: Quantum Channels, Error Correction, Multipartite Entanglement.

1 Introduction

The quantum theory of information furnishes a well established framework for characterizing communication via noisy quantum channels [1,2]. To correct errors, a certain amount of redundancy is added by encoding messages using codewords of a certain length n. The transmission rate is the ratio between the length of the message and the length of the codeword. A rate is said to be reliable if, in the limit of long codewords, the errors can be corrected reaching an arbitrary high fidelity. The maximum reliable transmission rate is, by definition, the channel capacity.

There exist several concepts of channel capacities in the framework of quantum information theory, depending on whether classical (bit) or quantum (qubit) information has to be sent (and if pre-shared entanglement is available). Here we consider the case in which quantum information carriers are used to encode and distribute classical information. The approach to the problem of determining the capacity of a quantum channels is in a certain extent similar to its classical counterpart. However, qualitative differences arise once one takes in account the phenomenon of *quantum entanglement*. Indeed, the possibility of encoding information by means of *entangled codewords* is peculiar of the quantum setting and has no classical analogue. Following a conservative approach, one can encode messages into codewords which are not entangled. The maximum reliable rate achievable with unentangled codewords is called the Holevo capacity. The presence of entanglement has led to a milestone problem in quantum information, i.e. the problem of determining whether the Holevo capacity is additive or not. Additivity of the Holevo capacity would imply that entanglement cannot be useful for achieving higher transmission rates. It has been recently proven [3] that the Holevo capacity is *not* additive, hence leading to the actual possibility of achieving higher transmission rates by means of entangled codewords.

---

* Corresponding author.

A. Sergienko, S. Pascazio, and P. Villoresi (Eds.): QuantumCom 2009, LNICST 36, pp. 26–33, 2010.

Earlier studies on this subject have focussed on *memoryless* quantum channel, characterized by the property that the noises affecting different channel uses are identical and independent. However, in many physical settings the assumption of memoryless noise may appear rather artificial and one is naturally led to consider quantum channel with memory [4]. In a memory channel, the noises acting at different channel uses are in general nor identical neither independent. For the case of quantum channels with memory, it has been explicitly shown that the use of entangled codewords can indeed increase the transmission rate in the context of continuous Gaussian channels, like lossy bosonic ones [5]. Here we are going to consider another kind of bosonic Gaussian channel, those with classical additive noise [6], and investigate the usefulness of entangled codewords in the presence of memory effects. For such channels, the attention has been initially focussed on short codewords [7,8], and recently the channel capacity has been computed in some cases [9]. Here, after introducing a suitable model of noise correlation, we explicitly show how multipartite entangled codewords can enhance the rate of reliable communication over arbitrary number of channel uses.

2   The Model

We study a model of bosonic Gaussian quantum channel with memory. The physical system in which information is encoded is a quantum harmonic oscillator (e.g. a spatial, spectral or polarization mode of the electromagnetic field). The channel model is the so-called additive noise channel [6]. Upon n uses of the quantum channel, a set of canonical variables $\mathbf{R} := (q_1, \ldots q_n, p_1, \ldots p_n)$ experience the Heisenberg picture transformation

$$\mathbf{R} \mapsto \mathbf{R} + \Xi, \tag{1}$$

where $\Xi := (X_1, \ldots X_2, Y_1, \ldots Y_n)$ is a vector of $2n$ real-valued stochastic variables. The noise vectors are taken to be Gaussian distributed. We consider a multivariate noise distribution of the form

$$P(\Xi) \simeq \exp\left[-\frac{1}{2}\left(\Xi\, \mathbb{V}^{-1}\, \Xi^{\mathsf{T}}\right)\right]. \tag{2}$$

In [9] the capacity of such a channel has been computed for a certain choice of the covariance matrix, describing a Markovian noise, showing that the optimal rate is achieved without the use of entangled codewords. Here we slightly modify the model by introducing a certain asymmetry in the noise acting on the quadratures q_k, p_k. We expect that the lack of symmetry corresponds to entangled optimal codewords (see comments in [7] and [9]). The noise covariance matrix is chosen as follows

$$\mathbb{V} = \begin{pmatrix} \mathbb{V}^X & \mathbb{O} \\ \mathbb{O} & \mathbb{V}^Y \end{pmatrix}, \tag{3}$$

where the $n \times n$ covariance matrices $\mathbb{V}^X$, $\mathbb{V}^Y$ have components

$$\mathbb{V}^X_{hk} = \sigma\mu^{|h-k|}, \tag{4}$$

$$\mathbb{V}^Y_{hk} = \sigma\delta_{hk}, \tag{5}$$

with $\mu \in [0,1]$, $\sigma \geq 0$, and we assume $\mathbb{V}^X_{hk} = \sigma\delta_{hk}$ for $\mu = 0$.

Moving to the Schroedinger picture, we describe quantum states of n bosonic modes by means of the Wigner function. The memory channel under consideration transforms a n mode Wigner function $W^{(n)}(\mathbf{R})$ according to:

$$W^{(n)}(\mathbf{R}) \mapsto \int \prod_{k=1}^{n} dX_k dY_k \, P(\boldsymbol{\Xi}) \, W^{(n)}(\mathbf{R} - \boldsymbol{\Xi}). \tag{6}$$

Let us recall that a state of n bosonic modes is said to be Gaussian if its Wigner function is Gaussian in the canonical variables $\mathbf{R}$. Analogously, a quantum channel is said to be Gaussian if it transforms Gaussian states to Gaussian states. From Eq. (6), using the fact that the noise variables are distributed according to the Gaussian in Eq. (2), it follows that the channel under consideration is indeed a Gaussian channel.

Clearly, in this model the memory effects in the quantum channel come from the correlations among the noise variables X_k. The limiting case of memoryless channel is obtained for $\mu = 0$, when the multivariate distribution in Eq. (2) factorizes as the product of independent and identical Gaussian distributions, each with variance σ. We can indeed consider the parameter σ as quantifying the amount of noise present at each channel use and $\mu \in [0, 1]$ as a memory parameter. In order to avoid unphysical results, it is customary to impose suitable constraints on the maximum energy carried by the encoding codewords. In the bosonic setting, it is natural to impose a constraint in the number of field excitations in average per mode, i.e.

$$\left\langle \frac{1}{n} \sum_{k=1}^{n} \frac{q_k^2 + p_k^2}{2} \right\rangle \leq N + \frac{1}{2}, \tag{7}$$

where the average is over codewords at the channel input. For any given σ and N, we can introduce the signal-to-noise ratio $SNR = N/\sigma$.

3   Block Encoding/Decoding Schemes

The classical capacity of a quantum channel is the maximum rate of reliable communication, where the maximum is over all possible ways of encoding classical information into quantum states and all possible measurements that can be performed at the channel output to reconstructed the encoded signal. Here we consider the maximum rate that can be achieved with a given choice of the measurement to be performed at the channel output, i.e. heterodyne detection. We optimize the transmission rates over a class of block encoding and decoding schemes, and compare their performances for transmitting classical information as function of the memory parameter and of the length of the blocks.

For any n, we consider codewords of $2n$ symbols, described by the vector $\mathbf{C} := (A_1, \ldots A_n, B_1, \ldots B_n) \in \mathbb{R}^{2n}$. Codewords are taken to be distributed according to a multivariate Gaussian distribution with zero mean and covariance matric $\mathbb{V}_c$. Codewords of length $2n$ are encoded in Gaussian states of n bosonic modes, corresponding to a block of successive channel uses. For encoding the codeword $\mathbf{C}$ we use a Gaussian state described by the Wigner function

$$W_{\mathbf{C}}^{(n)}(\mathbf{R}) \simeq \exp\left[-\frac{(\mathbf{R} - \mathbf{C})\mathbb{V}_{in}^{-1}(\mathbf{R} - \mathbf{C})^{\mathsf{T}}}{2} \right]. \tag{8}$$

These states can be obtained by applying displacement operators on a fiducial Gaussian state with zero mean and covariance matrix $\mathbb{V}_{in}$. This kind of Gaussian encoding can be proven to be optimal for several Gaussian channels, due to the recently proven *minimum output entropy conjecture* [10].

Regarding the decoding part, we consider ideal heterodyne measurement to be used to decode the classical information at the output of the quantum channel. For an encoding block of length n, heterodyne measurement is performed at the corresponding n output modes. For a given encoded codeword $\mathbf{C}$, the state of the n output modes is described by the Wigner function

$$W_{\mathbf{C},out}^{(n)}(\mathbf{R}) \simeq \exp\left[-\frac{(\mathbf{R} - \mathbf{C})(\mathbb{V}_{in} + \mathbb{V})^{-1}(\mathbf{R} - \mathbf{C})^{\mathsf{T}}}{2}\right] \tag{9}$$

(this equation has been derived inserting (2), (8) into (6)), and the ensemble state, averaged over all possible input codewords, is described by the Wigner function

$$W_{out}^{(n)}(\mathbf{R}) \simeq \exp\left[-\frac{\mathbf{R}(\mathbb{V}_{in} + \mathbb{V} + \mathbb{V}_c)^{-1}\mathbf{R}^{\mathsf{T}}}{2}\right]. \tag{10}$$

For each of the n output modes, heterodyne detection provides a joint measurement of both the quadratures. From Eq. (10) it follows that the probability of obtaining the vector $\mathbf{c} := (a_1, \ldots a_n, b_1, \ldots b_n)$ as output of an ideal measurement is given by the Gaussian distribution

$$P(\mathbf{c}) \simeq \exp\left[-\frac{\mathbf{c}(\mathbb{V}_{in} + \mathbb{V} + \mathbb{V}_c + \mathbb{I}/2)^{-1}\mathbf{c}^{\mathsf{T}}}{2}\right], \tag{11}$$

where the term $\mathbb{I}/2$, proportional to the unit matrix, accounts for the uncertainty principle (i.e. it is the noise introduced by the heterodyne measurement). Analogously from Eq. (9) the conditional probability of measuring $\mathbf{c}$, given that the codeword $\mathbf{C}$ was sent, is the Gaussian

$$P(\mathbf{c}|\mathbf{C}) \simeq \exp\left[-\frac{(\mathbf{c} - \mathbf{C})(\mathbb{V}_{in} + \mathbb{V} + \mathbb{I}/2)^{-1}(\mathbf{c} - \mathbf{C})^{\mathsf{T}}}{2}\right]. \tag{12}$$

The Shannon entropy of the decoded codeword $\mathbf{c}$, measured in bits, is hence given by the expression

$$H(\mathbf{c}) = \frac{1}{2}\log_2\left[\det(\mathbb{V}_{in} + \mathbb{V} + \mathbb{V}_c + \mathbb{I}/2)\right]. \tag{13}$$

Analogously, the conditional entropy is

$$H(\mathbf{c}|\mathbf{C}) = \frac{1}{2}\log_2\left[\det(\mathbb{V}_{in} + \mathbb{V} + \mathbb{I}/2)\right], \tag{14}$$

independently on the value of the encoded codeword $\mathbf{C}$.

In conclusion we can write the rate of transmission, measured in bits per channel use, for a block encoding of length n and heterodyne decoding. The transmission rate is given by the mutual information per channel use

$$F_n = \frac{I(\mathbf{c}; \mathbf{C})}{n} = \frac{H(\mathbf{c}) - H(\mathbf{c}|\mathbf{C})}{n} = \frac{1}{2n} \log_2 \left[\frac{\det (\mathbb{V}_{in} + \mathbb{V} + \mathbb{V}_c + \mathbb{I}/2)}{\det (\mathbb{V}_{in} + \mathbb{V} + \mathbb{I}/2)} \right]. \quad (15)$$

4 Optimal Transmission Rates

The aim of this section is to maximize, for any given n, the transmission rate in Eq. (15) by optimizing over all possible covariance matrices $\mathbb{V}_c$ and $\mathbb{V}_{in}$. We will find that for $n > 1$ the optimal encoding strategy involves states which are entangled among the bosonic modes belonging to the same encoding block.

First of all, let us consider the case $n = 1$. In this case the encoding blocks are made of one bosonic mode, hence only separable states are used to encode the codewords. The noise covariance matrix is 2×2, and reads:

$$\mathbb{V} = \begin{pmatrix} \sigma & 0 \\ 0 & \sigma \end{pmatrix}. \quad (16)$$

The quantum state used to encode information has covariance matrix

$$\mathbb{V}_{in} = \begin{pmatrix} i_q & i_d \\ i_d & i_p \end{pmatrix}, \quad (17)$$

where, to satisfy the uncertainty principle, the condition $i_q i_p - i_d^2 \geq 1/4$ has to imposed. Analogously, the covariance matrix of the codewords is

$$\mathbb{V}_c = \begin{pmatrix} c_q & c_d \\ c_d & c_p \end{pmatrix}. \quad (18)$$

The maximization on the encoding schemes is under the constraint

$$\frac{i_q + i_p + c_q + c_p}{2} \leq N + 1/2. \quad (19)$$

Using the Lagrange method one obtains that the maximum rate is reached in correspondence of the optimal values $i_d^{opt} = c_d^{opt} = 0$, and $i_q^{opt} = i_p^{opt} = 1/2$, $c_q^{opt} = c_p^{opt} = N$. The corresponding optimal rate is

$$R_1 = \max_{i_q, i_p, i_d, c_q, c_p, c_d} F_1 = \log_2 \left(\frac{N + \sigma + 1}{\sigma + 1} \right). \quad (20)$$

Let us move to the case of encoding/decoding blocks of length $n > 1$. To optimize the transmission rate we proceed along the same line of [9]. First notice that, for any n, there exists a $n \times n$ orthogonal matrix $\mathbb{T}$ which diagonalizes the covariance matrix $\mathbb{V}^X$, i.e.

$$\sum_{h,k=1}^{n} \mathbb{T}_{jh} \mathbb{V}_{hk}^{X} \mathbb{T}_{lk} = \sigma_j \, \delta_{jl}. \quad (21)$$

We define the *collective* noise variables $\tilde{\Xi} = (\tilde{X}_1, \ldots \tilde{X}_n, \tilde{Y}_1, \ldots \tilde{Y}_n)$, where $\tilde{X}_j :=$ $\sum_k \mathbb{T}_{jk} X_k$, $\tilde{Y}_j := \sum_k \mathbb{T}_{jk} Y_k$. This new set of noise variables are hence distributed according to

$$\tilde{P}(\tilde{\Xi}) \simeq \exp\left[-\frac{1}{2}\left(\tilde{\Xi}\,\tilde{V}^{-1}\,\tilde{\Xi}^{\mathsf{T}}\right)\right], \tag{22}$$

where $\tilde{V} = \mathrm{diag}(\sigma_1, \ldots \sigma_n, \sigma, \ldots \sigma)$. We analogously define the collective field variables $\tilde{R} = (\tilde{q}_1, \ldots \tilde{q}_n, \tilde{p}_1, \ldots \tilde{p}_n)$, where $\tilde{q}_j := \sum_k \mathbb{T}_{jk} q_k$, $\tilde{p}_j := \sum_k \mathbb{T}_{jk} p_k$. Notice that the latter is a canonical (i.e. symplectic) transformation, moreover it preserves the form of the energy constraint which reads

$$\left\langle \frac{1}{n}\sum_{j=1}^{n} \frac{\tilde{q}_j^2 + \tilde{p}_j^2}{2} \right\rangle \leq N + \frac{1}{2}. \tag{23}$$

Finally, we define the codewords $\tilde{C} = (\tilde{A}_1, \ldots \tilde{A}_n, \tilde{B}_1, \ldots \tilde{B}_n)$ with $\tilde{A}_j := \sum_k \mathbb{T}_{jk} A_k$, $\tilde{B}_j := \sum_k \mathbb{T}_{jk} B_k$. In terms of the collective field variables the encoding Gaussian states have Wigner function of the form

$$\tilde{W}_{\tilde{C}}^{(n)}(\tilde{R}) \simeq \exp\left[-\frac{(\tilde{R} - \tilde{C})\tilde{V}_{in}^{-1}(\tilde{R} - \tilde{C})^{\mathsf{T}}}{2}\right], \tag{24}$$

and the codewords are distributed according to

$$\tilde{P}(\tilde{C}) \simeq \exp\left[-\frac{1}{2}\left(\tilde{C}\,\tilde{V}_c^{-1}\,\tilde{C}^{\mathsf{T}}\right)\right]. \tag{25}$$

We conjecture that, in analogy with the $n = 1$ case, the optimal encoding is reached for jointly diagonal covariance matrices. We hence consider the following parametrization: $\tilde{V}_{in} = \mathrm{diag}(i_{q_1}, \ldots i_{q_n}, i_{p_1}, \ldots i_{p_n})$, $\tilde{V}_c = \mathrm{diag}(c_{q_1}, \ldots c_{q_n}, c_{p_1}, \ldots c_{p_n})$. Using Eq. (15) we have

$$F_n = \sum_{j=1}^{n} \frac{1}{2n} \log_2\left[\frac{(i_{q_j} + c_{q_j} + \sigma_j)(i_{p_j} + c_{p_j} + \sigma)}{(i_{q_j} + \sigma_j)(i_{p_j} + \sigma)}\right]. \tag{26}$$

The latter expression has to be maximized under the energy constraint

$$\sum_{j=1}^{n} \frac{i_{q_j} + c_{q_j} + i_{p_j} + c_{p_j}}{2n} \leq N + \frac{1}{2}. \tag{27}$$

The optimal transmission rate over an encoding block of length n is denoted $R_n = \max_{i_{q_j}, c_{q_j}, i_{p_j}, c_{p_j}} F_n$. For generic values of the parameters N, σ, μ it can be computed numerically. Figure 1 shows the ratio R_n/R_1 as function of n for several values of the memory parameter μ. The ratio expresses the gain in the transmission rate reached by encoding classical information on blocks of length n. As it is shown in the next section the optimal encoding scheme on blocks of length n involves Gaussian states which are entangled over n bosonic modes.

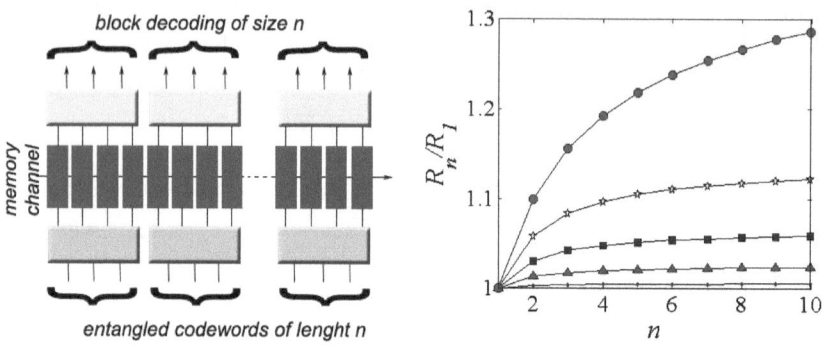

Fig. 1. On the left: block-encoding and decoding for the memory channel. On the right: the gain in the transmission rates using entangled codewords of length n, for several values of the memory parameter: dots: $\mu = 0.2$, triangles: $\mu = 0.4$, squares: $\mu = 0.6$, stars: $\mu = 0.8$, circles: $\mu = 1$. The value of the noise parameter is $\sigma = 1$, the signal-to-noise ratio is $SNR = 3$.

5   Discussion

In terms of the collective field variables $\tilde{\mathbf{R}}$ the covariance matrix of the optimal Gaussian state is diagonal of the form $\tilde{\mathbb{V}}_{in}^{opt} = \mathrm{diag}(i_{q_1}^{opt}, \ldots i_{q_n}^{opt}, i_{p_1}^{opt}, \ldots i_{p_n}^{opt})$. The optimal values $i_{q_j}^{opt}$, $i_{p_j}^{opt}$ can be in general determined numerically. It turns out that the optimal state is pure, satisfying $i_{q_j}^{opt} i_{p_j}^{opt} = 1/4$. Let us now move to the description in terms of the fields variables $\mathbf{R} = (q_1, \ldots q_n, p_1, \ldots p_n)$, where q_k, p_k are the canonical variables describing the kth channel use. This is the natural representation to study the entanglement between different channel uses. The optimal covariance matrix is

$$\mathbb{V}_{in}^{opt} = \begin{pmatrix} \mathbb{T}^{\mathsf{T}} & \mathbb{O} \\ \mathbb{O} & \mathbb{T}^{\mathsf{T}} \end{pmatrix} \tilde{\mathbb{V}}_{in}^{opt} \begin{pmatrix} \mathbb{T} & \mathbb{O} \\ \mathbb{O} & \mathbb{T} \end{pmatrix}. \tag{28}$$

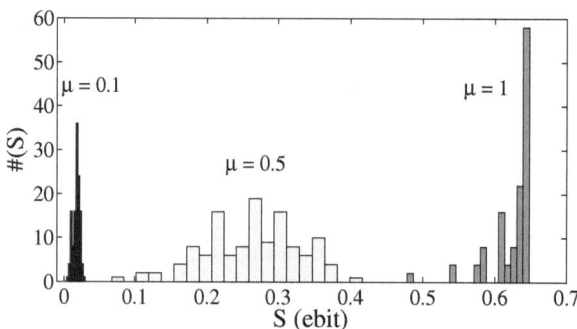

Fig. 2. The plot shows the distribution of the entanglement entropy (measured in *ebit*) over balanced bipartitions for the optimal encoding state of length $n = 10$. The histograms are for several values of the memory parameter μ and $\sigma = 1$, $SNR = 3$.

The optimal state turns out to be entangled among different channel uses. To study the entanglement in the found optimal state it is sufficient to look at the covariance matrix V_{in}^{opt}. For a block encoding of length n, we characterize the multipartite entanglement among different channel uses by means of the distribution of bipartite entanglement among all possible bipartition of the n modes into two subsets of length d, $n - d$ [11]. We estimate the amount of bipartite entanglement using the von Neumann entropy of the reduced state of one of the two subsets (i.e. the entanglement entropy). We hence consider the distribution of the entanglement among all possible subsets of a given length d. As an illustrative example Fig. 2 shows the distribution of bipartite entanglement for $d = [n/2]$ (i.e. balanced bipartitions, $[\cdot]$ indicates the integer part) and for different values of the memory parameter.

6 Conclusion

We have studied the optimal block encoding strategy for a bosonic Gaussian memory channel with additive classical noise, where the decoding measurement is heterodyne detection. A block of length n allows the use of entangled states of n bosonic modes as codewords. We found that the optimal encoding, making use of Gaussian states, is by means of multipartite entangled states. Entanglement in the optimal multipartite encoding states has been characterized by the distribution of the entanglement entropy over all possible balanced bipartition. The optimal codeword states turns out to be separable only for $\mu = 0$ (or $n = 1$). For an encoding block of length n the mean (and other quantifiers, e.g. the mode) of the distribution of the entanglement entropy increases with the memory parameter. For a given value of the memory parameter, the rate of reliable communication increases monotonically with the length of the encoding block.

References

1. Nielsen, M.A., Chuang, I.L.: Quantum Computation and Quantum Information. Cambridge University Press, Cambridge (2000)
2. Holevo, A.S.: IEEE Trans. Inf. Th. 44, 269 (1998); Schumacher, B., Westmoreland, M.D.: Phys. Rev. A 56, 131 (1997)
3. Hastings, M.B.: Nature Physics 5, 255 (2009)
4. Kretschmann, D., Werner, R.F.: Phys. Rev. A. 72, 062323 (2005)
5. Ruggeri, G., Soliani, G., Giovannetti, V., Mancini, S.: Europhys. Lett. 70, 719 (2005); Pilyavets, O., Zborovskii, V., Mancini, S.: Phys. Rev. A 77, 052324 (2008); Lupo, C., Pilyavets, O., Mancini, S.: New J. of Phys. 11, 063023 (2009)
6. Holevo, H.S., Sohma, M., Hirota, O.: Phys. Rev. A. 59, 1820 (1999)
7. Cerf, N.J., Clavareau, J., Macchiavello, C., Roland, J.: Phys. Rev. A 72, 042330 (2005)
8. Ruggeri, G., Mancini, S.: Quant. Inf. & Comp. 7, 265 (2007)
9. Lupo, C., Memarzadeh, L., Mancini, S.: Phys. Rev. A 80, 042328 (2009)
10. Lloyd, S., et al.: arxiv:0906.2758
11. Facchi, P., Florio, G., Pascazio, S.: Int. J. Quantum Inf. 5, 97 (2007)

High-Speed Single-Photon Detection Using 2-GHz Sinusoidally Gated InGaAs/InP Avalanche Photodiode

Naoto Namekata, Shunsuke Adachi, and Shuichiro Inoue

Institute of Quantum Science, Nihon University, 1-8-14 Kanda-Surugadai, Chiyoda-ku, Tokyo 101-8308, Japan
nnao@phys.cst.nihon-u.ac.jp

Abstract. We report a telecom-band single-photon detector for high-speed quantum key distribution systems. The single-photon detector is based on a sinusoidally gated InGaAs/InP avalanche photodiode. The gate repetition frequency of the single-photon detector reached 2 GHz. A quantum efficiency of 10.5 % at 1550 nm was obtained with a dark count probability per gate of 6.1×10^{-7} and an afterpulsing probability of 3.4 %.

Keywords: Single-photon detector, sinusoidal gate operation, telecomm-band.

1 Introduction

Single-photon detector (SPD) is the most important component to realize quantum key distribution (QKD) [1]. A gated InGaAs/InP avalanche photodiode (APD) is a practical solution for the SPD [2,3,4]. However, conventional one cannot operate at high gate repetition frequency, since afterpulses occur with a high probability. Therefore, there are many efforts to suppress the afterpulse by means of the avalanche signal detection with a low avalanche gain [5,6]. We have supposed the sinusoidally gate operation of InGaAs/InP APD (SG-APD) [5] that achieved much higher gate repetition frequency than that of conventional gate operations. the detector has been already applied to the QKD experiment and achieved key generation rates in orders of megahertz [7]. In SG-APD, it is realized to suppress a transient pulse noise due to the capacitive response of the APD to an applied gate voltage pulse. As a result, the avalanche multiplication gain can be reduced to $10^5 \sim 10^6$ (two orders of magnitude lower than that in the conventional gate operation), which contributes to suppression of afterpulsing. Ultimately, the avalanche multiplication gain enough to discriminate an avalanche signal can be reduced to the value defined only by the thermal noise. In the gating regime, discrimination of an extremely weak avalanche signal at the thermal noise limit of detector circuit has not been realized yet, since it is required to output no extra noise despite the fact that the huge gate voltage pulse is applied to the APD, described above. In this report, we demonstrated

A. Sergienko, S. Pascazio, and P. Villoresi (Eds.): QuantumCom 2009, LNICST 36, pp. 34–38, 2010.

that using the sinusoidal gating scheme at the thermal noise limit of a detection circuit, the SG-APD can be operated at the 2 GHz gate repetition frequency with a low afterpulsing probability.

2   Sinusoidally Gated InGaAs/InP Avalanche Photodiode

A diagram of our SPD is shown in Fig. 1. The tested InGaAs/InP APD is AGD-25-SE-1-T8 (Princeton Lightwave). The APD is cooled to -50 °C by a Peltier cooler driven by PID (proportional-integral-derivative) controller. In order to supply an AC voltage superposed on the DC reverse bias voltage V_{DC} to the APD, we used a gated passive quenching circuit (GPQC) [3]. The sinusoidal voltage at a frequency of ω_g was produced by the signal generator (SG) and used as the gate voltage after amplification by the high-power amplifier (HP-AMP). Here the amplified sinusoidal voltage passed through the band-pass-filter (BPF) (center frequency: $f = \omega_g$) to reject amplified sideband noise, harmonics, and non-harmonics, which contributes to reduce the noise level of the GPQC output. The GPQC output signal passed through three band elimination filters (BEFs) whose center (elimination) frequencies f were set to ω_g. Then the signal was amplified by an inverting broadband amplifier (B-AMP) whose gain and bandwidth were 40 dB and 3 GHz, respectively. The total elimination ratio of the BEFs was 100 dB at ω_g. The BEFs distorted the avalanche pulse, since the BEFs eliminated the ω_g component and gave a large phase shift around ω_g. This distortion causes a wider time jitter in the avalanche signal discrimination. Therefore, to fix the pulse form, the avalanche signal was passed through a low-pass-filter (LPF) with a cutoff frequency of $\sim 1.5\omega_g$ before entering the discriminator. The transferred

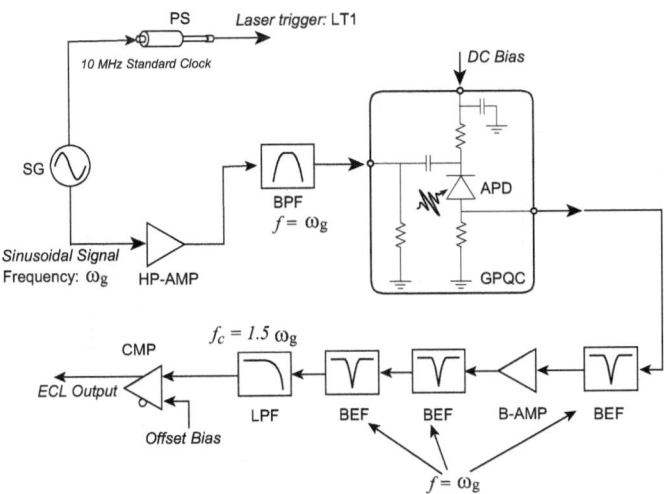

Fig. 1. Diagram of Sinusoidally gated avalanche photodiode

gate signal at ω_g was sufficiently rejected by the BEFs and the LPF, and the avalanche signal was distilled well. The rms voltage noise V_n of the LPF output was approximately 5 mV when the 23 dBm sinusoidal voltage at $\omega_g = 1$ GHz was used as a gate. Taking the gain and noise figure of the B-AMP into account, the V_n is close to the thermal noise limit $V_{th} = \sqrt{4kTRB}$ of a detector circuit, where k, T, R, and B are the Boltzmann constant, the absolute temperature of the output resister in the circuit, the resister's value, and bandwidth, respectively. A threshold voltage for the avalanche signal discrimination was set to -25 $\sim$ -30 mV, which indicates that avalanche signals containing only 10^4 electrons (before amplification) can be discriminated. The avalanche signal was discriminated by the ultra-high-speed comparator (CMP) circuit that accepts subnanosecond pulses. We finally obtained the logic output in the emitter-coupled-logic (ECL) level.

3   Detector Performance

We evaluated the SPD performance for 50 ps weak laser pulses at 1550 nm. The SPD performances were evaluated changing the V_{DC}. Figure 2 shows the dark count probability per gate P_d and the afterpulsing probability P_a as functions of quantum efficiency η at a gate repetition frequency of 2 GHz. P_d is in orders of 10^{-7} when η is less than 13 %, more specifically the SG-APD was achieved a detection efficiency of 10.5 % at 1550 nm with a dark count probability of 6.1×10^{-7} and an afterpulsing probability of 3.4 %. The dark count probability and afterpulsing probability is considerably lower than the other reported APD-based SPDs, while our APD-based SPD achieved the highest repetition frequency.

Figure 3(a) shows the time histogram of detection events when a gate repetition frequency ω_g was 2 GHz. The peak at 135.9 ns corresponds to the illuminated

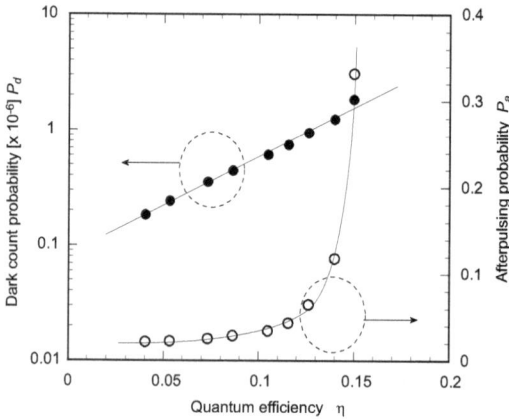

Fig. 2. Detector performance

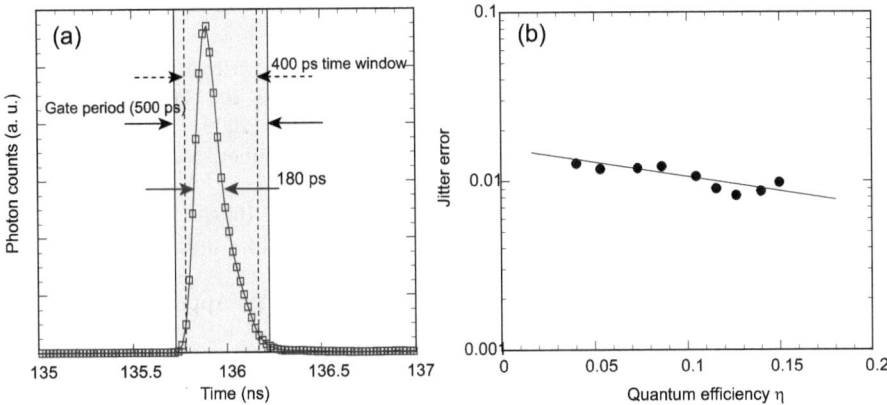

Fig. 3. Jitter characteristics. (a)Time histogram of detection events. (b) Probability for the jitter error as a function of the quantum efficiency η.

gate. The time jitter distribution has a FWHM of 180 ps, which is shorter than the gate period (500 ps). However, the peak is not completely separated from adjacent peak. This fact means that a fraction of the detection events is registered in an incorrect gate, which obviously causes bit errors. To reduce the error counts due to the time jitter, we employed a 400 ps time window to obtain the detector clicks. The probability P_e of error caused by the time jitter is plotted as a function of the quantum efficiency in Fig. 3(b). Increasing η, corresponding to increasing the reverse bias voltage applied to APD, P_e slightly reduces, since the response speed of APD made higher as the reverse bias voltage is increased. Although P_e can be reduced by employing a shorter time window, the net detection efficiency also reduces in this case, which indicates that ω_g higher than 2 GHz does not guarantee to obtain a higher detection rate.

4   Conclusion

In conclusion, we have developed a single-photon detector at 1550-nm using a sinusoidally gated InGaAs/InP APD. A gate repetition frequency of 2 GHz was achieved with high quantum efficiency, low dark count probability, and low afterpulsing probability. The single-photon detector can be easily applied to practical high-speed QKD systems.

Acknowledgement

This research was partially supported by the Grant-in-Aid for Scientific Research of Ministry of Education, Culture, Sports, Science and Technology, Japan (MEXT).

References

1. Gisin, N., Ribordy, G., Tittel, W., Zbinden, H.: Rev. Mod. Phys. 74, 145 (2002)
2. Yoshizawa, A., Tsuchida, H.: Jpn. J. Appl. Phys. Pt.1, 40, 200 (2001)
3. Namekata, N., Makino, Y., Inoue, S.: Opt. Lett. 27, 954 (2002)
4. Ribordy, G., Gisin, N., Guinnard, O., Stucki, D., Wegmuller, M., Zbinden, H.: J. Mod. Opt. 51, 1381 (2004)
5. Namekata, N., Sasamori, S., Inoue, S.: Opt. Express. 14, 10043 (2006)
6. Yuan, Z.L., Kardynal, B.E., Sharpe, A.W., Shields, A.J.: Appl. Phys. Lett. 91, 041114 (2007)
7. Namekata, N., Fujii, G., Honjo, T., Takesue, H., Inoue, S.: Appl. Phys. Lett. 91, 011112 (2007)

Local Transformation of Two EPR Photon Pairs into a Three-Photon W State Using a Polarization Dependent Beamsplitter

Toshiyuki Tashima[1], Tetsurho Wakatsuki[1], Şahin Kaya Özdemir[1,2], Takashi Yamamoto[1], Masato Koashi[1], and Nobuyuki Imoto[1]

[1] Graduate School of Engineering Science, Osaka University,
Toyonaka, Osaka 560-8531, Japan
tashima@qi.mp.es.osaka-u.ac.jp
[2] Department of Electrical and Systems Engineering,
Washington University in St. Louis, St. Louis, MO 63130 USA
ozdemir@ese.wustl.edu

Abstract. We have theoretically proposed and experimentally demonstrated that two EPR photon pairs can be transformed into a three-photon W state by local operation and classical communication (LOCC). The fidelity of the final state to the ideal W state was 0.778 ± 0.043. The obtained expectation value of the witness operator for distinguishing between the three-photon W state and bi-separable states was -0.111 ± 0.043.

Keywords: Multipartite entanglement, W state, LOCC.

1 Introduction

Entanglement has been used for key theoretical and experimental progresses in quantum information science. Unlike bipartite entanglement, where Einstein-Podolsky-Rosen (EPR) pairs of qubits act as a universal resource to prepare any bipartite state by local operation and classical communication (LOCC), there is no N-partite ($N \geq 3$) entanglement which can be used as a universal resource to prepare N-partite states due to the fact that there are distinct classes of multipartite entangled states which cannot be converted into each other by stochastic local operation and classical communication (SLOCC). Greenberger-Horne-Zeilinger (GHZ) and W states are well-known examples of such distinct classes [1].

It is of significance to study how different classes of multipartite entangled states can be prepared among distantly located parties sharing EPR pairs using only LOCC. Previously, preparation of GHZ states from two EPR pairs by LOCC was experimentally demonstrated [2]. Moreover, it was shown that a pseudo tripartite W state could be prepared from a tripartite GHZ state [3] although W and GHZ are distinct classes of entangled states. Such an approximate conversion

A. Sergienko, S. Pascazio, and P. Villoresi (Eds.): QuantumCom 2009, LNICST 36, pp. 39–45, 2010.

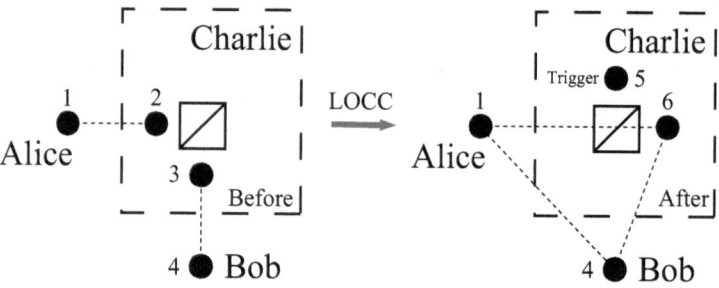

Fig. 1. Concept of local transformation from two EPR photon pairs to the W state using a polarization dependent beamsplitter. Two EPR photon pairs are shared by Alice-Charlie and Bob-Charlie.

has the drawback of trade-off between fidelity and success probbaility, i.e., unit fidelity cannot be achieved with a non-zero success probability.

Here we study a transformation of two EPR photon pairs into a three-photon W state by LOCC with unit fidelity and show an experimental demonstration [4]. When each of Alice-Charlie (modes 1 and 2) and Charlie-Bob (modes 3 and 4) share one EPR pair, the required transformation is performed by a local operation using the two photons on Charlie's side as in Fig. 1. The local operation uses a polarization dependent beamsplitter (PDBS) which has suitable transmission/reflection characteristics for horizontally and vertically polarized photons.

2 Theoretical Analysis

2.1 Optimal Method

We assume that four photons in state $|\,\mathrm{EPR}\rangle_{12}|\,\mathrm{EPR}\rangle_{34} = (|\,\mathrm{HHHH}\rangle_{1234} + |\,\mathrm{HHVV}\rangle_{1234} + |\,\mathrm{VVHH}\rangle_{1234} + |\,\mathrm{VVVV}\rangle_{1234})/2$ are distributed such that Alice has the photon in mode 1, Bob has mode 4, and Charlie has modes 2 and 3. Charlie sends his two photons to a PDBS, whose output modes are labeled as 5 and 6 in Fig. 1. The PDBS transforms the H- and V-polarized photons as

$$\hat{a}_{2\mathrm{H}}^{\dagger} = \sqrt{1-\mu}\,\hat{a}_{5\mathrm{H}}^{\dagger} - \sqrt{\mu}\,\hat{a}_{6\mathrm{H}}^{\dagger}, \quad \hat{a}_{3\mathrm{H}}^{\dagger} = \sqrt{\mu}\,\hat{a}_{5\mathrm{H}}^{\dagger} + \sqrt{1-\mu}\,\hat{a}_{6\mathrm{H}}^{\dagger}, \qquad (1)$$

and

$$\hat{a}_{2\mathrm{V}}^{\dagger} = \sqrt{1-\nu}\,\hat{a}_{5\mathrm{V}}^{\dagger} - \sqrt{\nu}\,\hat{a}_{6\mathrm{V}}^{\dagger}, \quad \hat{a}_{3\mathrm{V}}^{\dagger} = \sqrt{\nu}\,\hat{a}_{5\mathrm{V}}^{\dagger} + \sqrt{1-\nu}\,\hat{a}_{6\mathrm{V}}^{\dagger} \qquad (2)$$

where $\hat{a}_{j\mathrm{H}}^{\dagger}$ ($\hat{a}_{j\mathrm{V}}^{\dagger}$) denotes the creation operator of H (V)-polarized photon in the j-th mode of PDBS, and μ (ν) is the transmission coefficient for H (V)-polarization. Using the relations given in Eqs. (1) and (2), we find that the action of the PDBS on the four possible input states, $|\,1_{\mathrm{H}}\rangle_{2}|\,1_{\mathrm{H}}\rangle_{3} = \hat{a}_{2\mathrm{H}}^{\dagger}\hat{a}_{3\mathrm{H}}^{\dagger}|\,vac\rangle_{23}$, $|\,1_{\mathrm{H}}\rangle_{2}|\,1_{\mathrm{V}}\rangle_{3} = \hat{a}_{2\mathrm{H}}^{\dagger}\hat{a}_{3\mathrm{V}}^{\dagger}|\,vac\rangle_{23}$, $|\,1_{\mathrm{V}}\rangle_{2}|\,1_{\mathrm{H}}\rangle_{3} = \hat{a}_{2\mathrm{V}}^{\dagger}\hat{a}_{3\mathrm{H}}^{\dagger}|\,vac\rangle_{23}$ and $|\,1_{\mathrm{V}}\rangle_{2}|\,1_{\mathrm{V}}\rangle_{3} = \hat{a}_{2\mathrm{V}}^{\dagger}\hat{a}_{3\mathrm{V}}^{\dagger}|\,vac\rangle_{23}$ with $|\,vac\rangle$ denoting the vacuum state, transforms them into

$$|1_H\rangle_2|1_H\rangle_3 \rightarrow \sqrt{2\mu(1-\mu)}\,|2_H\rangle_5|0\rangle_6 + (1-2\mu)|1_H\rangle_5|1_H\rangle_6$$
$$-\sqrt{2\mu(1-\mu)}\,|0\rangle_5|2_H\rangle_6,$$
$$|1_H\rangle_2|1_V\rangle_3 \rightarrow \sqrt{\nu(1-\mu)}\,|1_V1_H\rangle_5|0\rangle_6 - \sqrt{\nu\mu}\,|1_V\rangle_5|1_H\rangle_6$$
$$+\sqrt{(1-\mu)(1-\nu)}\,|1_H\rangle_5|1_V\rangle_6 - \sqrt{\mu(1-\nu)}\,|0\rangle_5|1_V1_H\rangle_6,$$
$$|1_V\rangle_2|1_H\rangle_3 \rightarrow \sqrt{\mu(1-\nu)}\,|1_V1_H\rangle_5|0\rangle_6 + \sqrt{(1-\nu)(1-\mu)}\,|1_V\rangle_5|1_H\rangle_6$$
$$-\sqrt{\mu\nu}\,|1_H\rangle_5|1_V\rangle_6 - \sqrt{\nu(1-\mu)}\,|0\rangle_5|1_V1_H\rangle_6,$$
$$|1_V\rangle_2|1_V\rangle_3 \rightarrow \sqrt{2\nu(1-\nu)}\,|2_V\rangle_5|0\rangle_6 + (1-2\nu)|1_V\rangle_5|1_V\rangle_6$$
$$-\sqrt{2\nu(1-\nu)}\,|0\rangle_5|2_V\rangle_6. \tag{3}$$

Keeping only the cases leading to coincidence detection where a photon is present in each of the modes 1, 4, 5 and 6, we find that the state after the PDBS is given by

$$\frac{1}{2}\left[(1-2\mu)|HHH\rangle_{146} + \sqrt{(1-\mu)(1-\nu)}|HVV\rangle_{146} - \sqrt{\mu\nu}|VHV\rangle_{146}\right]|H\rangle_5$$
$$+\frac{1}{2}\left[(1-2\nu)|VVV\rangle_{146} - \sqrt{\mu\nu}|HVH\rangle_{146} + \sqrt{(1-\mu)(1-\nu)}|VHH\rangle_{146}\right]|V\rangle_5. \tag{4}$$

If Charlie has detected an H-polarized photon or a V-polarized photon in mode 5, he announces it and switches the polarization mode 6 as $|H\rangle_6 \leftrightarrow |V\rangle_6$. At this point, the three parties share the following states

$$\frac{1}{2}\left[(1-2\mu)|HHV\rangle_{146} + \sqrt{(1-\mu)(1-\nu)}|HVH\rangle_{146} - \sqrt{\mu\nu}|VHH\rangle_{146}\right] \tag{5}$$

and

$$\frac{1}{2}\left[(1-2\nu)|VVH\rangle_{146} - \sqrt{\mu\nu}|HVV\rangle_{146} + \sqrt{(1-\mu)(1-\nu)}|VHV\rangle_{146}\right], \tag{6}$$

respectively for H- and V-polarization detection in mode 5. With local filtering, the states in Eqs. (5) and (6) can be transformed into the ideal polarization encoded state $|W_3\rangle \equiv (|HHV\rangle + |HVH\rangle + |VHH\rangle)/\sqrt{3}$. The probabilities of obtaining these cases are

$$p_H \equiv \frac{3}{4}\min\{(2\mu-1)^2,\ (1-\mu)(1-\nu),\ \mu\nu\} \tag{7}$$

and

$$p_V \equiv \frac{3}{4}\min\{(2\nu-1)^2,\ (1-\mu)(1-\nu),\ \mu\nu\}. \tag{8}$$

In the following, we consider the optimization of μ and ν to obtain the highest success probability. For a fixed value of μ, $\min\{(1-\mu)(1-\nu),\ \mu\nu\}$ takes its maximum value $\mu(1-\mu)$ for $\nu = 1 - \mu$. The maximum value of $\min\{(2\mu - $

$1)^2$, $\mu(1 - \mu)\}$ is $1/5$ for $\mu = (5 \pm \sqrt{5})/10$. Therefore, the maximum success probability is $p_H = 3/20$ for $\mu = (5 + \sqrt{5})/10$ and $\nu = (5 - \sqrt{5})/10$ or vice versa. For this choice of μ and ν, p_V also takes its maximum value $3/20$. It is noted that for these optimal choices of μ and ν, the states in Eqs. (5) and (6) are already W states and the local filtering is not necessary.

2.2 Experimental Method

In our experiment, we used a sub-optimal choice of the PDBS parameters, $\mu = (7 + \sqrt{17})/16$ and $\nu = 1/2$. One of the reasons for this choice is that the two-photon interference for the V polarization is observed with a high visibility, which makes the alignment easier and gives us a clue about how well the two photons from different pairs are overlapped at the PDBS. Setting $\nu = 1/2$ in Eq. (4), we obtain

$$\frac{1}{2}\left[(2\mu - 1)|\,\mathrm{HHH}\rangle_{146} + \sqrt{\frac{1-\mu}{2}}|\,\mathrm{HVV}\rangle_{146} - \sqrt{\frac{\mu}{2}}|\,\mathrm{VHV}\rangle_{146}\right]|\,\mathrm{H}\rangle_5$$
$$+\frac{1}{2}\left[-\sqrt{\frac{\mu}{2}}|\,\mathrm{HVH}\rangle_{146} + \sqrt{\frac{1-\mu}{2}}|\,\mathrm{VHH}\rangle_{146}\right]|\,\mathrm{V}\rangle_5, \qquad (9)$$

where we see that if Charlie has detected a V-polarized photon in mode 5, three parties share a bi-separable state $(\sqrt{\mu}|\,\mathrm{HV}\rangle_{14} + \sqrt{1-\mu}|\,\mathrm{VH}\rangle_{14})|\,\mathrm{H}\rangle_6/2\sqrt{2}$. On the other hand, if Charlie has detected an H-polarized photon in mode 5 and Alice introduces a phase shift locally, they will end up with a W-like state from which $|\,\mathrm{W}_3\rangle$ can be prepared by equalizing the weights of the components with local filtering. From Fig. 2, we see that the success probability is optimized when $(\mu - 1/2)^2 = (1 - \mu)/8$ or $(\mu - 1/2)^2 = \mu/8$. These equations, respectively, give

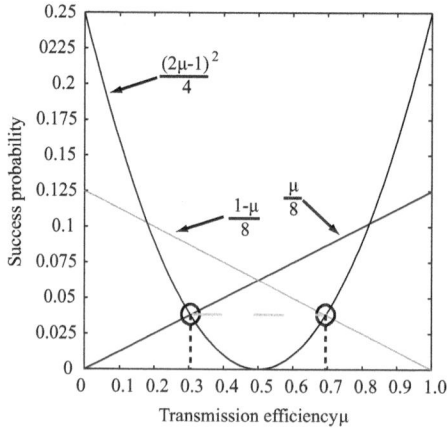

Fig. 2. Success probability of each terms by detecting H-polarized photon on Charlie's side. The circle gives $\mu = (9 - \sqrt{17})/16$ and $\mu = (7 + \sqrt{17})/16$.

$\mu = (7 + \sqrt{17})/16$ and $\mu = (9 - \sqrt{17})/16$, both of which lead to the same success probability. Substituting $\mu = (7 + \sqrt{17})/16$ in Eq. (9), and keeping only the terms leading to coincidence detection triggered by an H-photon detection in mode 5, the post-selected state in modes 1, 4 and 6 becomes

$$\frac{\sqrt{9 - \sqrt{17}}}{8\sqrt{2}} \left[|\,\mathrm{HHV}\rangle + |\,\mathrm{HVH}\rangle + \frac{\sqrt{7 + \sqrt{17}}}{\sqrt{9 - \sqrt{17}}} |\,\mathrm{VHH}\rangle \right]_{146} , \qquad (10)$$

after Charlie compensates the phase shift locally and changes the polarization of his photon in mode 6. We see that the component $|\,\mathrm{VHH}\rangle$ is the only one with a V-photon in mode 1; thus its weight can be equalized to the others by introducing polarization dependent losses in mode 1. Then the final state becomes

$$\sqrt{3(9 - \sqrt{17})/128}|\,\mathrm{W}_3\rangle_{146} \qquad (11)$$

implying that local transformation of two EPR photon pairs into a $|\,\mathrm{W}_3\rangle$ is achieved with unit fidelity at a success probability of $3(9 - \sqrt{17})/128 \sim 11.4\%$.

3   Experimental Demonstration

The experimental setup is shown in Fig. 3 (a). Two EPR photon pairs were prepared by using parametric down-conversion at a pair of type-I phase-matched β-barium borate (BBO) crystals stacked together with optical axes orthogonal to

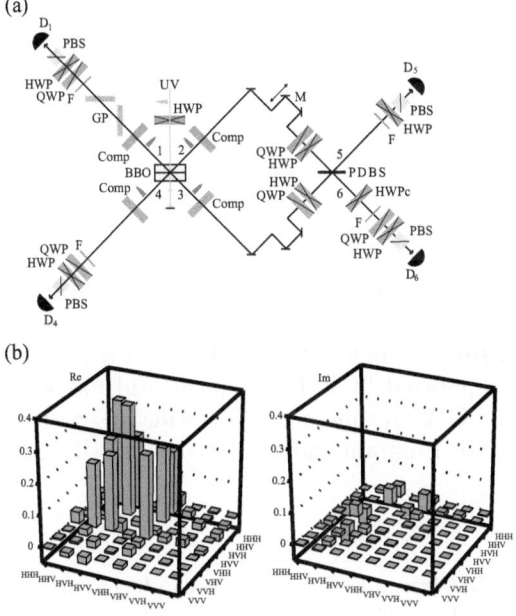

Fig. 3. (a) Proposed experimental setup. (b) Real and imaginary parts of the reconstructed density matrix of the experimentally obtained W state.

Table 1. Peres-Horodecki criterion, Concurrence and Entanglement of Formation (EOF) of the prepared marginal bipartite states

	Peres-Horodecki criterion	Concurrence	EOF
modes 1-4	-0.091 ± 0.026	0.322 ± 0.073	0.244 ± 0.066
modes 1-6	-0.143 ± 0.030	0.421 ± 0.066	0.263 ± 0.065
modes 4-6	-0.123 ± 0.027	0.415 ± 0.068	0.195 ± 0.065

each other. Then one photon from each of the EPR photon pairs was mixed at the PDBS. Whenever it was confirmed that an H-polarized photon was in mode 5, a three-photon W state was prepared in the rest of output modes (1, 4 and 6).

The density matrix of the three-photon state was estimated as in Fig. 3 (b) using quantum state tomography [5]. The fidelity of the converted state to the ideal three-photon W state was 0.778 ± 0.043. The obtained expectation value of the witness operator for distinguishing between the three-photon W state and biseparable states (including separable states)) [6] was -0.111 ± 0.043. This negative value proves that the prepared state is not a biseparable state. Since marginal bipartite states of a W state should be entangled, we have tested all pairwise combinations for the prepared W state. We evaluated the entanglement of the marginal bipartite states using the Peres-Horodecki criterion [7,8], concurrence [9] and entanglmement of formation (EOF) [10] which, respectively, have the values of -0.206, 0.667 and 0.55 for marginal entangled states of an ideal three-photon W state. The results of these analysis for the experimentally prepared W-state are given in Table. 1, which clearly shows the existence of entanglement in these marginal bipartite states.

4   Conclusion

We have theoretically proposed and experimentally demonstrated a scheme in which two EPR pairs are converted into a three-photon W-state using LOCC. The required transformation is provided by a polarization dependent beamsplitter, phase compensator and polarization dependent losses. Through these results, it is now possible to generate arbitrary three-qubit states of W and GHZ classes via LOCC starting from a single resource of two EPR pairs.

This work was supported by JSPS Grant-in-Aid for Scientific Research(C) 20540389 and by MEXT Grant-in-Aid for Scientific Research on Innovative Areas 20104003, Global COE Program and Young scientists(B) 20740232.

References

1. Dür, W., Vidal, G., Cirac, J.I.: Three qubits can be entangled in two inequivalent ways. Phys. Rev. A 62, 062314 (2000)
2. Resch, K.J., Walther, P., Zeilinger, A.: Full Characterization of a Three-Photon Greenverger-Horne-Zeilinger State Using Quantum State Tomography. Phys. Rev. Lett. 94, 070402 (2005)

3. Walther, P., Resch, K.J., Zeilinger, A.: Local Conversion of Greenberger-Horne-Zeilinger States to Approximate W States. Phys. Rev. Lett. 94, 240501 (2005)
4. Tashima, T., Wakatsuki, T., Ozdemir, S.K., Yamamoto, T., Koashi, M., Imoto, N.: Local Transformation of Two Einstein-Podolsky-Rosen Photon Pairs into a Three-Photon W State. Phys. Rev. Lett. 102, 130502 (2009)
5. James, D.F.V., Kwiat, P.G., Munro, W.J., White, A.G.: Measurement of qubits. Phys. Rev. A. 64, 052312 (2001)
6. Bourennane, M., Eibl, M., Kurtsiefer, C., Gaertner, S., Weinfurter, H., Guhne, O., Hyllus, P., Brus, D., Lewenstein, M., Sanpera, A.: Experimental Detection of Multipartite Entanglement using Witness Operators. Phys. Rev. Lett. 92, 087902 (2004)
7. Peres, A.: Separability Criterion for Density Matrices. Phys. Rev. Lett. 77, 1413 (1996)
8. Horodecki, M., Horodecki, P., Horodecki, R.: Separability of mixed states: necessary and sufficient conditions. Phys. Lett. A 223, 1 (1996)
9. Coffman, V., Kundu, J., Wootters, W.K.: Distributed entanglement. Phys. Rev. A 61, 052306 (2000)
10. Wootters, W.K.: Entanglement of Formation of an Arbitrary State of Two Qubits. Phys. Rev. Lett. 80, 2245 (1998)

Entanglement Degree Characterization of Spontaneous Parametric-Down Conversion Biphotons in Frequency Domain

Giorgio Brida[1], Valentina Caricato[1,2], Marco Genovese[1], Marco Gramegna[1], Mikhail V. Fedorov[3], and Sergey P. Kulik[4]

[1] I.N.RI.M. - Istituto Nazionale di Ricerca Metrologica, Turin, Italy
[2] Department of Physics, Politecnico of Turin, Turin, Italy
v.caricato@inrim.it, valentina.caricato@polito.it
[3] A.M.Prokhorov General Physics Institute, Russian Academy of Science, Moscow, Russia
[4] Faculty of Physics, M.V.Lomonosov Moscow State University, Moscow, Russia

Abstract. We present an experiment addressed to verify the validity of the entanglement quantifier R, defined as the ratio between single counts and coincidence distribution width for pure biphoton states generated by short pump pulses. We demonstrate that this ratio can be efficient to estimate entanglement degree in frequency domain. Our work has been performed with a femto second pulsed laser pump addressed to a LiIO$_3$ crystal generating correlated biphotons through collinear degenerate type-I SPDC. To study the dependence of R on the length of the crystal we performed two sequence of measurements for each crystal firstly with a 10 mm and then with a 5 mm.

Keywords: Entanglement, biphotons, SPDC.

1 Introduction

In the framework of quantum mechanics and quantum optics, entanglement is a powerful resource for developing quantum technologies [1], such as quantum communication, q-calculus, q-imaging, q-metrology, etc. For this reason it is very important to define a precise and easy-implementable method for characterizing entanglement properties. In literature there are several measures that quantify entanglement of quantum bipartite states both in discrete and continuous variables [2]; we want to cite as some examples the Schmidt rank, entropy, concurrence, etc. From a theoretical point of view the density matrix that describes a precise state of bipartite system is useful to evaluate all these quantifiers. On the other side from an experimental point of view a quantum tomography procedure (either complete or reduced) would be necessary [3]. In the case of high-dimensional systems this procedure requires a number of measurements that increases quadratically with the dimension of Hilbert space. Following these considerations, an innovative and alternative method is to analyze and develop

A. Sergienko, S. Pascazio, and P. Villoresi (Eds.): QuantumCom 2009, LNICST 36, pp. 46–55, 2010.

links between entanglement and measurable quantities related to the quantum state, defining precise parameters. For pure biphoton states one of these parameters is the Fedorov's parameter [4] R_q (where q denotes an arbitrary space, e.g. frequency or spatial one), defined as the ratio of the single-particle and coincidence distributions widths in q-space. The advantage of this parameter is that can be rather easily measured at variance with all other entanglement quantifiers. A qualitative approach to R_q follows from entropy : high entanglement leads to a better knowledge about composite bipartite system (narrower coincidence distribution) and a worse knowledge about individual subsystem(s) (wider single-particle distribution). This clear physical meaning makes R_q to be an extremely useful tool for entanglement control. It has been proved [4] that for bipartite states, described by a double-Gaussian wave function, the parameter R coincides exactly with the Schmidt number K. Also, if we consider special classes of non-double-Gaussian wave functions like those describing Spontaneous Parametric Down-Conversion (SPDC) their values remain quite close as well. This is true both for continue and discrete variables. Recently the R parameter has been successfully applied for demonstrating a very strong entanglement anisotropy in spatial distributions of biphotons [5].

1.1   Aim of the Paper

In this proceeding we present in detail an experimental work devoted to verify the operational entanglement quantifier R applied to the study of two-photon states entangled in frequency domain [6]. We want to demonstrate that the measure of R is an efficient alternative to other quantifiers like for example (in specific experiments) the measure of the visibility of interference pattern that can provide some knowledge about entanglement [7]. It is known that two-photon states belong to a multi-dimensional Hilbert space and can posses an extremely high entanglement degree (up to several hundreds) that makes them very perspective in applications of quantum information and quantum communication. It is important to emphasize that in this work our consideration are restricted to purely spectral entanglement. This means that we consider only photons propagating along the pump axis. In our experiment this condition has been realized inserting some very thin slits (50 μm width) in front of detectors in the focal plane of photons emitted from a $LiIO_3$ crystal. An opposite case, where the photon frequencies are fixed but directions of their propagation can change, has been investigated earlier [5].

2   Theoretical Considerations

We consider the pump described by a sequence of short Fourier-limited pulses, which provides conditions for generating pure biphoton states characterized by a wave function. We are interesting in type-I degenerate collinear SPDC where the wave function Ψ, depending only on frequencies of emitted signal and idler photons ω_1 and ω_2, is given by [8,9]

$$\Psi(\nu_1, \nu_2) \propto \exp\left(-\frac{(\nu_1 + \nu_2)^2 \tau^2}{8 \ln 2}\right) \times \text{sinc}\left\{\frac{L}{2c}\left[A(\nu_1 + \nu_2) - B\frac{(\nu_1 - \nu_2)^2}{\omega_p}\right]\right\}(1)$$

The first term in the wave function is a Gaussian exponent that describes the pump spectral amplitude. The pump-pulse duration is τ, L is the length of the crystal, ν_1 and ν_2 are deviations of frequencies of the signal and idler photons $\omega_{1,2}$ from the central frequencies $\omega_1^{(0)} = \omega_2^{(0)} = \omega_p/2$, $|\nu_{1,2}| \ll \omega_p$, ω_p being the central frequency of the pump spectrum. A and B are the temporal walk-off and dispersion constants

$$A = c\left(k_p'(\omega)\big|_{\omega=\omega_p} - k_1'(\omega)\big|_{\omega=\omega_p/2}\right) = c\left(\frac{1}{v_g^{(p)}} - \frac{1}{v_g^{(o)}}\right)$$

$$B = \frac{c}{4}\,\omega_p\,k_1''(\omega)\big|_{\omega=\omega_0/2}\,. \tag{2}$$

$v_g^{(p)}$ and $v_g^{(o)}$ are the group velocities of the pump and ordinary waves, and k_1 and k_p are the wave vectors of signal and pump photons.

We want to stress that in a more general case both linear and quadratic terms in the argument of the sinc-function in (1) are important and none of them can be eliminated. The sinc-function cannot be substituted by any Gaussian function and, for this reason, the wave function (1) is considered as a non-double-Gaussian one. We use the wave function (1) to estimate the theoretical coincidence and single-particle biphoton spectra. It is important to notice that these spectra are significantly different in the cases of short and long pump pulses. It has been defined the control parameter separating the regions of short and long pulses and it is expressed by [8]

$$\eta = \frac{\Delta\nu_{1\,\text{sinc}}}{\Delta\nu_{1\,\text{pump}}} \approx 2\frac{c\tau}{AL} = \frac{2\tau}{L/v_g^{(p)} - L/v_g^{(o)}}\,. \tag{3}$$

i.e., it changes linearly with the pump-pulse duration and decreases if the length of the crystal increases. For a fixed length of the crystal pump pulses are short if $\eta \ll 1$ and long if $\eta \gg 1$ and, typically, $\eta \sim 1$ at $\tau \sim 1$ ps. The two limits $\eta \ll 1$ and $\eta \gg 1$ correspond to a pump spectral amplitude expressed as a function of $\nu_{1,2}$, much narrower and much wider than the sinc-function in (1). The parameter (3) contains factors that can be easily controlled in the experiment of this paper, namely the length of the crystal and the pump pulse duration. The choose of a particular crystal with specific dispersion properties can act on the longitudinal walk-off effect expressed in the term A. Let's focus our attention on the short pump pulses $\eta \ll 1$: the FWHM of the coincidence and single-particle spectra were found analytically to be given by [8]

$$\Delta\omega_c = \frac{5.56\,c}{AL}, \quad \Delta\omega_s = \sqrt{\frac{2A\ln(2)\,\omega_p}{B\tau}}\,. \tag{4}$$

The first formula follows directly from (1) whereas the second one requires for its derivation a preliminary integration of $|\Psi(\nu_1, \nu_2)|^2$, for example over ν_2 [8]. For long pump pulses $\eta \gg 1$, the coincidence and single-particle spectral widths have the form [8]

$$\Delta\omega_c = \frac{4\ln 2}{\tau}, \quad \Delta\omega_s = \sqrt{\frac{2.78\, c\, \omega_0}{L B}}. \tag{5}$$

We use equations (4) and (5) to find the expression of the parameter $R(\eta)$ in regions of short and long pulses as :

$$R(\eta) = \frac{\Delta\omega_s}{\Delta\omega_c}. \tag{6}$$

In the intermediate region ($\eta \sim 1$) the $R(\eta)$ can be determined by a quadratic interpolation [8]

$$R(\eta) \approx \sqrt{R_{\text{short}}^2 + R_{\text{long}}^2} = 0.75\frac{A}{\sqrt{B}}\sqrt{\frac{L}{\lambda_0}}\sqrt{\eta^2 + \frac{1}{\eta}} = 55\sqrt{\eta^2 + \frac{1}{\eta}}. \tag{7}$$

The function $R(\eta)$ (7) is plotted in Fig. 1 together with the Schmidt number $K(\eta)$. The latter was calculated numerically by Silberhorn and Mauerer [10] who demonstrated that, in the region of short and long pulses, the phase of the wave function does not affect the Schmidt number.

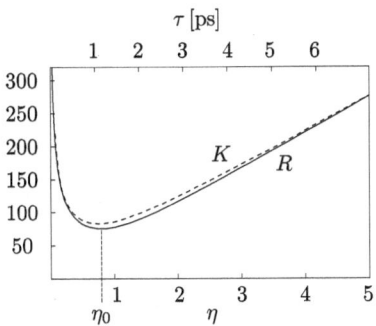

Fig. 1. Quantifiers calculated analytically (R [8]) and numerically (K [10]) for LiIO$_3$ crystal of a length $L = 0.5$ cm and the pump wavelength $\lambda_p = 400$ nm

It is visible from the picture that the functions $R(\eta)$ and $K(\eta)$ are very close each other, and this confirms the validity of the experimentally measurable parameter R for quantifying the degree of entanglement of biphoton states characterized by a non-double-Gaussian wave function of the form (1). From the previous theoretical considerations the degree of spectral entanglement reaches very high values in the whole range of pump-pulse durations when we are sufficiently far from the minimum localized at $\eta \sim 1$ or $\tau \sim 1$ ps.

3   Experimental Set Up and Results

The experiment here described was performed in fs pulsed regime with a Mode-Locked Titanium-Sapphire laser at a working wavelength of $\lambda_{IR} = (795.0 \pm$

0.1) nm with a $\Delta\lambda_{IR} = (5.9 \pm 0.1)$ nm. After doubling in frequency by second harmonic generation we obtained a pump of $\lambda_p = (397.5 \pm 0.2)$ nm and $\Delta\lambda_p = (1.8 \pm 0.1)$ nm, corresponding to a pulse duration of $\tau = (186 \pm 30)$ fs. We estimated the beam waist to be smaller than 2 mm. The pump spectrum observed on a spectrometer analyzer showed an IR component @ 795 nm still present in the pump so we inserted in the beam two quartz prisms oriented at the Brewster's angle to reduce the fundamental residual as much as possible. In order to study the dependence of the $R(\eta)$ parameter on the crystal length L, we performed two different sequences of measurements, one addressing the pump beam to a 10 mm LiIO$_3$ the other to a 5 mm crystal. With the correct phase matching angle the horizontally polarized beam generated two vertically polarized photons through the non linear optic process known as type I collinear SPDC. A couple of UV mirrors was inserted after the crystal to eliminate the UV pump, and biphotons produced were divided by a non polarizing beam-splitter and sent to two photodetection apparatuses consisting of red glass filters and two single photo-detectors (SPAD: Perkin-Elmer SPCM-AQR-15). The photodetectors are characterized by a photon efficiency variable with the wavelength. Our working range was from 700 nm to 900 nm (Fig. 2).

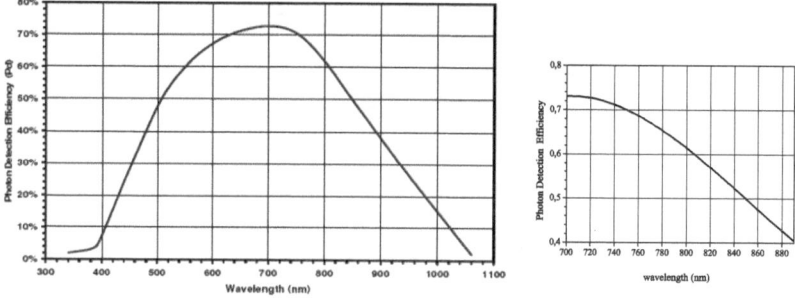

Fig. 2. Photon Detection Efficiency of Perkin-Elmer detectors and a detail in the range from 700 nm to 900 nm

In front of each detector it was placed a monochromator with a variable spectral resolution. A lens with a focal distance of 20 cm and with a diameter of 2.5 cm was fixed 135 cm far from the crystal in order to focalize the light inside each monochromator and a couple of lenses with the same focal distance were placed on the output of the monochromators. It is worth to mention that the lenses placed at this large distance, with this small diameter, have been worked as selective tool on the spatial modes of biphotons allowing to collect a single spatial mode in the monochromators.

All experiments were performed as a sequence of corresponding measurements of spectral distributions both in single counts and coincidences. When the best phase matching conditions had been found the measure of the coincidence distribution was performed fixing one of the two monochromators at the central

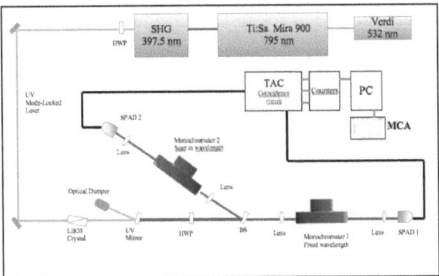

Fig. 3. Experimental set up: a duplicated titanium sapphire laser beam pumps a LiIO₃ crystal generating collinear SPDC. After a Half Wave Plate (HWP) the biphotons are split on a beam-splitter (BS) and fed to SPAD detectors, whose output feeds Time-to Amplitude Converter (TAC), counters and Multi Channel Analyzer (MCA).

wavelength of SPDC (795 nm) and scanning the other one in a range around this value. The same procedure has been applied for the single counts distribution.

3.1 Measure with a 10 mm LiIO₃ Crystal

The first series of measurements was performed with a 10 mm LiIO₃ crystal. To measure the coincidence distribution we have used monochromators with 0.2 nm resolution for scanning the signal wavelength. For each wavelength we got three measurements of the number of counts for real and accidental coincidence distributions. Each measurement was performed in an acquisition time of 100 s and we selected on the multichannel analyzer a window of 670 channels (i.e. 1,64 ns) for both distributions. After correcting the counts for the accidental ones, we evaluated the average and the standard deviation, then our measurements were corrected taking into account the efficiency of the detectors and finally the plot was normalized. Subsequently experimental points were fitted by a gaussian distribution in order to have for the distribution an expected value for the FWHM with its uncertainty. In this configuration we obtained a narrow peak with $\Delta\lambda_c = (0.29 \pm 0.03)$ nm (Fig. 4). By a comparison the pump width was 6.2 times larger than the coincidence spectrum.

The spectral distribution of single counts was performed using a monochromator in the transmission arm with a spectral resolution of 1 nm. For this measure we scanned each wavelength five times and we registered counts on one detector for a temporal window of 5 s. To evaluate the background we inserted a HWP cut for 400 nm before the pump in order to rotate the pump polarization from horizontal to vertical: in this way without rotating phase matching angle no correlated photons were produced. We subtracted again the background from the single counts and we evaluated the uncertainty with usual uncertainty propagation formula. Also in this case each value was corrected taking into account the efficiency of the detector for different wavelengths an the plot was finally normalized. We estimated the FWHM of the distribution from the experimental points and we associated the resolution of the CVI monochromator as the uncertainty

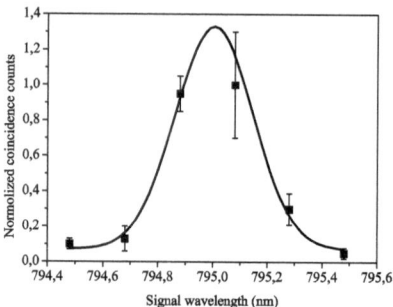

Fig. 4. Normalized coincidence distribution for 10 mm length LiIO₃ sample

on this measure, obtaining the single counts spectral width $\Delta\lambda_s = (101 \pm 1)$ nm (Fig. 5). It is worth to notice that data present a strong asymmetry of the right wing in the measured spectrum. This effect probably might be caused by loosing in spectral sensitivity of monochromator for long wavelength.

The distribution of single counts is 56 times larger than that of the pump, corresponding to an experimental ratio between widths of the two distributions $R_\omega = (349 \pm 43)$. The complete results are listed in the table 1.

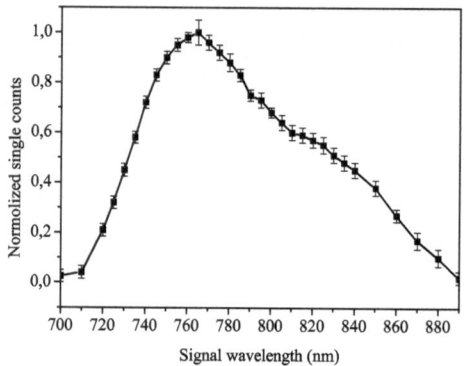

Fig. 5. Normalized single counts distribution for 10 mm length LiIO₃ sample

Table 1. Results for 10 mm LiIO₃ sample

Theory Model	Experimental results
Coincidence Distribution	
$\Delta\lambda_c = 0.32$ nm	$\Delta\lambda_c = (0.29 \pm 0.03)$
Single Counts Distribution	
$\Delta\lambda_s = 100$ nm	$\Delta\lambda_s = (101 \pm 1)$ nm
R-Quantifier	
$R_\omega \approx R_\lambda \approx 316$	$R_\omega \approx R_\lambda \approx (349 \pm 43)$

3.2   Measure with a 5 mm LiIO₃ Crystal

The same sequence of measurements, single counts and coincidence distribution, has been performed with a 5 mm length crystal obtaining for the coincidence distribution a $\Delta\lambda_c = (0.64 \pm 0.06)$ nm. The results are plotted in Fig. 6.

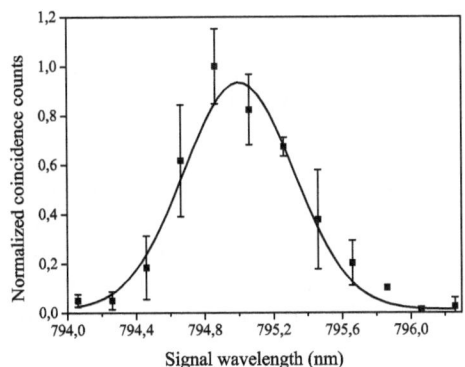

Fig. 6. Normalized coincidence distribution for 5 mm length LiIO₃ sample

The single counts distribution has now a width $\Delta\lambda_s = (115 \pm 1)$ nm. It is shown in Fig. 7 including the correction due to the detectors efficiency. This second set of measurements corresponds to $R_\omega = (179 \pm 18)$. It is important to notice that the pulse duration of the pump did not change respect to the previous measurements with the 10 mm crystal and we can observe that, in this case, the pump spectra is 2.8 times wider than the coincidence spectrum whereas the single counts distribution is 64 times larger than the width of the pump, in agreement with what expected from theory. Corresponding results are listed in table 2.

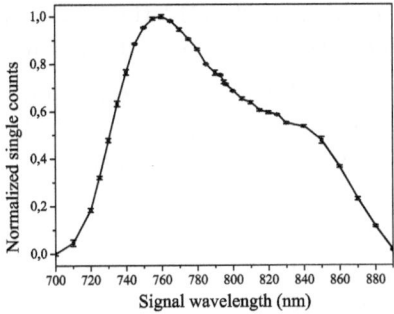

Fig. 7. Normalized single counts distribution for 5 mm length LiIO₃ sample

Table 2. Results for 5 mm $LiIO_3$ sample

Theory Model	Experimental results
Coincidence Distribution	
$\Delta\lambda_c = 0.63$ nm	$\Delta\lambda_c = (0.64 \pm 0.06)$
Single Counts Distribution	
$\Delta\lambda_s = 100$ nm	$\Delta\lambda_s = (115 \pm 1)$ nm
R-Quantifier	
$R_\omega \approx R_\lambda \approx 158$	$R_\omega \approx R_\lambda \approx (179 \pm 18)$

4   Conclusions and Future Plans

In this work we have experimentally verified the validity of the R-quantifier as an efficient tool to estimate entanglement degree in frequency domain for pure bipho-ton states generated by short pump pulses. More in detail, our work has been per-formed with a femto-second pulsed laser and the parameter we have changed to study the validity of this approach is the length of the $LiIO_3$ crystal. It is worth to notice that a variation in the η parameter can be realized by changing either the length L of the crystal or the pulse duration τ. However, working on the pulse duration in a wide range with the same laser results quite difficult because it leads to a loose of power and then decreases the production of biphotons after the crys-tal. Nevertheless, our results show a good agreement between the measured and predicted shapes of the curve shown at Fig. 1 and this completely confirms the adequacy of the R-quantifier approach. We observe that by doubling the sample length one doubles the entanglement degree for the short pump pulse regime of SPDC. Unfortunately, if we still decrease the length of the crystal we reduce too much the biphoton flux and reach the minimum of the $R(\eta)$ becomes quite diffi-cult. We remind that the method of measurement of R in frequency domain implies that other degrees of freedom of biphotons should be fixed. Concerning the spatial distribution of photon pairs emitted by crystal, these modes can be selected with the help of small pinholes. On the other side if we want to measure R in spatial variables fixing the frequencies of both photons, we need some very narrow-band filters [5]. We would like to mention also that a measurement of spectral widths can be done by transforming spectrum in time difference through a fiber [11]. Fi-nally we wish to acknowledge the work of [12] that is directly related to the subject of this work and was submitted right after we finished it.

Acknowledgments. This work has been partially supported by MIUR (PRIN 2007FYETBY), Regione Piemonte (E14),"San Paolo foundation", RFBR (08-02-12091) and NATO Grant (CBP.NR.NRCL 983251).

References

1. Genovese, M.: Research on hidden variable theories: A review of recent progresses. Phys. Rep. 413, 319–396 (2005)
2. Bengtsson, I., Zyczkowski, K.: Geometry of Quantum States. Cambridge Univ. Press, Cambridge (2006)

3. Zavatta, A., et al.: Tomographic reconstruction of the single-photon Fock state by high-frequency homodyne detection. Phys. Rev. A 70, 053821 (6 pages) (2004)

4. Fedorov, M.V., Efremov, M.A., Volkov, P.A., Eberly, J.H.: Short-pulse or strong-field breakup processes: a route to study entangled wave packets. J. Phys. B: At. Mol. Opt. Phys. 39, S467–S483 (2006)

5. Fedorov, M.V., Efremov, M.A., Volkov, P.A., Moreva, E.V., Straupe, S.S., Kulik, S.P.: Anisotropically and High Entanglement of Biphoton States Generated in Spontaneous Parametric Down-Conversion. Phys. Rev. Lett. 99, 063901 (4 pages) (2007)

6. Brida, G., Caricato, V., Genovese, M., Gramegna, M., Fedorov, M.V., Kulik, S.P.: Characterization of Spectral Entanglement of Spontaneous Parametric-Down Conversion Biphotons, http://arxiv.org/abs/0904.3009

7. Kim, Y., et al.: Temporal indistinguishability and quantum interference. Phys. Rev. A 62, 43820 (4 pages) (2000); Jeronimo-Moreno, Y., U'Ren, A.B.: Control, measurement, and propagation of entanglement in photon pairs generated through type-II parametric down-conversion. Phys. Rev. A 79, 033839 (14 pages) (2009)

8. Mikhailova, Y.M., Volkov, P.A., Fedorov, M.V.: Biphoton wave packets in parametric down-conversion: Spectral and temporal structure and degree of entanglement. Phys. Rev. A 78, 062327 (17 pages) (2008)

9. Keller, T.E., Rubin, M.H.: Theory of two-photon entanglement for spontaneous parametric down-conversion driven by a narrow pump pulse. Phys. Rev. A 56, 1534–1541 (1997)

10. Mauerer, W., Silberhorn, C.: Numerical Analysis of Parametric Downconversion. In: AIP Conf. Proc: Quantum Communication, Measurement and Computing (QCMC 2008), vol. 1110, pp. 220–223 (2009)

11. Brida, G., Chekhova, M.V., Genovese, M., Gramegna, M., Krivitsky, L.A.: Dispersion Spreading of Biphotons in Optical Fibers and Two-Photon Interference. Phys. Rev. Lett. 96, 143601 (4 pages) (2006); Brida, G., Genovese, M., Krivitsky, L.A., Chekhova, M.V.: Interference structure of two-photon amplitude revealed by dispersion spreading. Phys. Rev. A 75, 015801 (4 pages) (2007)

12. Avenhaus, M., Chekova, M.V., Krivitsky, L.A., Leuchs, G., Silberhorn, C.: Experimental verification of high spectral entanglement for pulsed waveguided spontaneous parametric down-conversion. Phys. Rev. A 79, 043836 (5 pages) (2009)

Matter-Matter Entanglement
for Quantum Communication

Julien Laurat

Laboratoire Kastler Brossel, Université P. et M. Curie, ENS and CNRS,
Case 74, 4 place Jussieu, 75252 Paris cedex 05, France
julien.laurat@upmc.fr

The distribution of entanglement between different parties enables quantum communication protocols, such as quantum key distribution or teleportation. Furthermore, control of entanglement between material systems is an essential capability for scalable architectures. This talk will focus on the experimental generation of such entanglement in different setups, in the regime of single photons or in the regime of continuous variables.

The first part will review experiments done recently in the single excitation regime. First, by following the seminal paper of Duan, Lukin, Cirac and Zoller ($DLCZ$) [1], entanglement between single collective excitations stored in two remote atomic ensembles can be generated. In the $DLCZ$ protocol, entanglement is created in a probabilistic but heralded way from quantum interference in the measurement process. The detection of a photon from one or the other atomic ensemble in an indistinguishable fashion results in an entangled state with one collective spin excitation shared coherently [2]. This entanglement has been used for the initial implementation of functional quantum nodes for polarization entanglement distribution involving the asynchronous preparation of two parallel pairs of entangled atomic ensembles [3]. Swapping of entanglement between entangled matter systems has also been investigated [4]. In this regime, a second way to generate entanglement between two atomic ensembles is to map into an entangled state of light [5]. A single photon is first split into two modes to generate photonic entanglement, which is then mapped to an entangled matter state in two ensembles by EIT. On demand, the stored entanglement is converted back into entangled photonic modes.

The two previous methods involve single excitations and single photons. An active direction concerns also the storage of noncalssical states in the regime of continuous variables. The talk will then focus on works recently developed, based on storage and retrieval of continuous variable of light into atomic ensembles [6,7]. For a vapor of cesium atoms, using EIT and Zeeman coherences, the case where a tunable single-sideband is stored independently of the other one to the case where the two symmetrical sidebands are stored using the same transparency window are compared. The optimal response frequency of the medium for storage can be adapted to the frequency to be stored by changing the control magnetic eld, keeping the EIT window rather narrow. Excess noise associated with spontaneous emission and spin relaxation is small, and quantum performance of the memory can be characterized by measuring the signal transfer

A. Sergienko, S. Pascazio, and P. Villoresi (Eds.): QuantumCom 2009, LNICST 36, pp. 56–57, 2010.

coefficient T and the conditional variance V and using the T-V criterion as a state independent benchmark. To generate entanglement between two ensembles in this regime, I will describe the squeezing source recently built [8] which enables to generate continuous variable entanglement between lateral sidebands. The principle of the on-going experiment will be described.

References

1. Duan, L.-M., Lukin, M.D., Cirac, J.I., Zoller, P.: Long-distance quantum communication with atomic ensembles and linear optics. Nature 414, 413 (2001)
2. Laurat, J., et al.: Heralded entanglement between atomic ensembles: preparation, decoherence, and scaling. Phys. Rev. Lett. 99, 180504 (2007)
3. Chou, C.-W., et al.: Functional quantum nodes for entanglement distribution over scalable quantum networks. Science 316, 1316 (2007)
4. Laurat, J., et al.: Towards experimental entanglement connection with atomic ensembles in the single excitation regime. New J. Phys. 9, 207 (2007)
5. Choi, K.S., Deng, H., Laurat, J., Kimble, H.J.: Mapping photonic entanglement into and out of a quantum memory. Nature 452, 67 (2008)
6. Cviklinski, J., et al.: Reversible quantum interface for tunable single sideband modulation. Phys. Rev. Lett. 101, 133601 (2008)
7. Ortalo, J., et al.: Atomic-ensemble-based quantum memory for sideband modulations. J. Phys. B 42, 11 (2009)
8. Burks, S., et al.: Vacuum squeezed light for atomic memories at the D2 cesium line. Opt. Express 17, 3777 (2009)

Manipulating Frequency Entangled Photons

Laurent Olislager[1], Johann Cussey[2], Anh Tuan Nguyen[3], Philippe Emplit[1],
Serge Massar[4], Jean-Marc Merolla[5], and Kien Phan Huy[5]

[1] Service OPERA-Photonique, CP 194/5, Université Libre de Bruxelles,
Avenue F.D. Roosevelt 50, 1050 Brussels, Belgium
[2] SmartQuantum SA, Espace Phœnix,
Route du Radôme, 22560 Pleumeur-Bodou, France
[3] Service de Physique et Electricité, Institut Meurice,
Haute Ecole Lucia de Brouckère, Avenue Emile Gryzon 1, 1070 Brussels, Belgium
[4] Laboratoire d'Information Quantique, CP 225, Université Libre de Bruxelles,
Boulevard du Triomphe, 1050 Brussels, Belgium
[5] Département d'Optique P.M. Duffieux, Institut FEMTO-ST,
Centre National de la Recherche Scientifique, UMR 6174,
Université de Franche-Comté, 25030 Besançon, France

Abstract. A parametric down conversion source pumped by a monochromatic laser will produce frequency entangled photon pairs. We demonstrate this by an experiment in which five-dimensional frequency entanglement is manipulated at telecommunication wavelengths using commercially available components such as electro-optic phase modulators and narrowband frequency filters. A theoretical intuition for this approach is developed by introducing the notion of *frequency bin* entanglement. We conclude by showing that using this method one can in principle violate the CHSH, the CGLMP, and a new – as yet unnamed – Bell inequalities.

Keywords: Quantum Optics, Entanglement, Frequency Bin.

1 Introduction

Entanglement is one of the most fascinating aspects of quantum mechanics, useful both for fundamental tests of physical principles and for applications such as Quantum Key Distribution (QKD). Many different kinds of photonic entanglement have been produced, including entanglement in polarization [1,2], momentum [3], angular momentum [4], time-energy [5,6,7] and its discretized version called *time bins* [8,9]. In the present work we show how energy – or equivalently frequency – entangled photons can be manipulated directly in the frequency domain.

Previous work on time-energy entanglement was mainly based on Franson's original proposal [10] which is based on three key ideas. First, a continuous pump laser produces time entangled photon pairs: the emission time of each photon is uncertain, but both photons are emitted simultaneously. Second, one uses measurements that resolve the time of arrival of the photons. This leads to

A. Sergienko, S. Pascazio, and P. Villoresi (Eds.): QuantumCom 2009, LNICST 36, pp. 58–65, 2010.
© Institute for Computer Sciences, Social-Informatics and Telecommunications Engineering 2010

the concept of *time bin*: two photons whose arrival time cannot be distinguished by the detectors belong to the same time bin. Third, different time bins are made to interfere by using unbalanced Mach-Zehnder interferometers. Together, these ideas provide a powerful platform to investigate and manipulate quantum entanglement, yielding seminal works such as long distance violation of Bell inequalities [7] and entanglement based QKD [11].

Here we use the same time-energy degree of freedom as in [5,6,7,8,9,10,11], but the way it is manipulated is very different. However at the conceptual level there is an instructive parallel between our approach and that of Franson. First, a narrowband pump laser produces frequency entangled photon pairs: the frequency of each photon is uncertain, but the sum of the frequencies is well defined. Second, we use narrowband filters that resolve the frequency of the detected photons. This leads to the concept of *frequency bin*: two photons whose frequency is so close that they cannot be distinguished are said to lie in the same frequency bin. Third, different frequency bins are made to interfere by using electro-optic phase modulators.

Our approach is inspired by, or related to, several earlier proposals for manipulating qubits in the frequency domain [12,13,14,15,16,17,18]. The experimental techniques we use follow closely those of QKD systems in which the quantum information is encoded in frequency sidebands of an attenuated coherent state [12,13,14,15]. Such systems allow efficient transmission of quantum information at telecommunication wavelengths and constitute a commercial alternative to time bin based QKD. The quantum state can be manipulated with either intensity [12,13] or phase [14] modulators. Here we transpose the setup of [14] to the entangled photon case, thereby showing how phase modulators can manipulate frequency entangled photons. In view of the proven success of this method for QKD, we believe this is a very promising approach for manipulating entangled photons, particularly in the context of long distance quantum communication at telecommunication wavelengths.

In the following we first describe our experiment and give the principle of our method. After presenting our experimental results, we discuss how, using this approach, one should be able to violate Bell inequalities.

2   Experimental Setup

Our experiment is schematized in Fig. 1. Photon pairs are generated by parametric down conversion in a 4 cm long Periodically Poled Lithium Niobate (PPLN) waveguide (HC Photonics). PPLN waveguides have emerged as the preferred photon pair source at telecommunication wavelengths because of their extremely high spectral brightness [19,20]. The narrowband pump (Sacher Lasertechnik, 20s-linewidth $\approx$ 2 MHz, λ_p = 776.1 nm, $P_p \approx$ 2 mW) is removed with a drop filter F insuring more than 30 dB isolation. The pairs, distributed around λ_0 = 1552.2 nm = $2\pi c/\omega_0$, are sent through a Polarization Controller (PC) and then separated with a polarization maintaining 3dB-coupler. Interesting cases occur when the photon pair is split: one photon is sent to Alice (A) and the other

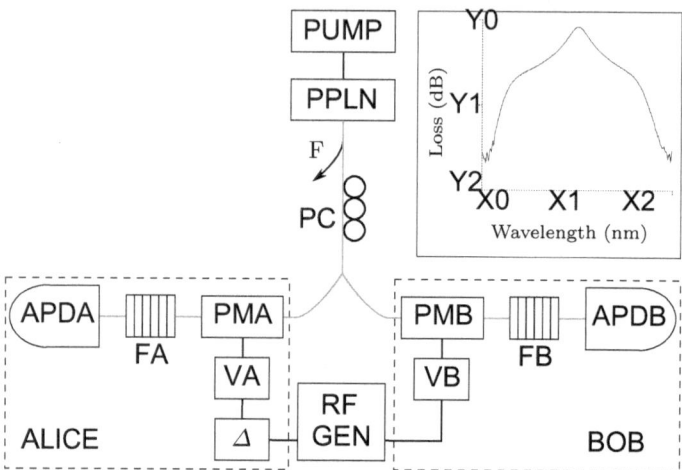

Fig. 1. Experimental setup. The quasi-monochromatic pump laser (PUMP) creates photon pairs in the periodically poled lithium niobate waveguide (PPLN) and is then removed by a filter (F). A fiber polarization controller (PC) is used to align the polarization of the photons with the active axis of electro-optic phase modulators (PM$_{A,B}$). The frequency bins are then selected by narrowband filters (F$_{A,B}$), whose typical transmission spectrum is shown in inset. The photons are detected by avalanche photo-diodes (APD$_{A,B}$). The phase modulators are driven by a 18 GHz radio frequency generator (RF GEN) whose output is controlled by variable attenuators (V$_{A,B}$) and a phase shifter (Δ).

to Bob (B). At the output of the coupler, the photons pass through electro-optic Phase Modulators (PM$_{A,B}$) (EOSPACE, 25 GHz bandwidth, 2.5 dB loss) to which are applied sinusoidally varying voltages at frequency $\Omega/2\pi = 18$ GHz (chosen to provide 20 dB isolation between each frequency bin), with amplitudes V$_{A,B}$ and phases α, β whose difference $\Delta = \alpha - \beta$ can be controlled. The induced time dependent optical phases $\phi_A(t) = a\cos(\Omega t - \alpha)$ and $\phi_B(t) = b\cos(\Omega t - \beta)$, where $a = \pi V_A/V_\pi$, $b = \pi V_B/V_\pi$, and V_π is the half-wave voltage of the modulators, lead to the unitary transformations

$$|\omega\rangle \rightarrow \sum_p |\omega + p\Omega\rangle U_p(c, \theta), \tag{1}$$

where $U_p(c, \theta) = J_p(c)e^{ip(\theta - \pi/2)}$, $c = a$ or b, $\theta = \alpha$ or β, and J_p is the pth-order Bessel function of the first kind. The photons are then sent through narrowband tunable filters (F$_{A,B}$) (AOS GmbH, FWHM $\approx$ 18 pm, 20 (40) dB isolation at 9 (18) GHz, tuning range $\approx$ 400 pm) centered on angular frequencies $\omega_{A,B}$ – in our case taken near the degeneracy point ω_0. The use of such narrowband filters together with a spectrally bright PPLN source of entangled photons has been reported previously in the context of four-photon experiments [21,20]. Finally the photons are detected by two Avalanche Photo-Diodes (APD$_{A,B}$) (id Quantique) and a time-to-amplitude converter performs a coincidence measurement. The maximum coincidence rate was approximately 0.5 Hz.

3   Theoretical Modelisation and Experimental Results

A parametric down conversion source pumped by a monochromatic beam produces an entangled state (subscripts A and B refer to Alice and Bob's photon)

$$|\Phi\rangle = \int d\omega |\omega_0 + \omega\rangle_A |\omega_0 - \omega\rangle_B, \qquad (2)$$

in which the total energy of the photon pair is well defined, but the energy of each photon is uncertain. The narrowband filters select the photons belonging to a small frequency interval. We view all the photons selected by a given filter as belonging to a specific *frequency bin*. The effect of the filters can be idealized by the state

$$|\Psi\rangle = \sum_k |\omega_0 + k\Omega\rangle_A |\omega_0 - k\Omega\rangle_B, \qquad (3)$$

which would correspond to the case of infinitely narrow filters equally spaced in frequency.

Note that the amount of entanglement in Eq. (2) would appear to be infinite. In practice it is approximately given by the ratio of the linewidth of the pairs to the linewidth of the pump laser and can in principle reach very high values, for instance 4 THz / 2 MHz $= 2\ 10^6$ in our case. Note also that for simplicity of notation we have not normalized Eqs. (2, 3). This does not affect our predictions as we are in fact interested in the *ratios* of the probabilities of finding photon A at one frequency and photon B at another frequency. Note finally that taking the Fourier transform of Eqs. (2, 3) would yield a description of the state in terms of time entanglement: the time of arrival of each photon is uncertain, but the difference between the time of arrival of Alice and Bob's photon is well defined.

According to Eq. (1), the phase modulators realize interferences between frequency bins separated by integer multiples of Ω. They thus play the same conceptual role as the Mach-Zehnder interferometers in Franson's scheme which realize interferences between different time bins. Using Eqs. (1, 3), one can readily compute the entangled state after the phase modulation

$$|\Psi\rangle \rightarrow \sum_{n,d} |\omega_0 + (n+d)\Omega\rangle_A |\omega_0 - n\Omega\rangle_B \ c_{nd}(a, b, \alpha, \beta), \qquad (4)$$

with $c_{nd}(a, b, \alpha, \beta) = \sum_p U_p(a, \alpha) U_{d-p}(b, \beta)$. Because of the symmetries of Eqs. (1, 3), the probability $\check{P}(n+d, -n|a, b, \alpha, \beta)$ of A and B detecting photons at angular frequencies $\omega_A = \omega_0 + (n+d)\Omega$ and $\omega_B = \omega_0 - n\Omega$ depends only on the sum of the frequencies $\omega_A + \omega_B = 2\omega_0 + d\Omega$ (but not on n) and on the phase difference Δ (but not on $\alpha + \beta$):

$$P(d|a, b, \Delta) = P(n+d, -n|a, b, \alpha, \beta) = |c_{nd}(a, b, \alpha, \beta)|^2. \qquad (5)$$

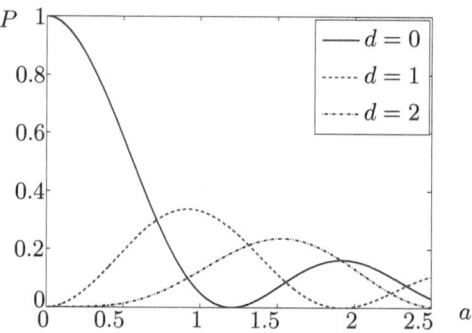

Fig. 2. Theoretical predictions for the probability $P(d|a, a, 0)$ of detecting photons with the sum of the frequencies given by $2\omega_0 + d\Omega$ for $d = 0$, 1, 2, when $a = b$ and $\Delta = 0$

As illustration, the predictions of Eq. (5) are plotted in Fig. 2 when $\Delta = 0$ as a function of the modulation amplitudes $a = b$ (taken to be equal) for $d = 0$, 1, 2. The amount of entanglement that is manipulated by the phase modulators, i.e. the number of frequency bins that interfere together, is approximately given by the number of values of d for which Eq. (5) takes a significant value, and increases when a, b increase.

In Fig. 3, we compare the prediction of Eq. 5 to our experimental results obtained when the phase Δ is scanned, with $a \approx b \approx 1.3$. This value is chosen so that $P(d = 0|a, a, 0)$ almost vanishes, leading to significant $d \neq 0$ contributions. The experimental visibility of interferences is approximately $V = 88\%$. We have also measured coincidence rates when $d = 1$ or 2. From these curves one can deduce that we are manipulating at least five-dimensional entanglement – otherwise the curve for $d = 2$ should be flat.

Two remarks about Fig. 3 are in order. First, we believe that noise and limited visibility are mainly due to drift of pump laser frequency and imperfect polarization control. Second, when changing Δ, because of reflections and interferences in the RF circuit, the amplitudes a and b also changed by up to 30%. This effect was carefully measured, and the curves in Fig. 3 take this into account. (This is why the theoretical curves are not perfectly symmetric with respect to $\Delta = \pi$).

4   Towards Bell Tests

We now discuss how our system can be used to realize quantum non locality experiments. Note that each measurement in principle produces an infinite number of possible outcomes, corresponding to the frequency bin in which the photon is found. (Of course in practice this is limited by the bandwidth of the photon pairs). For this reason it seems more appropriate to consider Bell inequalities adapted to high dimensional systems, such as the CGLMP inequality [22],

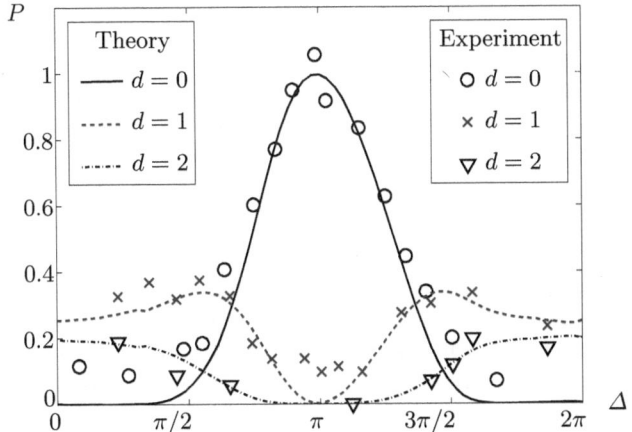

Fig. 3. Theoretical predictions (curves) and experimental measurements (symbols) of the coincidence probability $P(d|a, b, \Delta)$ when $a \approx b \approx 1.3$ and the phase Δ is scanned, for $d = 0, 1, 2$. Experimentally, P is the normalized coincidence rate, i.e. the ratio of true coincidences to accidental coincidences divided by this ratio for $d = 0$ when modulation is off. The acquisition time per point was constant, corresponding to a number of recorded coincidences approximately equal to $10^3 P$.

rather than Bell inequalities for two dimensional systems such as the Clauser-Horn-Shimony-Holt (CHSH) inequality [23]. We compare both cases below.

The three Bell expressions we will consider can all be written as a combination of four terms

$$S = C(A_1 B_1) + C(B_2 A_1) + C(B_1 A_2) - C(B_2 A_2), \tag{6}$$

where $C(A_i B_j)$, $C(B_j A_i)$ are the *generalized correlators* when Alice chooses measurement setting A_i and Bob chooses measurement setting B_j. By measurement setting we mean a choice of amplitude a (or b) and phase α (or β). The correlators are chosen so that $-1 \leq C(AB), C(BA) \leq +1$, and so that local hidden variable (LHV) theories all satisfy $S_{LHV} \leq 2$.

Let us first consider how the outcomes can be grouped to yield the CHSH expression. We associate with frequency bin $\omega_A = \omega_0 + (n + d)\Omega$ the variable $x = (-1)^{n+d}$, i.e. $x = +1 \ (-1)$ if $n + d$ is even (odd); and similarly we associate with frequency bin $\omega_B = \omega_0 - n\Omega$ the variable $y = (-1)^n$. The value of the correlator is taken to be the product $C_{CHSH} = xy$. In terms of the original variables n, d it can be written as $C_{CHSH}(AB) = P(d = 0 \bmod 2|AB) - P(d = 1 \bmod 2|AB) = C_{CHSH}(BA)$.

A second approach is to use the CGLMP inequality. The infinite dimensional limit of the inequality given in [22,24] corresponds to choosing $C_{CGLMP}(AB) = P(d \geq 0|AB) - P(d < 0|AB)$ and $C_{CGLMP}(BA) = P(d \leq 0|AB) - P(d > 0|AB)$.

A third approach is to define the correlators as

$$C_{EQ}(AB) = P(d = 0|AB) - P(d \neq 0|AB) = C_{EQ}(BA). \qquad (7)$$

Inserting these different choices of correlators in Eq. (6) yields three inequivalent Bell expressions. To our knowledge the one based on C_{EQ} has not been considered before in the literature, but it arose naturally from the symmetries of the present problem.

In Fig. 4 we compare how much the correlations Eq. (5) can violate each of these inequalities. We choose for simplicity the modulation amplitudes a_1, a_2, b_1, b_2 to be equal and numerically optimize the phases α_1, α_2, β_1, β_2. Surprisingly we find that the CGLMP expression yields the smallest violation even though it was introduced specifically for high dimensional systems. We attribute this to the fact that the set of measurements explored in the present work is too restricted. On the other hand the largest violation is obtained for the new inequality based on the correlators C_{EQ}.

We have not computed how much the data given in Fig. 3 would violate the above Bell inequalities. The reason is that, as mentioned above, when changing the phase α of Alice's phase modulator, we simultaneously changed the amplitude and phase of Bob's phase modulator. This violates the no-signalling conditions necessary for investigating quantum non locality. Nevertheless, we expect that our setup can significantly violate the above Bell inequalities. Indeed the amount of noise reported in Fig. 3 should yield a Bell expression approximately equal to the maximum value times the visibility $V(= 0.88)$, which for $a = 1.3$ is well above the LHV bound of 2.

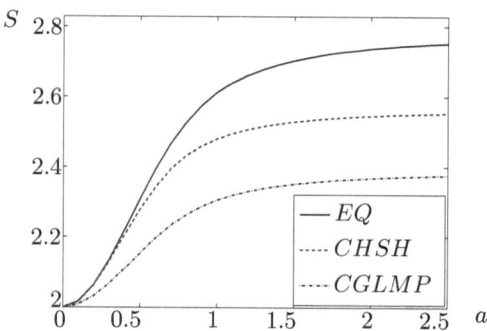

Fig. 4. Theoretical predictions for the violation of the CHSH, the CGLMP, and the EQ Bell inequalities as a function of the amplitude of the modulation a. The settings were chosen to be of the form $a_1 = a_2 = b_1 = b_2 = a$ and the phases $\alpha_{1,2}$ and $\beta_{1,2}$ were numerically optimized.

5 Conclusion

In summary we have demonstrated the manipulation of high dimensional frequency bin entanglement. In view of the proven success [12,13,14,15] of frequency encoding for long distance QKD, this seems a promising technique for quantum communication tasks. Future work will focus on studying the Bell inequalities discussed in the text, and demonstrating entanglement based QKD.

We acknowledge support from the European Union under project QAP (contract 015848), from the Belgian Science Policy under project IAP-P6/10 (Photonics@be), from the French Agence Nationale de la Recherche under project HQNET and from the Conseil Régional de Franche-Comté. This work also benefits from the Programme International de Coopération Scientifique PICS-3742 of the French Centre National de la Recherche Scientifique.

References

1. Aspect, A., Grangier, P., Roger, G.: Phys. Rev. Lett. 47, 460 (1981); 49, 91 (1982)
2. Kwiat, P.G., et al.: Phys. Rev. Lett. 75, 4337 (1995)
3. Rarity, J.G., Tapster, P.R.: Phys. Rev. Lett. 64, 2495 (1990)
4. Mair, A., Vaziri, A., Weihs, G., Zeilinger, A.: Nature 412, 313 (2001)
5. Kwiat, P.G., Steinberg, A.M., Chiao, R.Y.: Phys. Rev. A 47, R2472 (1993)
6. Brendel, J., Mohler, E., Martienssen, W.: Europhys. Lett. 20, 575 (1992)
7. Tittel, W., Brendel, J., Zbinden, H., Gisin, N.: Phys. Rev. Lett. 81, 3563 (1998)
8. Brendel, J., Gisin, N., Tittel, W., Zbinden, H.: Phys. Rev. Lett. 82, 2594 (1999)
9. Thew, R.T., Tanzilli, S., Tittel, W., Zbinden, H., Gisin, N.: Phys. Rev. A 66, 062304 (2002)
10. Franson, J.D.: Phys. Rev. Lett. 62, 2205 (1989)
11. Tittel, W., Brendel, J., Zbinden, H., Gisin, N.: Phys. Rev. Lett. 84, 4737 (2000)
12. Merolla, J.-M., Mazurenko, Y., Goedgebuer, J.-P., Porte, H., Rhodes, W.T.: Opt. Lett. 24, 104 (1999)
13. Merolla, J.-M., Mazurenko, Y., Goedgebuer, J.-P., Rhodes, W.T.: Phys. Rev. Lett. 82, 1656 (1999)
14. Bloch, M., McLaughlin, S.W., Merolla, J.-M., Patois, F.: Opt. Lett. 32, 301 (2007)
15. Cussey, J., Patois, F., Pelloquin, N., Merolla, J.-M.: Proceedings of the Optical Fiber Communication and National Fiber Optic Engineers Conference (2008)
16. Stefanov, A., Zbinden, H., Gisin, N., Suarez, A.: Phys. Rev. A. 67, 042115 (2003)
17. Huntington, E.H., Ralph, T.C.: Phys. Rev. A. 69, 042318 (2004)
18. Huntington, E.H., et al.: Phys. Rev. A. 71, 041802(R) (2005)
19. Tanzilli, S., et al.: Electron. Lett. 37, 26 (2001)
20. Halder, M., et al.: New J. Phys. 10, 023027 (2008)
21. Halder, M., et al.: Nat. Phys. 3, 692 (2007)
22. Collins, D., Gisin, N., Linden, N., Massar, S., Popescu, S.: Phys. Rev. Lett. 88, 040404 (2002)
23. Clauser, J.F., Horne, M.A., Shimony, A., Holt, R.A.: Phys. Rev. Lett. 23, 880 (1969)
24. Zohren, S., Gill, R.D.: Phys. Rev. Lett. 100, 120406 (2008)

Ground-State Entanglement Gives Birth to Quantum Energy Teleportation

Masahiro Hotta

Tohoku University, Sendai Miyagi-ken 980-8578, Japan
hotta@tuhep.phys.tohoku.ac.jp

Abstract. Ground-state entanglement induces emergence of negative-energy-density regions in quantum systems by squeezing zero-point oscillation, keeping total energy of the systems nonnegative. By use of the negativity of quantum energy density, protocols of quantum energy teleportation are proposed that transport energy to distant sites by local operations and classical communication. The energy is teleported without breaking any physical laws including causality and local energy conservation. Because intermediate subsystems of the energy transfer channel are not excited during the protocol execution, the protocol attains energy transportation without heat generation in the channel. We discuss the protocol focusing around qubit chains. In addition, we address a related problem of breaking ground-state entanglement by measurements.

Keywords: Entanglement, quantum teleportation, LOCC, quantum measurement.

1 Introduction

Recently protocols called quantum energy teleportation (QET) have been proposed which transport energy by local operations and classical communication (LOCC), respecting causality and local energy conservation. The protocols can be considered for various many-body quantum systems, including qubit chains [1,2], 1+1 dimensional massless Klein-Gordon fields [3], 1+3 dimensional electromagnetic field [4], and cold trapped ions [5]. The key point of the protocol is that there exists quantum correlation between local fluctuations of different sites in the ground state. The root of this correlation is the ground-state entanglement. By virtue of the correlation, a measurement result of local fluctuation in some site includes information about fluctuation in other sites. By selecting and performing a proper local operation based on the announced information, zero-point oscillation of a site far from the measurement site can be more suppressed than that of the ground state, yielding negative energy density. Here the origin of energy density is fixed such that the expectational value vanishes for the ground state. Such negative energy density appears due to quantum interference effects [6]. Even if we have a region with negative energy density in a system, we have other regions with positive energy density and the total energy of the system remains nonnegative. During the above local operation generating

A. Sergienko, S. Pascazio, and P. Villoresi (Eds.): QuantumCom 2009, LNICST 36, pp. 66–73, 2010.

negative energy density in the system, surplus energy is transferred from the quantum fluctuation to external systems and can be harnessed.

The organization of this report is as followed: In section 2, the relation between ground-state entanglement and emergence of negative energy density is explained. QET is realized by generating negative energy density at a distant site by LOCC. In section 3, a protocol of this QET is discussed for critical Ising spin chains. In section 4, a related problem of breaking ground-state entanglement by measurements are addressed. In section 5, recent results of QET analysis are summarized for other quantum systems.

2   Ground-State Entanglement and Negative Energy Density

The QET protocol is able to work by virtue of ground-state entanglement and emergence of negative energy density. In what follows, let us concentrate on qubit chain systems and explain the entanglement and the negative energy density. First of all, the Hamiltonian H is given by a site sum of energy density operators T_n, where n denotes site number. The origin of T_n can be shifted so as to satisfy

$$\langle g|T_n|g\rangle = 0 \tag{1}$$

without loss of generality. If each T_n is a local operator at site n satisfying $[T_n, T_{n'}] = 0$, all T_n can be simultaneously diagonalized. The ground state $|g\rangle$ becomes separable and an eigenstate for the lowest eigenvalue of each T_n. Clearly, in such a situation, T_n is nonnegative. However, the condition $[T_n, T_{n'}] = 0$ is not sustained for cases with interactions between qubits, and entangled ground states are generated. It is noted that a correlation function $\langle g|T_n O_m|g\rangle$ of a separable ground state $|g\rangle$ is given by $\langle g|T_n|g\rangle\langle g|O_m|g\rangle$ for a local operator O_m at site m apart far from n. On the other hand, in the case of the entangled ground state $|g\rangle$, this factorization relation does not hold in general:

$$\langle g|T_n O_m|g\rangle \neq \langle g|T_n|g\rangle\langle g|O_m|g\rangle. \tag{2}$$

This ground-state entanglement induces emergence of quantum states with negative energy density as follows. It turns out first that the entangled ground state $|g\rangle$ cannot be an eigenstate of T_n. The reason is following. If the eigenvalue equation $T_n|g\rangle = \tau|g\rangle$ with a real eigenvalue τ is satisfied, the above correlation function must be written as

$$\langle g|T_n O_m|g\rangle = \tau\langle g|O_m|g\rangle = \langle g|T_n|g\rangle\langle g|O_m|g\rangle,$$

where we have used $\langle g|T_n = \tau\langle g|$ and $\tau = \langle g|T_n|g\rangle$. This obviously contradicts Eq. (2). Therefore the entangled ground state $|g\rangle$ satisfying Eq. (2) is not an eigenstate of T_n. Next let us spectral-decompose the operator T_n as

$$T_n = \sum_{\nu,k_\nu} \epsilon_\nu(n)|\epsilon_\nu(n), k_\nu(n)\rangle\langle\epsilon_\nu(n), k_\nu(n)|,$$

where $\epsilon_\nu(n)$ are eigenvalues of T_n, $|\epsilon_\nu(n), k_\nu(n)\rangle$ are corresponding eigenstates, and the index $k_\nu(n)$ denotes the degeneracy freedom of the eigenvalue $\epsilon_\nu(n)$. Because $\{|\epsilon_\nu(n), k_\nu(n)\rangle\}$ is a complete set of orthonormal basis vectors of the total Hilbert space of the qubit chain, the ground state can be uniquely expanded as

$$|g\rangle = \sum_{\nu, k_\nu(n)} g_{\nu, k_\nu(n)} |\epsilon_\nu(n), k_\nu(n)\rangle,$$

where $g_{\nu, k_\nu(n)}$ are complex coefficients of the expansion. By use of this expansion, Eq. (1) gives an equation as follows:

$$\langle g|T_n|g\rangle = \sum_{\nu, k_\nu(n)} \epsilon_\nu(n) \left| g_{\nu, k_\nu(n)} \right|^2 = 0.$$

Clearly, this equation for $g_{\nu, k_\nu(n)}$ has no solution when the lowest eigenvalue $\epsilon_{\min}(n)$ of T_n is positive. The case with $\epsilon_{\min}(n) = 0$ is also prohibited for the equation because, if so, the entangled ground state $|g\rangle$ would become an eigenstate of T_n with its eigenvalue $\tau = 0$ and contradicts Eq. (2), as proven above. This means that $\epsilon_{\min}(n)$ must be negative. It is thereby verified that there exist quantum states $|\epsilon_{\min}(n), k_{\min}(n)\rangle$ with negative energy density due to the ground-state entanglement. Here it should be stressed that, because of Eq. (1), the eigenvalue of the ground state is zero:

$$H|g\rangle = 0,$$

and H is a nonnegative operator. Therefore, even if we have a region with negative energy density in a system, we have other regions with positive energy density so as to make the total energy of the system nonnegative. In the QET protocol, the negative energy density plays a crucial role as seen in the next section.

3   QET Protocol

By use of the negative energy density, protocols of QET can be constructed. In this section, a QET protocol for a critical Ising spin chain [2] is explained. The Hamiltonian is given by a sum of energy density operator T_n: $H = \sum_n T_n$. The operator T_n is given by

$$T_n = -J\sigma_n^z - \frac{J}{2}\sigma_n^x \left(\sigma_{n+1}^x + \sigma_{n-1}^x\right) - \epsilon, \tag{3}$$

where σ_n^z and σ_n^x are Pauli matrices at site n, J and ϵ are real constants. By fine-tuning ϵ, Eq. (1) is attained. The QET protocol is composed of the following three steps: (i) For the ground state $|g\rangle$, an energy sender A performs a local measurement of σ_A which is a one-direction component of the Pauli spin operator acting on A's qubit. Those eigenvalues of σ_A are $(-1)^\mu$ with $\mu = 0, 1$. Let us write the spectral decomposition of σ_A as

$$\sigma_A = \sum_{\mu=0,1} (-1)^\mu P_A(\mu),$$

where the operator $P_A(\mu)$ are projective operators onto the eigenspaces. In this measurement process, A must input positive amount of energy given by

$$E_A = \sum_{\mu=0,1} \langle g | P_A(\mu) H P_A(\mu) | g \rangle$$

to the qubit chain. (ii) A announces the measurement result μ to an energy receiver B by a classical channel. (iii) B performs a local unitary operation depending on the value of μ. The unitary operator is defined by

$$V_B(\mu) = I \cos \theta + i (-1)^\mu \sigma_B \sin \theta,$$

where σ_B is a one-direction component of the Pauli spin operator acting on B's qubit, and the above real parameter θ is fixed so as to extract the maximum energy from the chain. In this analysis, we assume that dynamical evolution of the system induced by H is negligible during short time interval t of the protocol: $\exp[-itH] \sim I$. Hence, the quantum state after step (iii) is written as follows.

$$\rho = \sum_{\mu=0,1} V_B(\mu) P_A(\mu) | g \rangle \langle g | P_A(\mu) V_B^\dagger(\mu).$$

Using this state, it can be shown that B extracts positive energy $+E_B$ on average from the qubit chain, accompanied by excitations with negative energy $-E_B$ in the qubit chain around B's site in step (iii). In fact, the expectational value of energy after step(iii) is calculated [1] as

$$\mathrm{Tr}\,[\rho H] = E_A + \frac{\eta}{2} \sin(2\theta) + \frac{\xi}{2} (1 - \cos(2\theta)), \tag{4}$$

where ξ and η are given by

$$\xi = \langle g | \sigma_B H \sigma_B | g \rangle \geq 0,$$
$$\eta = i \langle g | \sigma_A [H,\ \sigma_B] | g \rangle.$$

The coefficient η is a two-point correlation function of (semi-)local operators of A and B, and turns out to be real. It is a key point that η does not vanish in general because of the ground-state entanglement. By taking a value of θ defined by

$$\cos(2\theta) = \frac{\xi}{\sqrt{\xi^2 + \eta^2}}, \quad \sin(2\theta) = -\frac{\eta}{\sqrt{\xi^2 + \eta^2}},$$

the minimum value of $\mathrm{Tr}\,[\rho H]$ with respect to θ is written explicitly as

$$\mathrm{Tr}\,[\rho H] = E_A - \frac{1}{2} \left[\sqrt{\xi^2 + \eta^2} - \xi \right].$$

From the viewpoint of local energy conservation, this result implies that, during the operation $V_B(\mu)$, positive amount of energy given by

$$E_B = E_A - \mathrm{Tr}\,[\rho H] = \frac{1}{2} \left[\sqrt{\xi^2 + \eta^2} - \xi \right] > 0 \tag{5}$$

is transferred from the qubit chain to external systems including the device system executing $V_B(\mu)$. In addition, it is possible to calculate analytically the value of E_B for the critical Ising spin chain as follows [2].

$$E_B = \frac{2J}{\pi}\left[\sqrt{1 + \left(\frac{\pi}{2}\Delta(|n_A - n_B|)\right)^2} - 1\right], \tag{6}$$

where $\Delta(n)$ is defined by

$$\Delta(n) = \left(\frac{2}{\pi}\right)^n \frac{2^{2n(n-1)}h(n)^4}{(4n^2 - 1)h(2n)}$$

with $h(n) = \prod_{k=1}^{n-1} k^{n-k}$. The asymptotic behavior of $\Delta(n)$ for large n is given by

$$\Delta(n \sim \infty) \sim \frac{1}{4}e^{1/4}2^{1/12}c^{-3}n^{-9/4}, \tag{7}$$

where the constant c is evaluated as $c \sim 1.28$. Due to the criticality of this model, E_B decays following not an exponential law but a power law ($\propto |n_A - n_B|^{-9/2}$) for large separation.

4   Breaking Ground-State Entanglement by Measurements

In section 3, we have shown that B obtains energy from the qubit chain by the QET protocol. However, even after the last step (iii) of the protocol, there exists residual energy E_A that A had to first deposit to the qubit chain. Let us imagine that A attempts to completely withdraw E_A by local operations after step (iii). If A succeeded in this withdrawing, the energy gain of B might have no cost. However, if so, the total energy of the qubit chain became equal to $-E_B$ and negative. Meanwhile, we know that the total energy of the qubit chain system must be nonnegative. Hence, A cannot withdraw energy larger than $E_A - E_B$ by local operations at site n_A. This means that, in the QET protocol, B has borrowed energy E_B in advance from the qubit chain on security of the deposited energy E_A. The main reason for A's inability to withdraw is because A's local measurement breaks the ground-state entanglement between A's qubit and all the other qubits. The post-measurement state is an exact separable state with no entanglement. If A wants to recover the original state of her qubit with zero energy density, A must recreate the broken entanglement. However, entanglement generation needs nonlocal operations in general. Therefore, A cannot recover the state perfectly by her local operations alone. This interesting aspect poses a residual-energy problem of the ground-state entanglement broken by measurements. Let us imagine that A stops the QET protocol soon after step (i) of the protocol, and attempts to completely withdraw E_A by local operations. By the same argument as the above, it is shown that this attempt never succeeds because A breaks the ground-state entanglement. Of course, for a long time interval beyond the short time scale that we have considered, local cooling is naturally expected to make residual energy in the qubit chain approaching zero by an assist of dynamical evolution induced by the nonlocal Hamiltonian H. However, in this short time interval, the dynamical evolution is not available. Therefore it is concluded that the residual energy in the qubit chain has its nonvanishing minimum value E_r with respect to A's local cooling processes in short time. In order to make the argument more concrete, let us consider a general local cooling operation of A after step (i) obtaining the measurement result μ. The operation is expressed by use of μ-dependent Kraus operators $M_A(\alpha, \mu)$ satisfying

$$\sum_\alpha M_A^\dagger(\alpha, \mu) M_A(\alpha, \mu) = I. \tag{8}$$

Then the quantum state after this local cooling by A is given by

$$\rho_c = \sum_{\mu, \alpha} M_A(\alpha, \mu) P_A(\mu) |g\rangle\langle g| P_A(\mu) M_A^\dagger(\alpha, \mu). \tag{9}$$

The minimum value E_r of the residual energy with respect to $M_A(\alpha, \mu)$ satisfying Eq. (8) is written as

$$E_r = \min_{\{M_A(\alpha, \mu)\}} \mathrm{Tr}\,[\rho_c H]. \tag{10}$$

Evaluation of E_r is performed analytically in the Ising spin chains [2] and given by

$$E_r = \left(\frac{6}{\pi} - 1\right) J > 0,$$

for the critical chain. Surprisingly, A is not able to extract this energy by any local operation in the short time, though it exists in front of A. Because of the nonnegativity of H, it is easily checked by resuming the QET protocol after the local cooling that E_r is lower bounded by the teleported energy E_B in Eq. (5). In addition, the paper [2] gives a stringent argument that the teleported energy in an extended protocol gives a more tight lower bound of residual energy E_r for general qubit chains.

Finally, a comment is added about recent numerical researches of the ground-state entanglement. As a quantitative entanglement measure, the negativity has been computed between separated blocks of qubit chains [7] (the logarithmic negativity for harmonic oscillator chains [8]) showing that at criticality this negativity is a function of the ratio of the separation to the length of the blocks and can be written as a product of a power law and an exponential decay. In our setting of QET, this suggests that change of the entanglement between A's block and B's block after A's local measurement has a similar rapid-decay dependence on the separation with a fixed block length. Thus it may be concluded that the entanglement between A's block and B's block itself is not essential for QET. Though the entanglement between the two blocks may be rapidly damped, E_B shows a power law decay ($\propto n^{-9/2}$) for large separation n, as seen in Eq. (6)

and Eq. (7). This implies in a sense that almost "classical" correlation between A's block and B' block is sufficient to execute QET for large separation, and is expected to be robust against environment disturbance, contrasting to the entanglement fragility. It should be emphasized, however, that this "classical" correlation is originally induced by the ground-state entanglement characterized by Eq. (2). If the ground state is separable, we have no correlation between the blocks.

5   QET for Other Systems

The QET protocols can be considered for other quantum systems. In [3], a protocol of QET for 1+1 dimensional massless scalar fields is analyzed. Though the nonrelativistic treatment for the qubit chain in section 3 is valid for short-time-scale processes of QET in which dynamical evolution induced by the Hamiltonian is negligible, in this relativistic case, the dynamical effect propagates with light velocity, which is the upper bound on the speed of classical communication. Thus, we generally cannot omit global time evolution. It is also noted that any continuous limit of zero lattice spacing cannot be taken for the protocols in the lattice QET models as long as measurements in the protocols are projective, which becomes an obstacle to obtaining a smooth limit. Therefore, in [3], A makes not a projective but instead a well-defined POVM measurement to the vacuum state of the field. After wavepackets with light velocity excited by A's measurement have already passed by the position of B, B extracts energy from the local vacuum state of the field by a unitary operation dependent on the measurement result announced by A. In [4], two QET protocols with discrete and continuous variables are analyzed for 1+3 dimensional electromagnetic field. In the discrete case, a 1/2 spin is coupled with the vacuum fluctuation of the field and measured in order to get one-bit information about the fluctuation. In the continuous case, a harmonic oscillator is coupled with the fluctuation and measured in order to get continuous-variable information about the fluctuation. In the discrete case, the amount of the extracted energy is suppressed by an exponential damping factor when the energy infused by the measurement becomes large. This suppression factor becomes power damping in the continuous case, and it is concluded that more information about the vacuum fluctuation is obtained by the measurement, more energy can be teleported. In [5], a protocol of QET is proposed for trapped ions. N cold ions, which are strongly bound in the y and z directions but weakly bound in an harmonic potential in the x direction, form a linear ion crystal. The first ion that stays at the left edge of the crystal is the gateway of the QET channel where energy is input. The N-th ion that stays at the right edge of the crystal is the exit of the QET channel where the teleported energy is output. Two internal energy levels of the gateway ion are selected and regarded as energy levels of a probe qubit to measure the local phonon fluctuation. The probe qubit is strongly coupled with the phonon fluctuation in the ground state during short time via laser field and is projectively measured. In the measurement models, the kinetic energy of the gateway ion increases after the measurement, but the kinetic energy of other ions and the potential energy of all the ions remain unchanged. The obtained information is announced through a classical channel from the gateway point to the exit point. The speed of the information transfer can be equal to the speed of light in principle, which is much faster than that of the phonon propagation in the ion crystal. The phonons excited at the QET gateway do not arrive at the exit point yet when the information arrives at the exit point. However, by using the announced information, we are able to soon extract energy from the exit ion. Experimental verification of the QET mechanism has not been achieved yet for any system, and is a quite stimulating open problem.

Acknowledgments

I would like to M. Ozawa and A. Furusawa for fruitful discussions. This research is partially supported by the SCOPE project of the MIC and the Ministry of Education, Science, Sports and Culture of Japan, No. 21244007.

References

1. Hotta, M.: J. Phys. Soc. Japan. 78, 034001 (2009)
2. Hotta, M.: Phys. Lett. A 372, 5671 (2008)
3. Hotta, M.: Phys. Rev. D78, 045006 (2008); Hotta, M.: Controlled Hawking Process by Quantum Information, arXiv:0907.1378
4. Hotta, M.: Quantum Energy Teleportation with Electromagnetic Field: Discrete vs. Continuous Variables, arXiv:0908.2674
5. Hotta, M.: Phys. Rev. A 80, 042323 (2009)
6. Ford, L.H.: Proc. R. Soc. (London) A 346, 227 (1978); Birrell, N.D., Davies, P.C.W.: Quantum Fields in Curved Space. Cambridge Univ. Press, Cambridge (1982)
7. Wichterich, H., Molina-Vilaplana, J., Bose, S.: Phys. Rev. A. 80, 010304(R) (2009)
8. Marcovitch, S., Retzker, A., Plenio, M.B., Reznik, B.: Phys. Rev. A 80, 012325 (2009)

Network Games with Quantum Strategies

Giannicola Scarpa

Dipartimento Informatica e Applicazioni
Università di Salerno, Italy
giannicolascarpa@gmail.com

Abstract. Recently, Physics and Computer Science have both contributed to the field of Game Theory. The first by introducing quantum information concepts into the strategy set of the players and the second by giving methods to estimate the efficiency of equilibria and formalizing "network games". This work will aim to be a first step towards the merging of those existing ideas, by studying the behavior of players in a quantum network. It will focus on two classes of network games: formation and congestion games, showing that in some cases "classical" examples of inefficiency fail if players use quantum strategies.

Keywords: Game theory, network games, EWL protocol, quantum strategies, efficiency of equilibria, selfish routing, network formation.

1 Introduction

In game theory, a game is the study of competitive situations, formalized by a set of players (or agents) each one declaring a strategy chosen among a set of possible moves, and each one receiving a payoff based on the overall results. The outcome of the game can be predicted using the Nash Equilibrium concept, that is to say when no player can increase his utility by deviating from his initially declared strategy, while the other players' moves remain fixed. A mixed-strategy NE is similar to that, but instead of choosing a single (pure) strategy players declare a probability distribution on the strategy set.

The EWL protocol, proposed by Eisert et al. in [12], allows players to represent strategies using qubits, and sets up a classically impossible coordination creating *entanglement* between players' qubits. It was limited to 2x2 games (2 players, 2 strategies), but extensions to n-player case were discussed by Benjamin and Hayden in [14], and by many authors in following studies.

Starting from those results, this work extends studies about Network Games by Roughgarden and Tardos in [4], by using and extending EWL protocol and representing the same set of games in a quantum-network scenario. To see the different performances of the quantum and classical versions of the games, efficiency measures will be used: *Price of Stability (PoS)*, the ratio between the *best* possible equilibrium cost and the optimal cost, and *Price of Anarchy (PoA)*, which compares the *worst* possible equilibrium cost with the optimum.

A. Sergienko, S. Pascazio, and P. Villoresi (Eds.): QuantumCom 2009, LNICST 36, pp. 74–81, 2010.

2   EWL Protocol

The EWL protocol was initially used by Jens Eisert et al. for the discussion of the "Quantum Prisoner Dilemma". It was later used for more general purposes by many others (e.g. [14] and [15]).

For a 2x2 game, the EWL scheme is as follows:

- Initially we have 2 qubits in the state $|\psi\rangle = |0\rangle \otimes |0\rangle = |00\rangle$
- A referee creates entanglement using *entangling gate*

$$\hat{J} = \frac{1}{\sqrt{2}} \begin{bmatrix} 1 & 0 & 0 & i \\ 0 & 1 & i & 0 \\ 0 & i & 1 & 0 \\ i & 0 & 0 & 1 \end{bmatrix}, \tag{1}$$

leaving $|\psi\rangle$ in the *maximally entangled* state

$$|\psi'\rangle = \hat{J}|\psi\rangle = \frac{1}{\sqrt{2}}(|00\rangle + i|11\rangle). \tag{2}$$

After that, qubits are sent to the respective players, which are forbidden to communicate.

- Each player i applies his own 1-qubit unitary operation s_i and successively sends the qubit back, leaving the state in

$$|\psi''\rangle = (s_1 \otimes s_2)|\psi'\rangle \tag{3}$$

- Finally the referee can determine the outcome by applying $\hat{J}^\dagger$ operation (conjugate transpose of $\hat{J}$) and measuring the qubits in their standard basis.

3   Local Formation Games

Network formation games deal with n selfish agents building a network.

In the *local* case, each player $u \in N$ is a node in a graph, and can decide to build up to $n - 1$ direct arcs from himself to any other player. The strategy set for u, S_u, then contains all possible subsets of the $n - 1$ incident arcs to u, and strategies can be represented by a binary vector $s_u \in S_u$. The strategy concatenation $s = (s_1, \ldots, s_n)$ forms the outcome network, $G(s)$. The goal for each agent is to have a connected $G(s)$, while taking as low as possible the sum of the distances from the other agents plus the cost α of arcs built, thus minimizing the function:

$$c_u(s) = \alpha n_u(s) + \sum_{v \in N} dist_s(u, v). \tag{4}$$

The social cost of the solution s is:

$$C(s) = \sum_{u,v \in N} dist_s(u, v) + \alpha|E| \tag{5}$$

where $|E|$ is the number of arcs[1] in $G(s)$.

---

[1] Counting only one α for each arc is a valid assumption, because in an equilibrium solution there is no arc (u, v) that is paid by both interested nodes.

Tardos proves in [4] that PoS can be as large as $\frac{4}{3}$ and PoA is directly dependent on the maximum diameter of the network. Using quantum strategies and entanglement in an extended EWL setting, the Price of Stability turns out to be always 1, and the maximum diameter of an equilibrium is reduced by a $\sqrt{2}$ factor.

3.1   Classical and Quantum Version

As said before, it comes natural to represent players' strategies with strings:

$$s_i = (e_1, e_2, \ldots, e_{i-1}, e_{i+1}, \ldots, e_{n-1}, e_n) \tag{6}$$

of $n-1$ bits. To use the EWL protocol with this setting, we can create entanglement on qubits representing the same arc (u, v) in both s_u and s_v, implementing many 2-players quantum games: qubits $e_v \in s_u$ and $e_u \in s_v$ will be in an entangled state.

In the case with only 2 players, there is the following payoff matrix (C= contribute to the arc , R=refuse to contribute):

	C	**R**
C	$(\alpha + 1), (\alpha + 1)$	$1, (\alpha + 1)$
R	$(\alpha + 1), 1$	∞, ∞

The classical Nash Equilibria are (C, R) and (R, C), however with the EWL protocol these are not equilibria anymore, as it is not convenient for a player to play a pure strategy. For every unitary operation a player can chose to play, in fact, the opponent can build a perfect counter-strategy that maximizes his payoff, and vice-versa[2].

Considering Quantum mixed-strategy Nash Equilibrium of the type[3]

$$A^1 \sim \begin{bmatrix} 1 & 0 \\ 0 & 1 \end{bmatrix} \qquad\qquad A^2 \sim \begin{bmatrix} -i & 0 \\ 0 & i \end{bmatrix} \tag{7}$$

$$B^1 \sim \begin{bmatrix} 0 & 1 \\ -1 & 0 \end{bmatrix} \qquad\qquad B^2 \sim \begin{bmatrix} 0 & -i \\ -i & 0 \end{bmatrix} \tag{8}$$

makes each player the unique builder of the arc with probability $\frac{1}{2}$, with a expected cost $E[c_i] = \frac{\alpha}{2} + 1$. What is changed with respect to classical equilibria is that now the agent who pays for the arc is selected randomly, and there is a decrease of the average cost of "putting an arc into game" (that was classically $\alpha + 1$). This is the key change for the PoS of the multiplayer case.

In fact, in [4], Tardos shows that the efficiency of equilibria change in function of α. The only case in which $PoS > 1$ is when $1 < \alpha < 2$, when the complete graph is optimal but the best equilibrium is a star-graph. This happens because

---

[2] The situation is like the Prisoner's Dilemma, see [12].

[3] The symbol "$\sim$" means that there are infinite equivalent equilibria obtained with appropriate rotations on the strategies, preserving the mutual optimality property.

no agent has incentive to pay $2 < \alpha + 1 < 3$ for building a direct arc, instead of just using the existing link and paying $dist(u,v) = 2$.

This does not hold using EWL protocol, as an agent u with strategy cost

$$c_u(s) = \alpha n_u(s) + dist_s(u,w) + \sum_{v \in N \setminus \{w\}} dist_s(u,v) \qquad (9)$$

and $dist_s(u,w) \geq 2$, can "put into game" (u,w) with a new solution s', expecting

$$E[c_u(s')] = \alpha(n_u(s) + 1) + 1 + \sum_{v \in N \setminus \{w\}} dist_{s'}(u,v) \qquad \leq c_u(s). \qquad (10)$$

The complete graph is a Nash Equilibrium with quantum strategies for $\alpha \leq 2$, while nothing changes for $\alpha \geq 2$. This means that with quantum strategies the PoS is always equal to 1.

About the PoA, Tardos shows in [4] that it depends on the maximum diameter[4] of a graph in an equilibrium solution, that is classically at most $2\sqrt{\alpha}$. In a quantum setting, suppose two nodes u and v are at distance $dist(u,v) \geq 2k$ for some k. Putting the arc (u,v) in competition, node u would pay $\frac{\alpha}{2}$ in expectation, while reducing the distance from nodes in the second half of the shortest-path $u - v$ of $(2k-1) + (2k-3) + \cdots + 1 = k^2$. The construction of the direct arc is convenient if $dist(u,v) > 2\sqrt{\frac{\alpha}{2}}$, and that's why the maximum diameter of a graph in a Nash Equilibrium with Quantum Strategies is at most $2\sqrt{\frac{\alpha}{2}} = \sqrt{2\alpha}$.

4   Global Formation Games

Does the introduction of EWL protocol define a new game with at least the same performance of the classical counterpart? It turns out that this is not true, and the following is an example.

The *global* case of formation game shows players having source-target pairs and willing to connect those by building arcs, each of them having an associated cost. If more than on player decide to build the same arc, the cost is equally shared. The *PoA* in this setting is up to $\mathcal{H}_k$, the k-th harmonic number, as shown by Tardos and Wexler in [4]. Here it is convenient to use an extended EWL protocol that, given strategy strings, creates entanglement between *all* the qubits representing the same arc. Let's show with a simple example an instance where the classical game performs better than the quantum one (figure 1). The strategy of a player is communicated to the referee using a string (e_{pb}, e_{pvt}) of 2 qubits, one for the private path and one for the public (the shared) one. The two e_{pb} qubits are in an entangled state, while the e_{pvt} ones remain independent.

Player 2 finds convenient to buy the e_{pb} link together with Player 1, who is not interested in it because the loss of $\frac{\epsilon}{2}$. Because of the entanglement, Player 1

---

[4] The length $max_{(u,v)}d(u,v)$ of the "longest shortest path" between any two graph vertices (u,v) of a graph, where $d(u,v)$ is a graph distance.

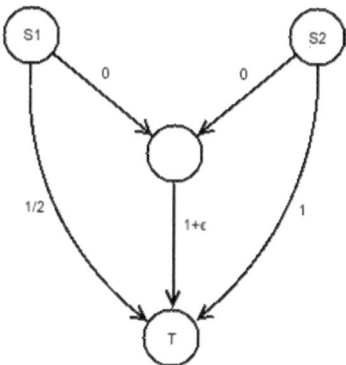

Fig. 1. A simple two-player Global Formation instance

cannot avoid to play the public link, the best he can do for his defense is to play the mixed strategies:

$$S_0 \sim \begin{bmatrix} 1 & 0 \\ 0 & 1 \end{bmatrix} \qquad\qquad S_1 \sim \begin{bmatrix} i & 0 \\ 0 & -i \end{bmatrix}, \qquad (11)$$

who act as randomizers for the choice of Player 2. Player 1 will confirm (and follow) or negate the strategy of Player 2, thus ending in both or neither buying the arc with equal probability. Expected cost for Player 1 is then:

$$c_1[s] = \frac{1}{2} \cdot \frac{1}{2} + \frac{1}{2} \cdot \left(\frac{1}{2} + \frac{1+\epsilon}{2}\right) = \frac{3+\epsilon}{4}. \qquad (12)$$

Now Player 2 has 50% probability of losing e_{pb}, so to avoid an infinite expected cost he has to build also e_{pvt}, with expected cost:

$$c_1[s] = \frac{1}{2} \cdot 1 + \frac{1}{2} \cdot \left(1 + \frac{1+\epsilon}{2}\right) = \frac{5+\epsilon}{4}. \qquad (13)$$

This is the only class of mixed Nash Equilibria, maximizing expectations for both players, and there are no pure NE. Let's consider the two classical cases:

- If both players would play only the e_{pvt} arcs, Player 2 could increase his payoff by playing Strategy S_1 on e_{pb} and leaving his e_{pvt} to 0. This way he could force Player 1 to follow him on the public link, paying only $\frac{1+\epsilon}{2}$ instead of 1.
- If players would use only e_{pb}, Player 1 could use the same S_1 strategy to "undo" the decision of Player 1, and use only his e_{pvt} arc, saving $\frac{\epsilon}{2}$.

PoS for this instance is then

$$PoS = \frac{\frac{3+\epsilon}{4} + \frac{5+\epsilon}{4}}{1+\epsilon} \simeq 2 \qquad (14)$$

while the classical was only $1.5 + \frac{1}{\epsilon}$. Extending the game to n players, PoS becomes $\mathcal{H}_k + \frac{1+\epsilon}{2}$.

5   Congestion Games

In a network congestion game, also called *selfish routing*, players have a source node, a target node and a unit of traffic to transfer, and they have to decide the route to take. The strategy set for each agent consists of all possible paths from source to target, and the utility is the latency time he faces, considering that arcs get congested in function of the number of players using them. The unit of traffic may be splittable or unsplittable and defined as non-atomic or atomic instances. It is possible to calculate an *optimal* solution that minimizes the sum of all latencies faced by the players. However, this solution is not always an equilibrium as some players may find profitable to change route and face less latency for themselves, increasing latency for others. Introducing quantum strategies with a slightly modified EWL protocol, classical examples of inefficiency in selfish routing, present Price of Anarchy and Stability of 1. This does not mean, however, that quantum equilibria are always optimal, as we can show a network inefficient with quantum strategies.

5.1   Quantum Version of Pigou's Example

Consider the network in Figure 2. We have a set of n players, each of them wanting to route a negligible amount of unsplittable traffic from s to t. In players' strategies there are only two paths who have a different congestion function: P_1 of constant cost $c(x) = 1$, and P_2 that has load dependent cost $c(x) = x/n$.

The optimal outcome for this problem is equally splitting the agents between the paths, thus achieving a social cost of

$$cost(OPT) = \frac{n}{2} \cdot 1 + \frac{n}{2} \cdot \frac{1}{2} = \frac{3}{4}n.$$

At Nash Equilibrium all players use the P_2 path, thus

$$cost(NE) = n \cdot \frac{n}{n} = n. \tag{15}$$

We can then calculate the Price of Stability:

$$PoS = \frac{cost(NE)}{cost(OPT)} = \frac{n}{\frac{3}{4}n} = \frac{4}{3}. \tag{16}$$

To use the EWL Protocol with n players, we can create a *Entanglement by couples* (if the number is odd, the last one is free). Fixed the behaviour of other $n - 2$ players, supposing that k of them (with $k < n - 2$) chose the P_2 path, payoff matrix for the two remaining players is:

	P_1	P_2
P_1	$(1,1)$	$(1, \frac{k+1}{n})$
P_2	$(\frac{k+1}{n}, 1)$	$(\frac{k+2}{n}, \frac{k+2}{n})$

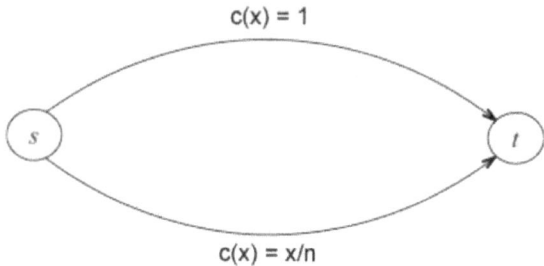

$$c(x) = 1$$

$$c(x) = x/n$$

Fig. 2. Pigou's example

Note that classically the dominant strategy P_2 is still valid, and considering entanglement we reach an equilibrium with the same mixed-strategies of equations 7 and 8. The individual expectation is:

$$E[c_j] = \frac{1}{2}\left(\frac{1 + \frac{k+1}{n}}{2}\right) + \frac{1}{2}\left(\frac{\frac{k+1}{n} + 1}{2}\right) = \frac{1 + \frac{k+1}{n}}{2}. \tag{17}$$

Because the remaining $n-2$ agents will face the same situation, this equilibrium has *exactly* one half of the agents taking the P_2 link, with an individual expected cost even lower than the classical:

$$E[c_j] = \frac{1 + \frac{k+1}{n}}{2} = \frac{1 + \frac{\frac{n}{2} - 1 + 1}{n}}{2} = \frac{1 + \frac{1}{2}}{2} = \frac{3}{4}. \tag{18}$$

It achieves optimal social cost and $PoS = 1$ (for this particular network):

$$cost(NE) = \sum_j c_j = n \cdot E[c_j] = \frac{3}{4}n = cost(OPT). \tag{19}$$

This quantum Pigou's example answers one important question: *are Nash Equilibria with Quantum Strategies and EWL protocol equivalent to classical correlated equilibria?* Obviously there is no correlated equilibrium for this example, because a player that is told to take the P_1 path has no incentive to obey. With quantum strategies, instead, players expect a better payoff by *taking the risk* to take the P_1 path. Finally, note that this special example does not imply that every Selfish Routing problem is optimal. In fact we can find a network where $PoS = \frac{8}{7}$, making $C(x) = \frac{2x}{n}$ in the P_2 path of Figure 2.

6   Conclusions

This work is an attempt to adapt the EWL protocol to a multiplayer game with unbounded players and to answers two natural questions about general Quantum Games, with counterexamples. It also shows how the introduction of quantum information technology in a large-scale network may lead to improvement in terms of congestion avoidance and link formation, and it is an invitation for future research and practical applications, such as quantum routers.

Acknowledgement

The work of the author has been supported in part by the European Commission through the FP6 program under contract FP6-1596 AEOLUS.

References

1. Nielsen, M.A., Chuang, I.L.: Quantum Computation And Quantum Information. Cambridge University Press, Cambridge (2000)
2. Einstein, A., Podolsky, B., Rosen, N.: Can quantum-mechanical description of physical reality be considered complete? Phys. Rev. 47(777) (1935)
3. Bell, J.S.: On the Einstein-Poldolsky-Rosen paradox. Physics 1(195) (1964)
4. Nisan, N., Roughgarden, T., Tardos, É., Vazirani, V.: Algorithmic Game Theory. Cambridge University Press, Cambridge (2007)
5. Roughgarden, T., Tardos, É.: How Bad is Selfish Routing? Journal of the ACM 49(2) (2002)
6. Pigou, A.C.: The Economics of Welfare. Macmillan, Basingstoke (1920)
7. Awerbuch, B., Azar, Y., Epstei, A.: The Price of Routing Unsplittable Flow. In: 37th Annual ACM Symposium on Theory of Computing, STOC (2005)
8. Christodoulou, G., Koutsoupias, E.: The price of anarchy of finite congestion games. In: 37th Annual ACM Symposium on Theory of Computing, STOC (2005)
9. Shapley, L., Monderer, D.: Potential Games. Games and Economic Behavior 14 (1996)
10. Chakrabarty, D., Mehta, A., Nagarajan, V., Vazirani, V.: Fairness and optimality in congestion games. In: Proceedings of the Sixth ACM Conference on Electronic Commerce, Vancouver, BC, pp. 52–57. ACM Press, New York (2005)
11. Meyer, D.A.: Quantum Strategies. Phys. Rev. Lett. 82, 1052–1055 (1999)
12. Eisert, J., Wilkens, M., Lewenstein, M.: Quantum Games and Quantum Strategies. Phys. Rev. Lett. 83, 3077–3080 (1999)
13. Eisert, J., Wilkens, M.: Quantum Games. Journal of Modern Optics 47(14) (2000)
14. Benjamin, S.C., Hayden, P.M.: Multi-Player Quantum Games. Phys. Rev. A 64 (2001)
15. Chen, K., Hogg, T., Beausoleil, R.: A Quantum Treatment of Public Goods Economics. Quantum Information Processing 1(6) (2002)

Optical Free-Space Communication on Earth and in Space Regarding Quantum Cryptography Aspects

Christian Fuchs* and Dirk Giggenbach

German Aerospace Center (DLR)
{christian.fuchs,dirk.giggenbach}@dlr.de

Abstract. Mobile FSO Communication links have gained significant attention over the last years due to their increasing maturity. They are used in point to point link scenarios where the high gain in power efficiency and datarate is crucial, compared to state-of-the-art RF links.

The Optical Communication Group of German Aerospace Center's Institute of Communication and Navigation is conducting research and practical investigations of mobile optical high-speed communication links between diverse platforms, e.g. optical links from aircrafts, UAVs or Satellites to Ground.

This paper will give an overview on the current and future work of the Optical Communication Group in this field, with additional regard to quantum communications.

Keywords: FSO, Free-Space Optical Communications, Quantum Cryptography.

1   Introduction

Free-space optical communications for mobile applications have become more and more popular during the last couple of years. Especially in applications that require high datarates and power efficiencies, FSO links are a good alternative to state-of-the-art RF links. An additional advantage is the fact that the optical spectrum is unregulated and thus no time-consuming licensing process is necessary.

The Optical Communication Group of German Aerospace Center's Institute of Communication and Navigation is conducting research for a broad variety of possible applications. This includes the demonstration of free-space optical links from stratospheric platforms to ground, aircraft to ground and satellites to ground. Furthermore, challenging topics as e.g. adaptive optics systems for highly turbulent atmospheric channels or deep space links for exploration probes are investigated.

As the design of classical FSO terminals is very similar to quantum communication terminals, it seems feasible to combine both into a single terminal. This

---

* Corresponding author.

A. Sergienko, S. Pascazio, and P. Villoresi (Eds.): QuantumCom 2009, LNICST 36, pp. 82–95, 2010.

approach would enable a number of additional applications, as e.g. the establishment of virtually tap proof communication links. This goal can be achieved by using the quantum communication subsystem for exchanging a secure key at the beginning of communication, while transmitting the actual data with a high rate over the classical communication channel.

2 High Altitude Platform-to-Ground Links

A very interesting and promising application for free-space optical links is the interconnection of airships placed in the stratosphere, so called High Altitude Platforms (HAPs). Equipped with RF-systems for the connection to users on ground and optical terminals for building up a network among them, HAPs offer a cost-effective solution for offering broadband data access in regions where only little terrestrial infrastructure exists or where it would be too expensive to build it up.

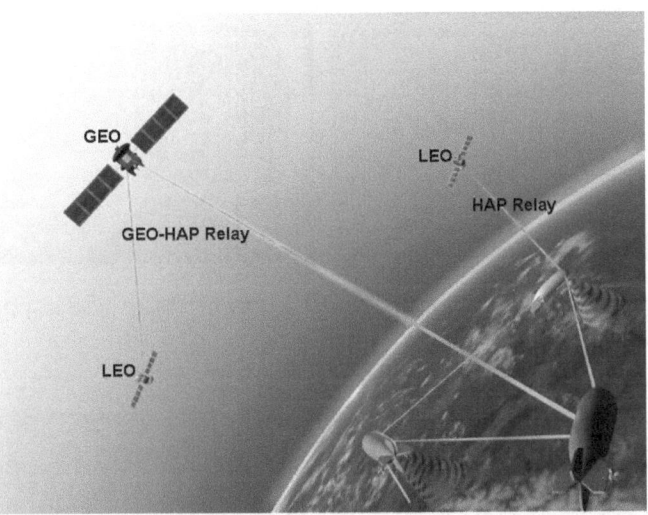

Fig. 1. Artist's Impression of a network including High Altitude Platforms (HAPs) in the stratosphere. Users can connect through weather-insensible RF communication links, while optical Inter-HAP links serve as backbone. Further applications are the use of HAPs as data relay for LEO- and GEO-satellites.

As the operating altitude of HAPs is above the height of clouds, no link blockings for the Inter-HAP-links can be expected due to clouds. At the same time, the RF user links can operate through the clouds, also during bad weather conditions.

Furthermore, HAPs can be used as data-relay for optical downlinks from satellites. As an example, a LEO satellite could dump it's data via a high-rate optical link to the HAP during the relatively short contact time of several minutes. Subsequently, the data could be forwarded to ground by means of a standard RF link that is not influenced by the weather underneath the HAP[1].

However, the development of optical terminals for HAPs is a challenging task. Virtually no air-pressure and very low temperatures at the operation altitude demand a sophisticated system design, especially if a high durability is required for the equipment.

DLR has developed an FSO terminal for stratospheric applications, and demonstrated its functionality in the framework of the EU funded project CAPANINA[2,3]. A downlink from a stratospheric ballon with a height of up to 25 km with a data rate of 1,25 Gbit/s over a distance of 64 km has been performed. Figure 2 shows pictures of the development. A periscope-type Coarse Pointing Assembly (CPA) was used for the terminal.

The validation trial was carried out at the ESRANGE facilities in Kiruna, Sweden. Figure 3 shows a picture of the ballon launch. The payload was mounted in a compartment at the bottom of the ballon.

Fig. 2. Freespace Experimental Laser Terminal (FELT) that has been developed for the CAPANINA Trial

Fig. 3. Launch of the stratospheric ballon carrying the FELT payload

Data was transmitted at rates of 622 Mbit/s and 1.25 Gbit/s. Figure 4 shows an eye pattern and a Bit Error Rate (BER) that has been measured during the trial. Virtually error free data communication could be achieved with both data rates.

Furthermore, the turbulent behaviour of the atmosphere has been characterized with several measurement devices. A detailed description of the experiments carried out can be found in [4].

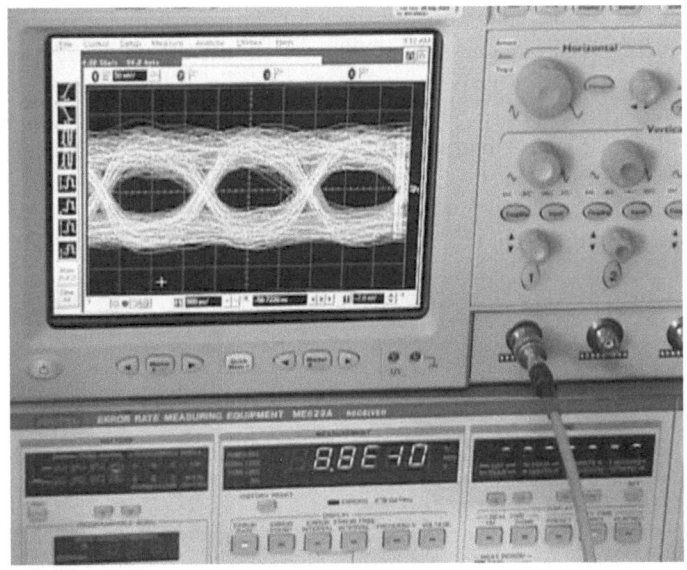

Fig. 4. Eye at the receiver and BER measured for a data rate of 622 Mbit/s during the CAPANINA trial

3   Aircraft-to-Ground Links

Optical Downlinks from aircraft have recently been demonstrated in DLR's ARGOS project. The ARGOS project (airborne wide area high altitude monitoring system) is meant to develop a platform to supply decision makers during mass- or even catastrophic events with real-time reconnaissance data of the current situation. The available sensors include a visible camera system and a synthetic aperture radar[5]. An optical downlink of the gathered data to a transportable optical ground station, placed at the operations center, is foreseen to obtain real-time data access.

In the frame of the project, an Airborne Optical Terminal for one of DLR's research aircraft has been developed. Its operability was shown over a link distance of up to 90 km. The data rate for the system was driven by project requirements, and has been chosen to the rather low value of 125 Mbit/s. Although this is a moderate data rate for FSO links, it is sufficient for the project requirements, and still greatly outperforms typical microwave links ($\approx$ 30...40 Mbit/s).

Figure 5 shows a picture of the terminal installed onboard the aircraft. Outside the aircraft, only the coude-type Coarse Pointing Assembly (CPA) is visible, while the rest of the terminal hardware is mounted inside the passenger cabin. Figure 6 shows the terminal in a laboratory and the Do228 aircraft during validation trials. Visible is the red marker laser, that was used for adjustment purposes. More information about the terminal design and its hardware can be found in [6].

Measurement results that were obtained for the beacon-uplink at the Airborne Terminal during trials in December 2008 are shown in Figure 7. The mean received power of approx. 15 nW is consistent with prior link budget calculations, however, the signal shows a remarkable dynamic range. This is due to the fact that virtually no aperture averaging effect is existent for the comparatively small 30 mm aperture of the AT.

The Pointing Error, that was calculated from the tracking camera's data, was most of the times below 1 mrad, with a mean of 266 μrad. This order of

Fig. 5. Left: Picture of the Do228 Aircraft with installed Optical Terminal. Right: Picture of Optical Ground Station (OGS) with 40cm Cassegrain Telescope for data reception and an RF link antenna for reception of GPS data from the aircraft.

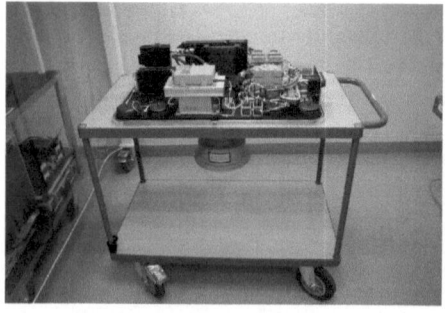

Fig. 6. Left: Optical Terminal in Lab. Right: View of Do228 aircraft during the trials. Visible is the red marker laser used for adjustment purposes.

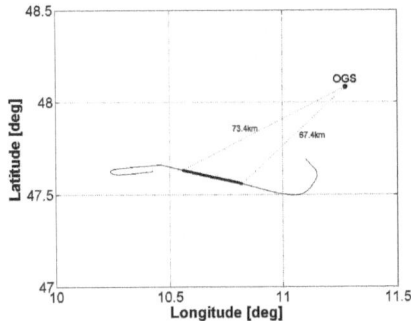 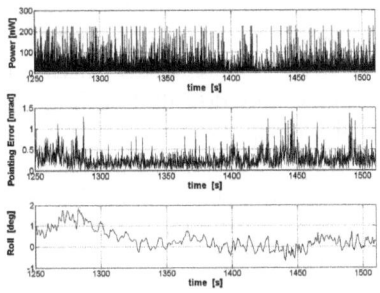

Fig. 7. Left: Flight Path example during validation flight. Right: Measurement data in the beacon-uplink for the thick painted line; Top: Received Power at Airborne Terminal, Middle: Pointing Error of Airborne Terminal, Bottom: Roll Angle of aircraft.

magnitude could be ascertained even during relatively fast and large changes of the aircraft's roll angle, visible in the bottom graph of Figure 7.

Considering the beam divergence of 2 mrad, and taking into account the forward error correction that is applied to the sent data, an error free communication link can be established with this setup.

However, the development of the terminal is still ongoing. Possible further improvements include the implementation of a Fine Pointing Assembly (FPA), consisting of a fast steering mirror and a fast tracking sensor. With this new subsystem, a very fast tracking of the impinging beacon light would be possible, resulting in a stable tracking during agile flight maneuvers. Furthermore, the system can compensate for vibrations of the aircraft that are transferred to the terminal hardware.

The new developments will be verified by further flight testing campaigns. The ultimate goal of the project is to demonstrate an optical link over a distance higher than 100 km with a datarate of over 1 Gbit/s.

4   Satellite-to-Ground Links

Also for the downlink of earth observation data gathered by LEO satellites, optical links to ground stations or High Altitude Platforms (HAPs) can be a solution for the increasing amount of downlink volume.

Typical RF Systems for high data rates in the X- and/or Ka-Band require tens to hundreds of Watts of electrical power and antenna diameters of several to tens of centimeters onboard the satellite. Furthermore, antenna dishes with several meters diameter are necessary on ground station side. It is believed that optical downlink terminals for LEO satellites can be built with lower power consumptions (< 50 W) and smaller telescope diameters (few centimers). Furthermore, ground station telescope diameters in the range from 10 cm to 40 cm are sufficient and available at reasonable prices.

The largest drawback of optical downlink technology, link blocking by clouds, can be mitigated with ground station diversity concepts. It has been shown that availabilities of almost 100 % can be achieved with four ground stations placed at beneficial locations throughout europe[7]. Further improvements of availability and downlink volume can be achieved with using HAPs as data relay platform[8,1].

Successful downlink experiments from JAXA's OICETS satellite have been carried out to Tokio during the KODEN trials[9,10]. Further experiments have been carried out by DLR during the KIODO trials to Oberpfaffenhofen close to Munich[11]. OICETS uses a wavelength in the 800 nm range for communication, what is a very common wavelength for quantum communication systems. The data-rate was around 50 Mbit/s.

In 2006, a total of eight trials have been performed to Oberpfaffenhofen, of which five were successful, while the others failed due to cloudy weather. Figure 8 shows the Optical Ground Station Oberpfaffenhofen (OGS-OP) with its 40cm cassegrain telescope during the trials.

As an example, Figure 9 shows the received power during KIODO Trial 3. It is visible that the power level increases while the fading gets lower during the trial. This is for the fact that the link distance through the atmosphere gets lower for higher elevation angles, resulting in a less turbulent communication channel and, of course, a higher power due to the shorter link distance. The visible signal breaks are characterstic of the laser terminal onboard OICETS[10], and don't appear due to mispointing of the telescope.

Figure 10 shows the Bit Error Rate results that have been obtained during a selection of the KIODO trials. It is visible that the Bit Error Rates get lower for higher elevation angles, as the received power increases and the scintiallation due to atmospheric turbulence gets lower.

Fig. 8. Optical Ground Station Oberpfaffenhofen during KIODO Trials in 2006

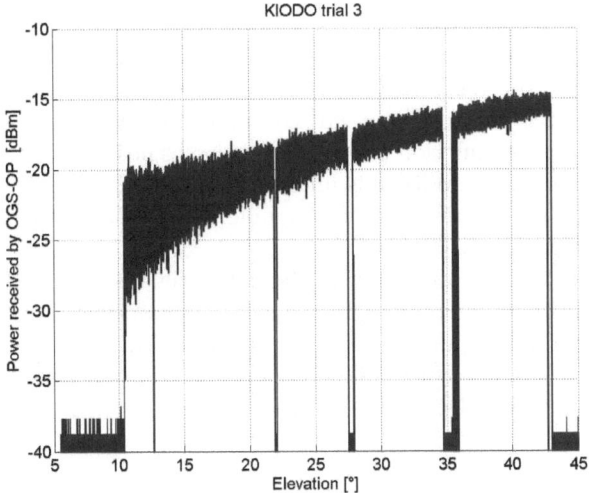

Fig. 9. Received Power during KIODO Trial 03[11]

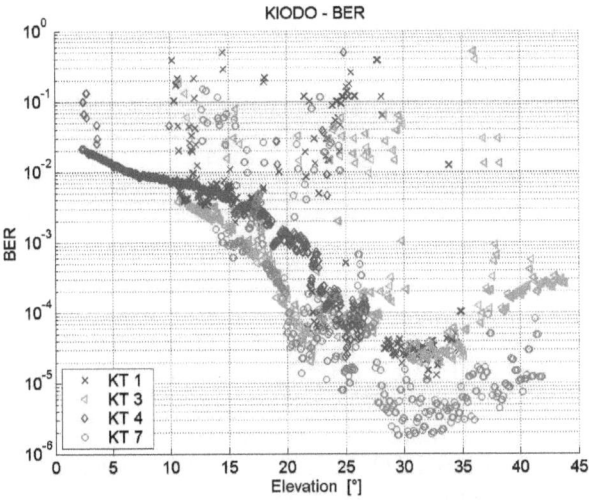

Fig. 10. Bit Error Rate Results for different elevation angles[11]

Further trials with OICETS were carried out during a seconds campaign in the summer of 2009. In addition to the previous trials, a shack-hartmann wavefront sensor was installed as additional measurement instrument. The gathered data will be used to evaluate the turbulent behaviour of the atmosphere, especially for low elevation angles (and thus high link distances). Furthermore, it will allow the simulation of adaptive optics systems and their theoretical performance in this very demanding scenario.

Besides this new measurement device, a data acquisition device with a sample-rate of five times the data-rate (250 MBit/s) was used to sample the signal at the output of the receiver frontend. By implementing the subsequent signal processing chain in a simulation program, it becomes possible to evaluate the Bit Error Rate results offline. With this information, the parameters of the receiver can be optimized for the purpose of designing an optimal receiver frontend for the turbulent atmospheric channel.

The data that was acquired during the 2009 trials is still under evaluation, however, it can already be stated that during all cloud-free links a signal from the satellite could be acquired and a data-communication link was be set up. However, again a number of links failed due to cloudy weather. Figure 11 show OGS-OP during the KIODO 2009 trials.

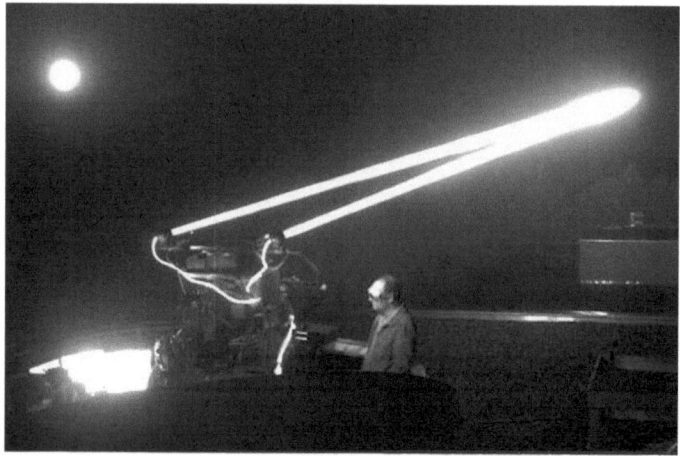

Fig. 11. Optical Ground Station Oberpfaffenhofen (OGS-OP) during the KIODO 2009 trials. Visible on the infrared picture is the beacon laser's backscatter due to haze.

The possibility of optical LEO downlinks for the transmission of e.g. earth sensing data using state-of-the art 1550 nm components is currently investigated at DLR. Figure 12 shows an example link budget for a downlink from a LEO satellite to the Optical Ground Station in Oberpfaffenhofen (OGS-OP) for two different aperture diameters with diffraction limited radiation from the satellite. Atmospheric attenuation based on [12] has been considered for clear sky conditions and moderate vulcanic activity.

It is visible that already with the small aperture of 10 mm onboard the satellite decent power levels can be achieved at the OGS, enabling data links with data rates from 100 Mbit/s up to more than 1 Gbit/s, depending on the sensitivity of the receiver. However, it is remarkable that the influence of the atmospheric

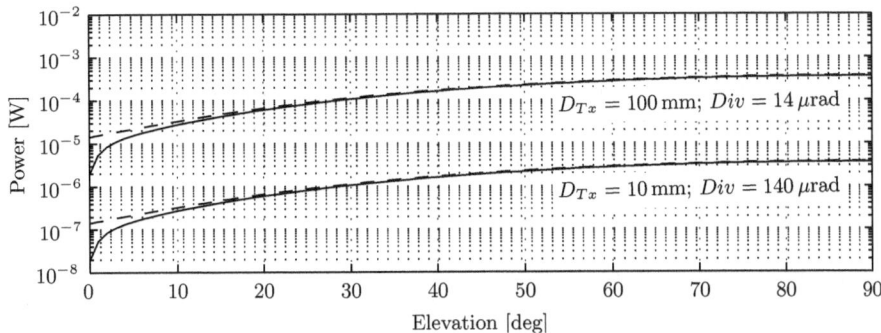

Fig. 12. Example Link-Budget for a diffraction limited optical LEO Downlink to OGS-OP with atmospheric attenuation (solid line) and without (dashed line) for different Tx-Apertures; Link Parameters: 3 dB Rx-System Loss, 40 cm Rx-Aperture, Wavelength 1550 nm, Orbit Height 510 km, Clear Sky, Moderate Vulcanic Activity

attenuation increases drastically for elevation angles close to 0, as for these cases the path through the atmosphere is very long and close to a horizontal link with worst-case attenuation conditions.

5   Further Developments

The research field of free-space optical communications offers still many open questions that are worth investigating them. One topic that is currently researched at DLR is e.g. an adaptive optics system fitted to the particular needs of FSO systems operating in the turbulent atmosphere. A system consisting of a wavefront sensor and a deformable mirror can be used to mitigate the influences of the atmosphere on the communication beam. A first lab-setup of this challenging development is intended to be operational in 2010.

Another interesting topic for further research is the application of FSO links for Deep-Space communications. With the inherent advantages of Optical Communications, as e.g. the reduced beam divergence, lower antenna size and lower power consumption, data rates from remote places in the solar system could be increased, by relaxing the power budget for Space Probe Missions at the same time.

Currently, development and qualification actions are carried out for the purpose of developing a small and lightweight optical terminal for the deployment on small- and micro- LEO satellites. It is foreseen to use the extremely precise attitude control systems used on these satellites to point the communication laser beam towards the ground station by rotating the satellite. Thus no bulky Coarse Pointing Assembly is necessary for the beam pointing, resulting in a very power-efficient and light-weight optical terminal with high data rates. In-Orbit validations of this promising development will be accomplished in the future.

6 Applicability of Quantum Cryptography to Standard FSO Links

In principle, the communication subsystems of the mentioned optical terminals can be modified for using them in quantum communication applications. A possible design of a combined optical transceiver for traditional FSO- and quantum communications using state-of-the-art technology has e.g. been proposed in [13]. Corresponding developments are already carried out - for instance, the *Space-QUEST* initiative is aiming at the installation of a quantum communication terminal onboard the ISS[14,15].

Common FSO terminals for communication purposes operate at a wavelength of 1550 nm. The atmospheric transmission at this wavelength is more benefical than for shorter wavelengths in the 800 nm region, as they are typical for quantum communication systems. This effect is illustrated in Figure 13 for the example of an aircraft-to-ground link.

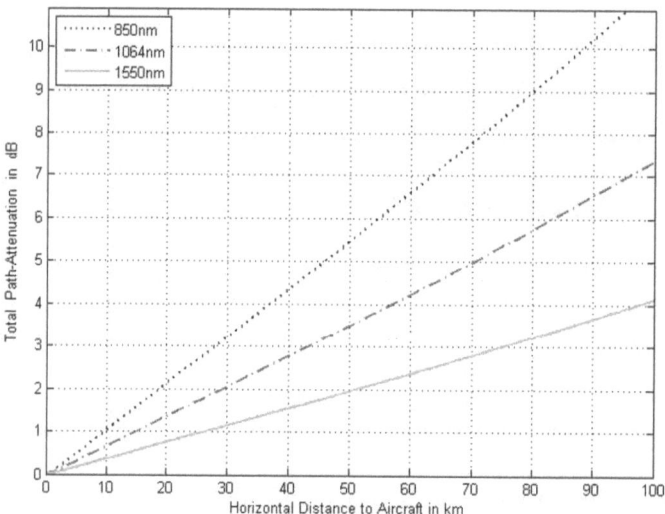

Fig. 13. Different atmospheric attenuations for an aircraft-to-ground link. It is visible that systems with longer wavelengths can take advantage of lower atmospheric attenuations[16].

For the relatively short link distances of optical aircraft-to-ground links, there is no need to design the communication system close to the diffraction limit. Instead, larger beam divergences can still fulfill the project requirements in terms of free-space loss and thus data-rate, with the advantage that the requirements for the tracking system can be relaxed and a less accurate tracking system is sufficient. Furthermore, many mature and low prized off-the-shelf components are available in the 1550 nm range.

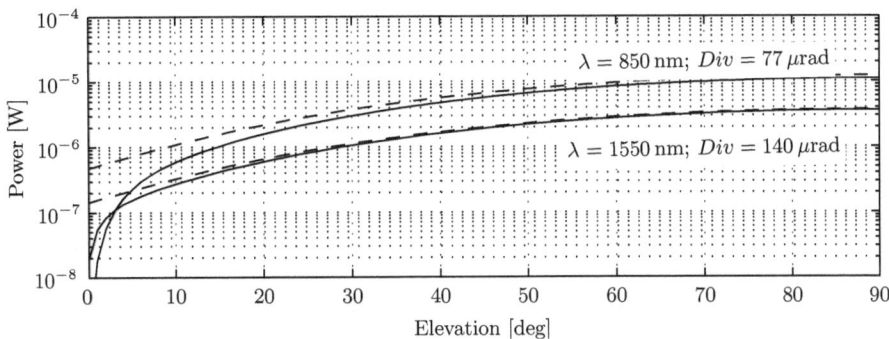

Fig. 14. Example Link-Budget for a diffraction limited optical LEO Downlink to OGS-OP with atmospheric attenuation (solid line) and without (dashed line) for different wavelengths; Link Parameters: 3 dB Rx-System Loss, 10 mm Tx-Aperture, 40 cm Rx-Aperture, Orbit Height 510 km, Clear Sky, Moderate Vulcanic Activity

However, for system designs close to or at the diffraction limit, as e.g. for satellite downlinks, the lower possible beam divergences of shorter wavelengths can compensate for the higher atmospheric attenuation. An example link budget for an optical downlink visualizing this effect is depicted in Figure 14. As for Figure 12, atmospheric attenuation based on [12] has been considered.

It is visible, that the better atmospheric attenuation for a 1550 nm system has only an effect for very low elevation angles. For any elevation angle over 5 degrees, the received power is higher for an 850 nm system.

Due to the mentioned component availability and -maturity, 1550 nm is still a good choice for communication systems. However, considering a combination of a classical FSO- and of a quantum communication terminal, the tracking could be accomplished with the 1550 nm wavelength. As the beam divergence for this system would be larger, tracking requirements would be somewhat relaxed. A quantum communication system in the 800 nm region could be included in the same system, as the splitting between the wavelengths for communication and quantum key exchange could be accomplished with ease and by means of standard components.

In such a combined system, the same aperture could be used for both systems. Thus, only one tracking system is necessary and the extension of a standard FSO communication terminal to a combined FSO-/Quantum communication terminal can be accomplished with relatively little effort. This procedure would allow the tap-proof exchange of an encryption key at the beginning of a data communication session, e.g. for a downlink from LEO satellites. Especially for the direct downlink of sensible data from LEO- or GEO satellites, such a kind of communication terminal would be a promising development.

7 Conclusions

Optical Free-Space Communications offer a solution for the continuously increasing demand of higher data rates in many applications. The German Aerospace Center's Institute of Communication and Navigation's Optical Communication Group pursues research concerning terminal design and characterization of the atmospheric influences on FSO links for many of these applications.

The feasibility of FSO links has been shown for a diverse spectra of applications. This includes the execution of validation trials from stratospheric platforms, aircrafts and satellites.

Furthermore, it seems feasible to comprise both standard FSO- and quantum communication subsystems into a combined optical terminal. This procedure would allow the establishment of tap-proof communication links over very high link distances, combining the advantages of both worlds: The tap-proof exchange of encryption keys with quantum communications, and the very high data-rates and power efficiencies of FSO links.

The next step towards an introduction of this combined technology would be the development of a technology demonstrator. Similar to the mentioned *Space-QUEST* initiative, the demonstator could be aimed at showing it's functionality with an actual LEO downlink from either the ISS, or another suitable LEO satellite.

Acknowledgement

The authors wish to thank all members of the Optical Communication Group for their efforts and dedication, without whose the activities mentioned above would not have been possible. Furthermore, the support of JAXA during the KIODO-trials is gratefully appreciated.

References

1. Knapek, M., et al.: Optical high-capacity satellite downlinks via high-altitude platform relays. In: Proceedings of the SPIE (2006)
2. Giggenbach, D., Horwath, J.: Optical free-space communications downlinks from stratospheric platforms - overview on stropex, the optical communications experiment of capanina. In: IST Summit Dresden (2005)
3. Horwath, J., et al.: Broadband backhaul communication for stratospheric platforms: The stratospheric optical payload experiment (stropex). In: Proceedings of the SPIE (2006)
4. Knapek, M., et al.: The DLR ground station in the optical payload experiment (stropex) - results of the atmospheric measurement instruments. In: Proceedings of the SPIE (2006)
5. Kurz, F., et al.: Near real time airborne monitoring system for disaster and traffic applications. In: ISPRS Hannover Workshop (June 2009)
6. Horwath, J., Fuchs, C.: Aircraft to ground unidirectional laser-comm. terminal for high resolution sensors. In: Free-Space Laser Communications Technologies XXI, San Jose, USA, January 2009, SPIE (2009)

7. Moll, F., Knapek, M.: Wavelength selection criteria and link availability due to cloud coverage statistics and attenuation affecting satellite, aerial, and downlink scenarios. In: Proceedings of the SPIE (2007)
8. Giggenbach, D., et al.: Optical satellite downlinks to optical ground stations and high-altitude platforms. In: IST Mobile & Wireless Communication Summit (2007)
9. Toyoshima, M., et al.: Ground-to-oicets laser communication experiments. In: Proceedings of the SPIE (2006)
10. Takayama, Y., et al.: Tracking and pointing characteristics of oicets optical terminal in communication demonstrations with ground stations. In: Proceedings of the SPIE (2007)
11. Perlot, N., et al.: Results of the optical downlink experiment kiodo from oicets satellite to optical ground station oberpfaffenhofen (ogs-op). In: Proceedings of the SPIE (2007)
12. Mayer, B., et al.: Dlr-internal electronic data base of atmospheric absorption coefficients. In: DLR internal (2002)
13. Pfennigbauer, M., et al.: Satellite-based quantum communication terminal employing state-of-the-art technology. Journal of Optical Networking (JON) 4 (2005)
14. Ursin, R., et al.: Space-quest: Experiments with quantum entanglement in space. Europhysicsnews 40(3) (2009)
15. Armengol, J.M.P., et al.: Quantum communications at ESA: Towards a space experiment on the ISS. In: 58th International Astronautical Congress, Hyderabad, India (2007)
16. Giggenbach, D.: Mobile optical high-speed data links with small terminals. In: Proceedings of the SPIE (2009)

Feasibility Analysis for Quantum Key Distribution between a LEO Satellite and Earth

C. Bonato[1,2], A. Tomaello[1], V. Da Deppo[1], G. Naletto[1], and P. Villoresi[1]

[1] Department of Information Engineering, University of Padova, Italy
CNR-INFM LUXOR Laboratory for Ultraviolet and X-ray, Padova, Italy
{tomaello,paolo.villoresi}@dei.unipd.it
[2] Huygens Laboratory, Leiden University,
P.O. Box 9504, 2300 RA Leiden, The Netherlands
bonato@molphys.leidenuniv.nl

Terrestrial QKD channels can connect two links with a maximum distance of few hundred kilometres. In the case of fibre links, this is due to the signal attenuation in the fibre; in the case of free-space link the losses are due to atmospheric turbulence and absorption. Free-space optical terminals exploiting satellite-based relays are the only resource that can enable global scale quantum key distribution, since single photon propagation is for the main part in vacuum with no turbulence or absorption, and just a small part of the path is through the atmosphere. Several proof-of-principle experiments have been carried out recently: among these the feasibility of single-photon exchange between a satellite and an optical ground station was demonstrated in 2008 [1].

Signal Attenuation. The main factor limiting the performance of free-space optical communication is atmospheric turbulence, both for terrestrial horizontal links or for links between ground and satellites. Turbulent eddies whose size is large compared to the size of the beam induce a deflection of the beam (beam wandering), while smaller-scale turbulent features induce beam broadening. In other words, observing a beam which propagates through turbulent atmosphere at different time instants, one can see a broadened beam randomly deflected in different directions. When integrating the observation over a time-scale longer than the beam-wandering characteristic time, the global effect is a broadening of the beam. For a Gaussian beam of waist w_0 and intensity I_0, the long-term intensity distribution is described by [3]:

$$< I(r, L) >= I_0 e^{-2r^2/w_{LT}^2}$$

where:

$$w_{LT}^2 = w_{ST}^2 + 2 < \beta^2 >$$

w_{LT} is the long-term beam width, w_{ST} is the short-term one and β is the instantaneous beam displacement from the unperturbed position.

A. Sergienko, S. Pascazio, and P. Villoresi (Eds.): QuantumCom 2009, LNICST 36, pp. 96–99, 2010.

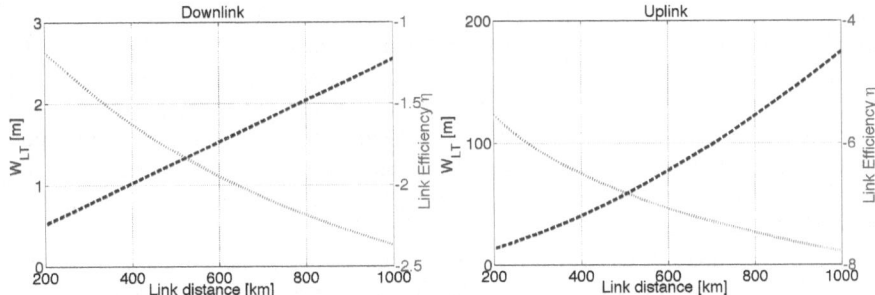

Fig. 1. Beam width w_{LT} and link efficiency for the uplink and the downlink

The results are shown in Fig. 1 for the uplink and the downlink. For the uplink, the beam first propagates through the turbulent atmosphere and then, aberrated, in vacuum, resulting in a large broadening (around 100 m diameter at 500 km). For the downlink, the beam propagates through turbulence only in the final stage, and the spreading is much less (around 1 m at 500 km). Therefore, the attenuation is much stronger in the uplink (more than 50 dB for a 30-cm diameter telescope) compared to the downlink (around 10 dB).

Background noise. As regards the expected background noise in the uplink, during day-time the main contribution is given by sunlight reflection on the Earth surface into the telescope field-of-view. We calculated this contribution to be between 10^7-10^9 photons per second (for a 1 nm of bandwidth). During night-time the main sources of noise are moonlight reflection from the Earth surface, which we calculated as six-orders of magnitude less than it is in day-time (around 10^1-10^3 photons per second) and light pollution from human activities.

We show that the signal-to-noise ratio is proportional to:

$$SNR = \frac{\epsilon_S}{\epsilon_N} \propto \frac{\eta_0}{w_{LT}^2 (IFOV)^2 \Delta\nu\Delta t}$$

where η_0 comprises the detection efficiency, the pointing losses and the atmospheric attenuation, (IFOV) is the telescope field of view and Δt is the detector gating time. In first approximation the SNR does not depend on the radius R of the receiving telescope. The results show that during day-time it is impossible to achieve a SNR higher than 1. During night-time a good SNR can be obtained both for the uplink (~15 dB) and the downlink (~20 dB), provided that a strong filtering is implemented.

Key generation rate. We calculated the expected key generation rates as a function of the link distance for different configurations (uplink, downlink) during night-time for different quantum key distribution protocol.

In most practical quantum communication experiments, single photons are implemented with weak coherent pulses, which have a non-zero probability of multi-photon emission. On such multi-photon pulses Eve could perform a photon-number-splitting attack (PNS)[4]. In the case of high-loss channels, like the ground-to-satellite one, multi-photon pulses are more likely to survive the channel attenuation and get to

Bob's detector than single-photon pulses. The probability of tagged bits in the key, for which Eve can have information without introducing any perturbation, is very high. In the case of the BB84 protocol, a worst-case estimate is taken on the fraction of tagged bits, assuming that all multi-photon pulses are correctly intercepted by Eve. In this case the only way to guarantee security is to reduce the probability of having multi-photon pulses, reducing the source mean photon number. This results in the impossibility to establish a BB84 uplink to a LEO satellite, while for the downlink the results are much better (see Fig. 2).

A better estimate of the fraction of tagged bits can be obtained using weak pulses with different mean photon numbers, the decoy-state technique [5]. Such technique mitigates the need to have a very low intensity source, so that a meaningful key generation rate can be achieved even in the uplink. Assuming a three-intensities decoy state protocol (vacuum, $\mu = 0.27$, $\mu' = 0.4$) a key generation rate of 10^{-6} can be obtained for the uplink to a satellite orbiting at 350 km. The cut-off distance for the uplink is around 300-400 km (depending on the QBER).

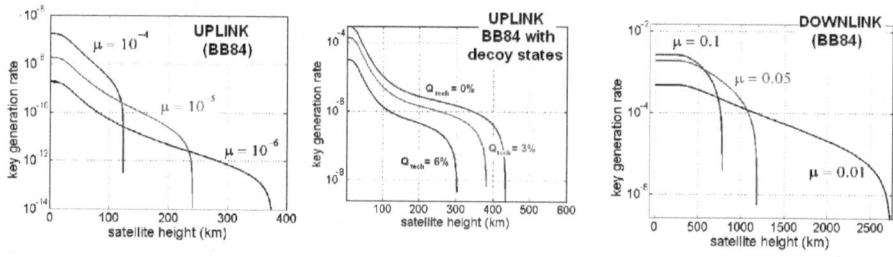

Fig. 2. Key generation rate for uplink (BB84 with and without decoy states) and downlink (BB84). For the uplink, it is possible to establish a QKD channel only using the decoy-state technique and the cut-off distance is around 300-400 km.

We analyzed also the possibility to establish an entanglement-based link between a LEO satellite and Earth. In this case the most important parameter is the SNR [5]: only achieving a 6:1 SNR Bell inequalities can be violated.. We show that a configuration with one local receiver and the other to or from a LEO satellite is feasible. The configuration with two downlinks [6] is also be feasible, but with very strict hardware requirements.

In conclusion, satellite technology can provide a rich environment for quantum information experiments. We believe that the dream of quantum key distribution in Space is possible and not far from being demonstrated.

Acknowledgments

The authors are glad to acknowledge many fruitful discussions with Prof. C. Barbieri, Prof. G. Cariolaro, Dr. F. Tamburini, Dr. I. Capraro and Dr. T. Occhipinti. This work has been carried out within the Strategic-Research-Project QUINTET of the Department of Information Engineering, University of Padova and the Strategic-Research-Project QUANTUMFUTURE of the University of Padova.

References

[1] Villoresi, P., et al.: Experimental verification of the feasibility of a quantum channel between space and Earth. New Journal of Physics 10, 033038 (2008)

[2] Bonato, C., et al.: Feasibility of satellite quantum key distribution. New Journal of Physics 11, 045017 (2009)

[3] Dios, F., et al.: Scintillation and beam-wander analysis in an optical ground station-satellite uplink. Appl. Opt. 43, 3866 (2004)

[4] Lo, H.-K., Ma, X., Chen, K.: Decoy state quantum key distribution. Phys. Rev. Lett. 94, 230504 (2005)

[5] Aspelmeyer, M., et al.: Long distance quantum communication with entangled photons using satellites. IEEE Sel. Top. In Quantum Electronics 9, 1541 (2003)

[6] Armengol, J., et al.: Quantum communications at ESA: towards a space experiment on the ISS. Acta Astronautica 63, 165 (2008)

Enhanced Free Space Beam Capture by Improved Optical Tapers

Tim Bartley[1,2], Bettina Heim[1,2], Dominique Elser[1,2], Denis Sych[1,2],
Metin Sabuncu[1,2], Christoffer Wittmann[1,2], Norbert Lindlein[1,2],
Christoph Marquardt[1,2], and Gerd Leuchs[1,2]

[1] Institute of Optics, Information and Photonics, University of Erlangen-Nuremberg,
Staudtstr. 7/B2, 91058 Erlangen, Germany
[2] Max Planck Institute for the Science of Light, Günther-Scharowsky-Str. 1,
Building 24, 91058 Erlangen, Germany
t.bartley1@physics.ox.ac.uk

Abstract. In our continuous variable quantum key distribution (QKD)
scheme, the homodyne detection set-up requires balancing the intensity
of an incident beam between two photodiodes. Realistic lens systems are
insufficient to provide a spatially stable focus in the presence of large
spatial beam-jitter caused by atmospheric transmission. We therefore
present an improved geometry for optical tapers which offer up to four
times the angular tolerance of a lens. The effective area of a photodiode
can thus be increased, without decreasing its bandwidth. This makes
them suitable for use in our free space QKD experiment and in free
space optical communication in general.

Keywords: Optical taper, free space communication, beam-jitter, atmospheric optics, random media, quantum key distribution.

1 Introduction

Quantum Key Distribution (QKD) (reviewed in e.g. [1,2]) concerns the exchange
of quantum states between two legitimate parties, conventionally named Alice
and Bob. From these states, secret key data can be distilled. Unlike classical
cryptography, for which the security is based on the unproven assumptions of
computational difficulty, in principle QKD can be unconditionally secure.

The first QKD protocol, BB84 [3], was demonstrated experimentally in
1992 [4]. Since then, numerous other protocols have emerged, e.g. [5,6,7,8].
Such schemes have been established over very large distances, using both fibres (250 km) [9] and free space (144 km) [10] as the quantum channel. In the
absence of quantum repeaters [11], free space optics (FSO) is required for worldwide quantum communication via satellites [12,13]. Furthermore, FSO facilitates
urban free space communication which would bypass the need and expense of
new fibres being laid. This is one solution to the "last mile" problem currently
faced by the telecommunications industry [14].

Recently, we have demonstrated experimentally the feasibility of using continuous variables (CV), rather than single photons, to facilitate QKD [15] in a

A. Sergienko, S. Pascazio, and P. Villoresi (Eds.): QuantumCom 2009, LNICST 36, pp. 100–107, 2010.
© Institute for Computer Sciences, Social-Informatics and Telecommunications Engineering 2010

real world free space environment with unrestrained daylight operation [16]. In this paper we explain why loss is of central importance to our scheme and we will present a new, practical approach to beam collection to reduce this loss. A characterisation of our set-up can be found in [16].

1.1   Noise and Attenuation

Noise and attenuation in the quantum channel are the limiting factors when determining the secure key rate. One of the central assumptions of any QKD protocol is that an adversary, Eve, has absolute control over the quantum channel. This means any and all attenuation in the channel are attributed to Eve, as well as any excess noise picked up during transmission.

In the continuous variable regime, attenuation gives Eve additional information and increases Bob's errors, both of which limit potential key rate [17]. In principle, it is always possible to generate a secret key even under high losses [18], although the rate becomes negligible if losses are too high.

While we have shown that it appears no polarisation excess noise is present in the channel [16], it is worth considering the effects of intensity noise on quantum states [19,20,21]. Imperfect detection means security analysis is more involved and further post processing is required.Security analysis under imperfect detection conditions, possibly caused by intensity noise, exists in the single photon regime [22]. It remains to be seen what the implications are for continuous variables (we do not address these issues here), suffice to say that detection efficiency should be optimised, especially in the CV regime.

2   Homodyne Detection

We measure the signal states using a balanced homodyne detection scheme, as shown in Fig. 1. Homodyne detection uses a local oscillator (LO) which interferes with the signal beam at a beam splitter. The difference of the two resulting photocurrents gives the amplitude of the signal state. Successful detection relies on splitting the incident intensity equally and amplifying the resulting photocurrent difference electronically (using an appropriate low-noise amplifier). This technique allows quantum noise to become visible using standard PIN photodiodes.

Conventionally, the local oscillator is generated locally by the receiver, e.g. in [23]. However, using polarisation variables, we are able to multiplex the signal and LO in the same spatial mode at the sender. This results in perfect mode matching at Bob's beam splitter (as well as numerous other benefits explained in [16]). However, since the LO is actually part of the detection system, the effect of loss is compounded.

2.1   Atmospheric Fluctuations

In standard FSO, an intensity-modulated beam is focussed on a single diode [14]. For a beam propagating from a source far away, all paths are considered paraxial, therefore only a lens is required to maximise detection efficiency.

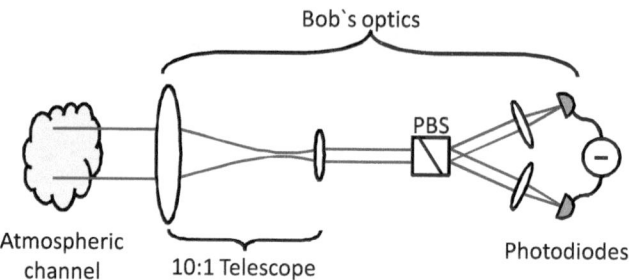

Fig. 1. Our free space homodyne detection. We use a Wollaston prism as the polarising beam splitter (PBS) in our set-up.

In our case, the homodyne detection set-up used to measure the polarisation states requires more optics. A telescope reduces the beam size before it is split at a polarising beam splitter (PBS) and focussed on the photodiodes, as shown in Fig. 1. These additional optics cause some paths of the beam to become non-paraxial in the presence of atmospheric beam jitter. This leads (in practice) to uncorrelated partial detection noise on each photodiode. However, any uncorrelated partial detection noise due to beam jitter leads to an imbalance across the diodes which may affect the homodyne detection [16]. It is therefore desirable to remove the spatial dependence of detection due to a jittering beam.

Compensating for Beam Jitter. In principle, the active area of the photodiode can be increased, such that it captures the entire jittering beam. However, diode area scales with capacitance, which in turn limits the speed of operation and thus key rate.

Another strategy is to focus using suitable lenses. However, imperfectly aligned aspheric lenses are susceptible to poor focussing of beams that are not incident normal to the surface to the lens (i.e. are not paraxial). Atmospheric beam jitter translates to angular deviations from the optical axis and the focus is no longer well defined. This effect is more pronounced in moving target implementations, such as surface to aircraft communications, e.g. [24]. Hence the motivation to design an optical component which offers angular as well as spatial tolerance in its transmission behaviour.

3   Improved Optical Tapers

Microscale optical tapers exist to couple beams between fibres and waveguides, see e.g. [25,26]. They typically operate in the single-mode to single-mode regime. While larger tapers have been suggested to operate in free space applications [27], to the best of our knowledge they have not been optimised and are not widely used. Using numerical ray trace analysis, we present an improved geometry of an optical taper.

3.1   Taper Geometry

The aim of our taper is to collect all light incident on a large aperture and transmit it onto a photodiode of much smaller size. We want to effectively increase the active area of the photodiode without decreasing its speed of operation. Furthermore, we require the transmittance of the taper to be both spatially invariant with respect to the incident beam and offer higher angular tolerance than lenses.

Truncated Parabolic Mirror. One solution to the problem of compressing a wide incident beam to one point is a parabolic mirror. We therefore base our geometry on a parabolic fully-reflective surface, as shown in Fig. 2 (left). The equation of a parabola is given by $z(r) = \alpha r^2$, where r is the radial extension and α is a constant. In our case, the parabola is truncated in the z-axis by the input and output apertures of radii r_1, r_2 at $z(r_1), z(r_2)$, respectively, such that the length of the taper is given by $l = z(r_1) - z(r_2)$. The constant $\alpha = \frac{l}{r_1^2 - r_2^2}$ is thus written in terms of these parameters.

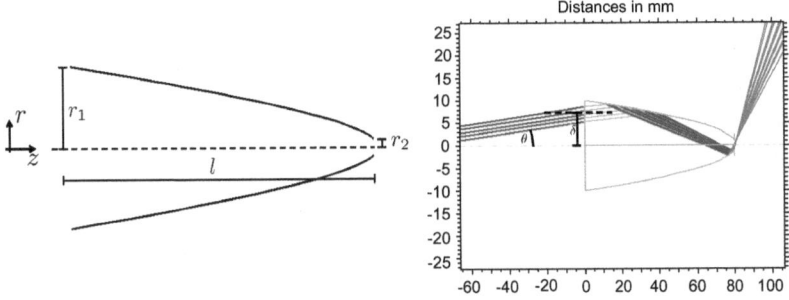

Fig. 2. Geometry and definitions of the parabolic taper (left). Example of ray trace of five rays (right). The colours of the rays change according to reflection/refraction events. Our actual simulations used over 10000 rays, from which transmission can be deduced by counting the resulting rays that pass through the aperture at the end of the taper. Software: RayTrace [28].

We require all the incident light to exit the taper, i.e. the focus z_f of the parabola lying outside the taper $(z_f > z(r_2))$. This imposes the condition that the gradient $r'(z)$ of the parabola (with respect to the z-axis) can never exceed unity, otherwise some paths within the taper would be back-reflecting. If we assume r_1 and r_2 are fixed by the size of beam jitter and diode area respectively, we therefore seek a taper of length l such that the condition $|r'(z)| < 1$ is fulfilled. Given the gradient $r'(z) = \frac{1}{2\alpha r}$ is maximal at $r = r_2$, upon substituting for α the condition above becomes

$$\left| \frac{r_1^2 - r_2^2}{2lr_2} \right| < 1 . \tag{1}$$

This means that the length l is limited by

$$l > \frac{\left(\beta^2 - 1\right) r_2}{2} , \tag{2}$$

where $\beta = \frac{r_1}{r_2}$ is the ratio of the input and output aperture radii of the taper.

3.2   Numerical Simulation

It is important to stress that unlike conventional tapers (e.g. [25,26]), we are operating in the highly multimode regime, rather than guiding single mode to single mode. The large size of the taper compared to the wavelength of the light permits the use of ray approximations rather than wave optics. Using the in-house numerical analysis programme RayTrace [28], the response of the taper to geometric rays can be simulated. An example of a trace is shown in Fig. 2 (right). Each incident ray represents an amount of energy governed by a Gaussian distribution centred on the middle ray. The incident rays can be parameterised by a radial and angular displacement vector (δ, θ), as used in ray transfer matrices. Using this technique, the transmission of the taper for different values of δ and θ can be calculated.

We simulate a beam of width 2 mm (similar to beam widths used in our QKD set-up) incident on a fully-reflecting taper of initial aperture 20 mm and final aperture of 2 mm, as shown in Fig. 2 (right). An absorbing plane with a 2 mm aperture is placed immediately after the taper. The transmission through this aperture represents the total intensity that would impinge on a photodiode of diameter 2 mm. Figure 3 (left) shows the transmission as a function of radial displacement δ (with $\theta = 0 \; \forall \; \delta$) for a taper compared to a lens of equal aperture. The "photodiode" is placed at the focus of the lens. As can be seen, both optical components serve to increase the effective diameter of the aperture to 17 mm.

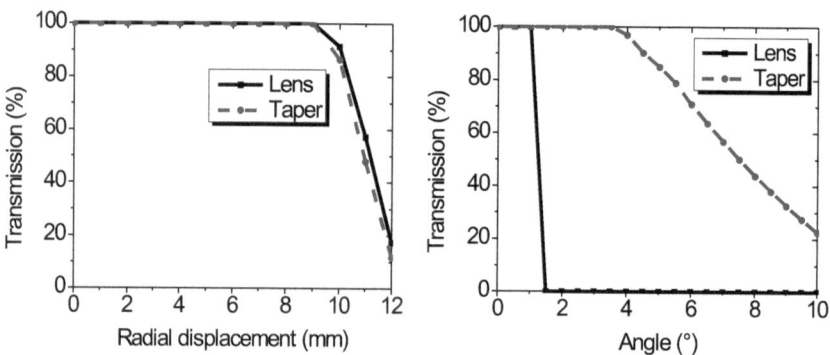

Fig. 3. Taper and lens transmission as a function of radial displacement (left) and angle (right). Both components show similar radial dependence. However, the lens is only fully transmittant up to angular deviations of $1°$, whereas the taper transmits fully up to $4°$.

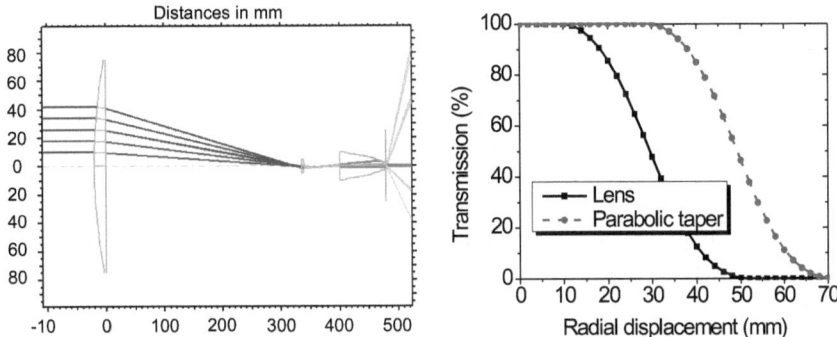

Fig. 4. Left: ray trace of the detection set-up including telescope. We see how angular deviations build up due to the optics. Right: transmission to a photodiode as a function of radial displacement at the initial telescope. The taper offers tolerance up to 30 mm, amounting to a three-fold improvement over a lens when coping with beam jitter.

The angular dependence is tested slightly differently. For a taper of non-linear geometry, whether or not a ray will be transmitted depends on both angle of incidence and point of entry into the taper. We therefore simulate a number of rays of equal amplitude spread across the input aperture. We test this at a number of different angles, with the results shown in Fig. 3 (right). The lens is only fully transmittant up to 1°, whereas the taper offers four times more angular tolerance, remaining fully transmittant up to 4° and offering at least 50 % efficiency up to 7°.

We also simulated how these general results translate to improvements in our specific set-up [16]. The receiver telescope and taper are shown in the ray trace in Fig. 4 (left). Here, we vary the displacement incident on the receiver telescope, as would be the case for a jittering beam. We do not need to consider angular variations since the beam propagating from a source far away is considered paraxial. However, radial displacement here is converted to angular deviations on the way to the photodiode, as shown in Fig. 4 (left). These angular deviations arise from the realistic case of slight misalignment of the telescope or lens aberrations. Explicitly quantifying the tolerance of axial misalignment of the telescope is difficult. Here we simulate lens positions within 1 mm of perfect alignment, which could be reasonably expected experimentally in our set-up. The improvement of the taper over a lens is shown in Fig. 4 (right). In our set-up, the taper offers considerably more tolerance than a lens, remaining fully transmittant up to radial deviations at the telescope of up to 30 mm.

4   Outlook

We have already begun looking at producing such tapers and characterising them experimentally. Their performance for a jittering beam can then be analysed in our existing 100 m free space channel. We then plan to implement such tapers in our 1.6 km point-to-point QKD link, which is currently under construction.

Conventional tapers are made out of glass and rely on total internal reflection to guide light. In our case, the rays may strike the glass/air boundary outside the range of the critical angle, such that some portion would be lost. This leads to the undesired effect of position-dependent transmission. To counter this, we coat our glass tapers with silver (offering high reflectivity for our wavelength of 800 nm), negating the critical angle condition for reflection. We also plan to coat the input and output apertures with an anti-reflective layer for our wavelength.

At the moment we study glass cores derived from the waste products of a fibre-drawing machine. As a result, they are currently not machined precisely to our specifications. However, constructing arbitrary taper geometries is possible in principle [29]. An alternative approach would be use a hollow taper with a highly-reflective inner surface.

Acknowledgements. The authors would like to thank Silke Rammler and Leyun Zang for useful discussions.

References

1. Gisin, N., Ribordy, G., Tittel, W., Zbinden, H.: Quantum Cryptography Rev. Mod. Phys. 74(1), 145 (2002)
2. Scarani, V., et al.: The Security of Practical Quantum Key Distribution. Rev. Mod. Phys. 81(3), 1301 (2009)
3. Bennett, C.H., Brassard, G.: Quantum Cryptography: Public Key Distribution and Coin Tossing. In: International Conference on Computers, Systems and Signal Processing, Bangalore, India (1984)
4. Bennett, C.H., Bessette, F., Brassard, G., Salvail, L., Smolin, J.: Experimental Quantum Cryptography. J. Cryptol. 5(1), 3 (1992)
5. Bennett, C.H.: Quantum Cryptography Using Any Two Nonorthogonal States. Phys. Rev. Lett. 68(21), 3121 (1992)
6. Ekert, A.K.: Quantum Cryptography Based on Bell's Theorem. Phys. Rev. Lett. 67(6), 661 (1991)
7. Ralph, T.C.: Continuous Variable Quantum Cryptography. Phys. Rev. A 61(1), 010303 (1999)
8. Silberhorn, C., Korolkova, N., Leuchs, G.: Quantum Key Distribution with Bright Entangled Beams. Phys. Rev. Lett. 88(16), 167902 (2002)
9. Stucki, D., et al.: High Rate, Long-Distance Quantum Key Distribution Over 250 km of Ultra Low Loss Fibres. New J. Phys. 11(7), 075003 (2009)
10. Schmitt-Manderbach, T., et al.: Experimental Demonstration of Free-space Decoy-state Quantum Key Distribution Over 144 km. Phys. Rev. Lett. 98(1), 010504 (2007)
11. Briegel, H.-J., Dür, W., Cirac, J.I., Zoller, P.: Quantum Repeaters: The Role of Imperfect Local Operations in Quantum Communication Phys. Rev. Lett. 81(26), 5932 (1998)
12. Villoresi, R., et al.: Experimental Verification of the Feasibility of a Quantum Channel Between Space and Earth. New J. Phys. 10(3), 033038 (2008)

13. Bonato, C., et al.: Feasibility of Satellite Quantum Key Distribution. New J. Phys. 11(4), 045017 (2009)
14. Majumdar, A.K., Ricklin, J.C. (eds.): Free-Space Laser Communications (Optical and Fiber Communications Reports), vol. 2. Springer, Berlin (2008)
15. Lorenz, S., Korolkova, N., Leuchs, G.: Continuous-Variable Quantum Key Distribution Using Polarization Encoding and Post Selection. App. Phys. B 79(3), 273 (2004)
16. Elser, D., et al.: Feasibility of Free Space Quantum Key Distribution with Coherent Polarization States. New J. Phys. 11(4), 045014 (2009)
17. Heid, M., Lütkenhaus, N.: Efficiency of Coherent-State Quantum Cryptography in the Presence of Loss: Influence of Realistic Error Correction. Phys. Rev. A 73(5), 052316 (2006)
18. Silberhorn, C., Ralph, T.C., Lütkenhaus, N., Leuchs, G.: Continuous Variable Quantum Cryptography: Beating the 3-dB Loss Limit. Phys. Rev. Lett. 89(16), 167901 (2002)
19. Dong, R., et al.: Experimental Entanglement Distillation of Mesoscopic Quantum States. Nature Phys. 4(12), 919 (2008)
20. Semenov, A.A., Vogel, W.: Quantum Light in the Turbulent Atmosphere. Phys. Rev. A 80(2), 021802 (2009)
21. Heim, B., et al.: Atomspheric Channel Characteristics for Quantum Communication with Continuous Polarization Variables. To appear in Applied Physics B, http://dx.doi.org/10.1007/s00340-009-3838-8
22. Fung, C.H.F., et al.: Security Proof of Quantum Key Distribution with Detection Efficiency Mismatch. Quantum Inf. Comput. 9(1), 131 (2009)
23. Lange, R., et al.: 142 km, 5.625 Gbps Free-Space Optical Link Based on Homodyne BPSK Modulation. In: Proceedings of SPIE, vol. 6105, p. 61050A (2006)
24. Horvath, J., Fuchs, C.: Aircraft to Ground Unidirectional Laser-Communication Terminal for High Resolution Sensors. In: Proceedings of SPIE, vol. 7199, p. 719909 (2009)
25. Burns, W., Abebe, M., Villarruel, C., Moeller, R.: Loss Mechanisms in Single-mode Fiber Tapers. J. Lightwave Technol. 4(6), 608 (1986)
26. Love, J.D., et al.: Tapered Single Mode Fibres and Devices. IEE Proceedings-J 138(5), 343 (1991)
27. Yun, G., Kavehrad, M.: Application of Optical Tapers to Receivers in Free Space/Atmospheric Optical Links. In: Military Communications Conference MILCOM 1990, vol. 3, p. 899 (1990)
28. Raytrace version 0.9 ©Universität Erlangen-Nürnberg (2008)
29. Birks, T., Li, Y.W.: The Shape of Fiber Tapers. J. Lightwave Technol. 10(4), 04432 (1992)

Entanglement Based Quantum Key Distribution Using a Bright Sagnac Entangled Photon Source

C. Erven[1], D. Hamel[1], K. Resch[1], R. Laflamme[1,2], and G. Weihs[1,3]

[1] Institute for Quantum Computing, University of Waterloo, Waterloo, ON, N2L 3G1, Canada
cerven@iqc.ca
http://www.iqc.ca/~cerven/
[2] Perimeter Institute, 31 Caroline Street North, Waterloo, ON, N2L 2Y5, Canada
[3] Institut für Experimentalphysik, Universität Innsbruck, Technikerstrasse 25, 6020 Innsbruck, Austria

Abstract. We report on improvements in an entangled free-space quantum key distribution (QKD) system by replacing the original non-collinear type-II spontaneous parametric down-conversion (SPDC) polarization entangled photon source with a new brighter Sagnac interferometric entangled photon source. While the SPDC source was integral to the initial setup of the system, it was limited in photon pair production rate and entanglement quality. Initial experiments with the new Sagnac source have already yielded substantially higher entangled photon rates and improved visibilities. In order to examine the integration of the new source with the QKD system, a local QKD experiment is performed where the source is pumped with 5 mW of power yielding an average raw key rate of 9,423 bits/s and an average final secret key rate of 2,695 bits/s, with an observed average QBER of 2.48%. Initial experiments distributing entangled photons over a single 1,305 m free-space link have seen entangled photon pair coincident detection rates as high as 3,000 cps. Extrapolating based on these initial numbers and previous experiments, we hope to obtain an average secret key rate of 715 bits/s for a two free-space link QKD experiment running the source at full power which will represent an order of magnitude increase over our previous experiments. An additional benefit of the new source is that it has a much narrower bandwidth which will aid in making the system compatible with daylight experiments. However, one drawback of the source is an appreciable double pair emission rate which initial experiments indicate.

Keywords: Quantum Cryptography, Free-Space Quantum Key Distribution, QKD, Entangled Photons, Sagnac Interferometric Source.

1 Introduction

Free-space quantum key distribution (QKD) has seen many experiments validate the possibility of its implementation in various real world scenarios; for example, see the more recent experiments performed by Marcikic *et al.* [1], Ursin *et al.* [2],

A. Sergienko, S. Pascazio, and P. Villoresi (Eds.): QuantumCom 2009, LNICST 36, pp. 108–116, 2010.

and Erven *et al.* [3]. Now that the feasibility of the basic idea has been demonstrated, experiments have shifted to improving the practical use of such systems. Some groups have started working on performing quantum key distribution with an orbiting satellite, such as the International Space Station, in order to achieve world wide quantum communication [4,5,6]. While other groups have focused on operating a free-space QKD system during daylight hours where high background light can make communication difficult [7]. Towards the same goal of improving free-space QKD for practical applications, we focus on improving the key generation rate of our system from previous experiments [3] through the use of a new brighter source of entangled photon pairs.

2   Experimental Setup

Our system performs the BBM92 entanglement based QKD protocol developed by Bennett *et. al.* [8] in 1992. Entangled photon pairs are distributed from a source to Alice and Bob who randomly measure a photon from each pair in one of two complementary bases. These measurement results become their generated secret key while the laws of quantum mechanics guarantee the security of the key. Precise integrity for the system is assured with the security proof by Ma *et. al.* [9] neglecting the need for authenticated classical communication, the loopholes opened from detector efficiency mismatch and double clicks, finite key statistics, and some simplifying assumptions made in the proof such as bit and phase errors being equal. We also require the squashing model [10,11,12] and the knowledge that the requirement for active polarization detection can be relaxed to include the passive scheme [13] in order to apply the security proof.

In previous experiments [3], entangled photon pairs were generated using a non-collinear type-II spontaneous parametric down-conversion (SPDC) source first reported by Kwiat *et al.* [14]. In this setup, a BBO non-linear optical crystal in a non-collinear configuration was used to produce entangled photon pairs. While the SPDC source was one of the first stable, high-intensity sources of polarization-entangled photon pairs developed and was integral in the initial setup of the experiment; it was limited to a local coincident entangled photon detection rate of 18,000 counts/sec(cps) with observed visibilities of 99.5% and 92% in the rectilinear (H/V) and diagonal (+45°/-45°) bases respectively. These visibilities corresponded to a baseline error rate for the system of 2.1% which increased the amount of error correction and privacy amplification needed, even without an eavesdropper present, and thus reduced the final secret key rate of the system significantly.

In order to improve the key rate of the system and make it more practical in a real-world scenario, we built a brighter Sagnac entangled photon source originally developed by Kim *et al.* [15] and optimized by Fedrizzi *et al.* [16] to replace the previous SPDC source. The Sagnac entangled photon source, shown in Fig. 1, utilizes a periodically poled KTP (PPKTP) non-linear optical crystal in a collinear configuration placed in an interferometer loop to generate photon pairs. Entangled photons are produced by sending 45° polarized light onto a dual

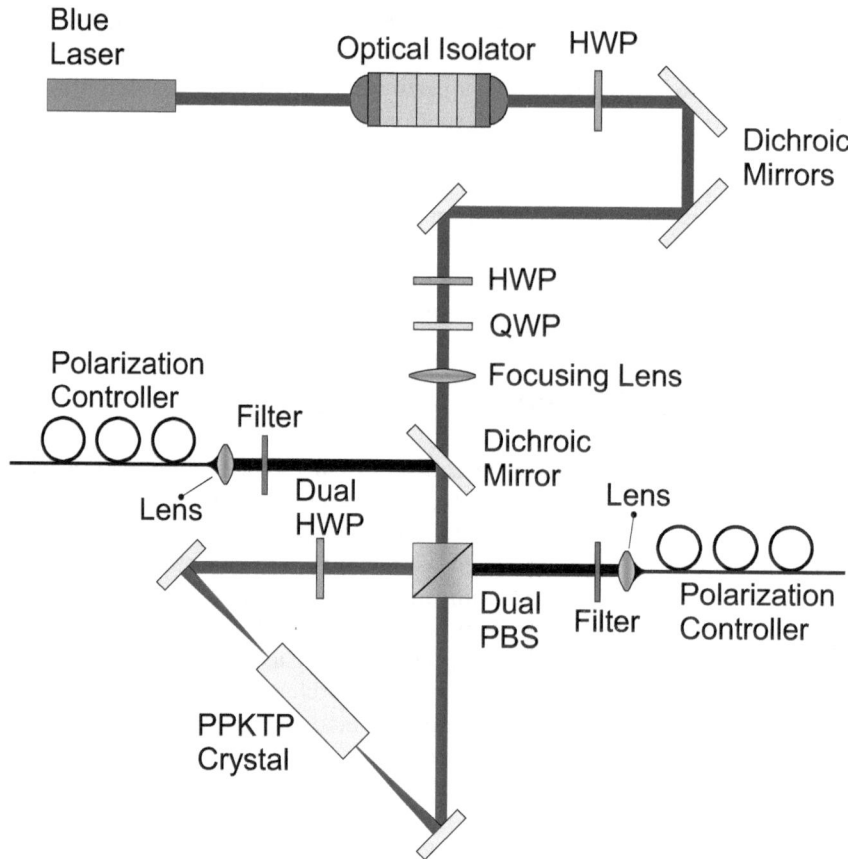

Fig. 1. Experimental schematic of the Sagnac interferometric entangled photon source. Entangled photon pairs are produced by bi-directionally pumping a PPKTP non-linear optical crystal which produces down-converted correlated photon pairs. The dual wavelength HWP and PBS are responsible for removing the path information of the photons, thus producing entangled photons, and for separating the pairs of photons into two paths to be sent to Alice and Bob.

wavelength polarizing beamsplitter (PBS) which has the effect of bi-directionally pumping the PPKTP crystal sitting in the middle of the interferometer loop. A dual wavelength half-waveplate (HWP) rotates the light in one arm by 90° and ensures that the blue laser light is properly polarized so that the crystal produces down-converted polarization correlated photon pairs in both directions around the loop. The dual wavelength HWP also rotates the polarization of the down-converted photons traveling counter-clockwise around the loop, this has the effect that after the down-converted photons are split on the dual wavelength PBS the path information has been erased. Thus, after the PBS, polarization entangled photons have been generated. The two beams of entangled photon pairs are collected directly from one port of the PBS and via a dichroic mirror responsible for splitting the down-converted photons from the blue pump laser at the second exit port.

For our source, we pump the PPKTP crystal with a 404 nm grating stabilized, power tuneable, laser with a maximum power of 50 mW from Toptica Photonics. We use a 10 mm × 1 mm × 1 mm PPKTP crystal made by Raicol. Local experiments with the new source yield single photon detection rates of 400,000 cps in either output path and coincident entangled photon detection rates of 40,000 cps at only 5 mW of pump power. This is currently the maximal detection rate which the data acquisition system can handle. For comparison, the previous source was pumped with 50 mW of power; thus, we have already doubled the number of entangled photon pairs produced at one-tenth the power level of the previous source. In order to get more entangled photon pairs from the source to counteract losses experienced in the polarization detection units and eventually over the free-space link we can still increase the pump power to 50 mW and also use a longer crystal.

Another advantage of the Sagnac source over the previous SPDC source is that it is extremely narrowband, with a bandwidth of 0.36 nm, which allows us to filter much narrower at the receiver stations and reduce the number of errors due to the detection of background light. In the experiment detailed in this paper we use 5 nm interference filters as well as 635 nm long pass filters (responsible for filtering the blue laser light). In previous experiments with the old SPDC source we were forced to use 10 nm filters in order to maintain appreciable detection rates.

Lastly, for the experiment detailed in this paper, we observed local visibilities of 99.2% and 96.1% in the rectilinear and diagonal bases respectively. These represent much improved visibilities over the ones observed with the previous source. Additionally, the visibilities seem to degrade less when they pass through the polarization detection optics which is most likely due to the improved polarization compensation accomplished with manual polarization controllers. The compensation is better since the bandwidth of the photons is much narrower than before allowing the polarization controllers to correct the random rotation of the fibres accurately for more photon pairs. While visibilities for a Sagnac source as high as 99.5% were observed by Fedrizzi et al. [16] in the diagonal basis, these were measured at very low power levels which greatly reduced the effects of accidental coincidences. Even so, our observed visibilities lead to a lower baseline error rate of 1.18% which coupled with the fact that the polarization compensation is more accurate means less error correction and privacy amplification are required thus producing a higher final key rate.

We implemented a local QKD system due to polarization stability issues with our free-space link which we are currently still debugging. Thus, Alice and Bob locally measured each half of their photon pairs while sitting next to the source connected to it with short optical fibres. They measured the photons with passive polarization detector boxes consisting of: a filter to reject background light, a 50/50 non-polarizing beamsplitter (BS) to perform the basis choice, a polarizing beamsplitter in the reflected arm of the BS to separate horizontally and vertically polarized photons, and a half waveplate and PBS in the transmitted arm of the BS to separate photons polarized at $+45°$ and $-45°$. Avalanche photodiode single photon detectors converted the photons into an electronic signal which was stamped with the polarization measured and a highly accurate time of arrival (accurate to 156.25 ps). This information was then transferred to Alice's and Bob's laptops and custom written software then performed the rest of the BBM92 protocol including entangled photon pair identification, sifting, error correction with an optimized Cascade algorithm [17,18], and privacy amplification with a 2-universal hash function [19].

3   Results

For the experiment detailed below, the source was operated at a power of 5 mW producing an average of 134,311 cps in Alice's detectors, 182,047 cps in Bob's detectors, and an average coincident entangled photon detection rate of 9,423 cps. As was mentioned before, this is the maximal detection rate which the current data acquisition system can handle. However, once we move to free-space experiments and link losses are taken into account this should still be adequate for detection while pumping the source with 50 mW of power.

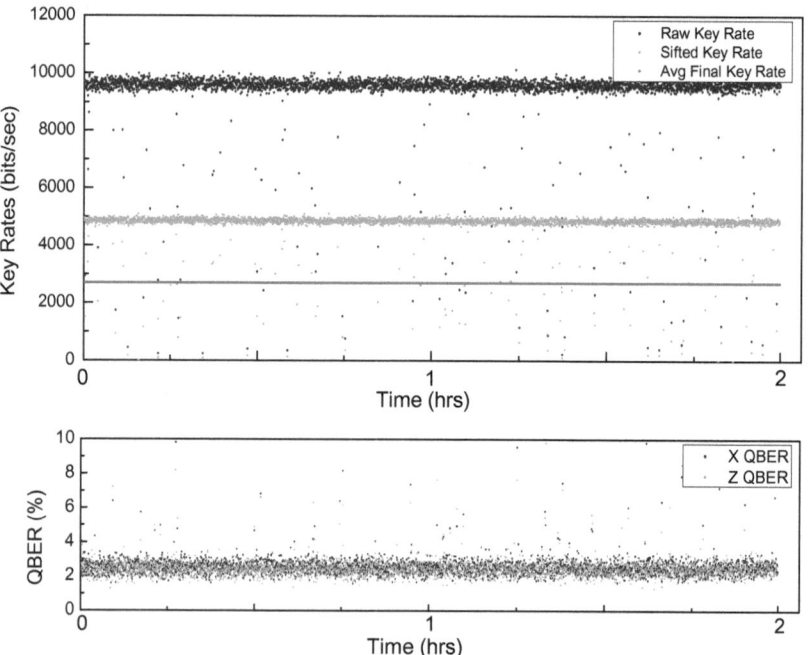

Fig. 2. Top Graph: Observed raw (top series, in blue), sifted (middle series, in green), and average final (bottom series, in magenta) key rates during the experiment. Bottom Graph: Observed X (in blue) and Z (in green) QBER's throughout the experiment.

Table 1. Reconstructed coincidence matrix for Alice and Bob from the experiment

		Alice				
		H	V	+	-	Total
	H	72,545	3,992,143	1,436,747	2,041,365	7,542,800
Bob	V	3,053,610	103,283	2,675,152	2,629,527	8,461,572
	+	1,049,401	2,246,669	88,197	3,986,896	7,371,163
	-	1,503,492	1,783,033	4,296,481	126,318	7,709,324
	Total	5,679,048	8,125,128	8,496,577	8,784,106	

The lower panel of Fig. 2 shows the observed QBER over the course of the experiment. As was discussed earlier, we see a lower average total QBER of 2.48% over the course of the experiment compared to previous experiments with the old SPDC source. The top panel of Fig. 2, shows the observed raw key rate (top series, in blue), sifted key rate (middle series, in green), and average final key rate (bottom series, in magenta) throughout the experiment. We observed an average raw key rate of 9,423 bits/s, an average sifted key rate of 4,765 bits/s, and an average final key rate of 2,695 bits/s. Table 1 shows the reconstructed coincidence matrix from Alice's and Bob's measurement data recorded during the experiment. The table also allows one to calculate the average visibilities of 95.13% and 94.95% in the rectilinear and diagonal bases from the experiment.

4   Discussion

As can be seen from the local experiment detailed above, the new Sagnac source shows much promise for producing entangled photon pairs with a lower error rate than before. Experiments with the source over one 1,305 m free-space link have yielded coincident detection rates for entangled photons of between 1,500 cps and 3,000 cps at 10 mW of pump power. While polarization stability issues in the 30 m singlemode fibre, which transports the entangled photons to rooftop sending telescopes, and the free-space link have so far made experiments with the full free-space link impossible; the initial numbers from free-space experiments make us very optimistic for good rates in a two free-space link experiment. Extrapolating from results in our previous experiment [3] where the transmission of one free-space link was ∼20% and the transmission of two free-space links was ∼9%, we are expecting to get coincident detection rates of approximately 850 cps at 10 mW of pump power and 2,500 cps at 50 mW of pump power for a two free-space link experiment. With an efficiency of ∼0.2860 secret key bits per raw key bit after sifting, error correction, and privacy amplification; we are hoping to get final key rates of ∼243 bits/s at 10 mW and ∼715 bit/s at 50 mW of pump power. This would represent an order of magnitude increase in our secret key rate over our previous experiments and would almost put the system in the kbits/s range, making it much more practical for real world applications.

One potential problem with the higher entangled photon rates generated with the new Sagnac source is the possibility of double pair emission events where two entangled photon pairs are created in the crystal within one coincidence window. Alice's and Bob's resulting measurements will no longer agree if they happen to measure photons from different pairs. Worse than this, the assumption that double clicks are negligible, needed to apply the security proof, will no longer be reasonable as the number of double pair emissions becomes appreciable. Indeed, preliminary experiments by one of us (D.H.) indicate that the visibilities degrade from 98.6% and 98.3% (H/V and +/-) to 85.0% and 84.4% (H/V and +/-) when increasing the laser power from 0.6 mW to 31.5 mW (measured after the optical isolator). This will be the subject of a forthcoming publication by some of the authors. In order to re-apply the security proof for the system, special care will be required to detect double clicks and assign each of them a random outcome. This will have the unfortunate consequence of increasing the error rate and shrinking the final key rate.

The narrow bandwidth of the new Sagnac source also leads to the additional benefit of helping towards daylight compatibility for the system. As was mentioned earlier, there are a number of groups [7] focused on operating a free-space QKD system during daylight hours. The main problem encountered with operating a free-space system during daylight hours is the extremely high background light levels. To combat this, there are really only 3 possible filtering techniques one can use: spatial, spectral, and temporal. Having a much narrower bandwidth than the previous source allows us to greatly improve the spectral filtering of the photons at the receiver stations. With a bandwidth of 0.36 nm we should be able to move from 10 nm to 1 nm filters to greatly reduce the background light without seeing a significant drop in the entangled photon detection rates. With slightly improved spatial and temporal filtering, the hope is that we shall be able to run the system in daylight conditions.

5   Conclusions

In conclusion, in order to improve on our previous experiments with an entangled free-space QKD system and make the system viable for real-world practical applications we have focused on improving the key generation rate of our system. In order to do this, we replaced the type-II SPDC entangled photon source, which was limited to an entangled photon pair detection rate of 18,000 cps with 50 mW of pump power detected locally, with a new brighter Sagnac entangled photon pair source, capable of producing local coincidence rates of 40,000 cps with just 5 mW of pump power. The narrow bandwidth of the source will also allow us to better filter out background light, thus helping to further reduce errors and the additional error correction and privacy amplification required. The narrow bandwidth will furthermore be useful in implementing daylight compatibility for the system. The one observed drawback of the source is a non-negligible double pair emission rate which will have to be carefully handled by the system so as not to expose a security loophole.

A local experiment examining the integration of the new source with the QKD system was detailed using 5 mW of pump power, which yielded a raw key rate of 9,423 bits/s and a final secret key rate of 2,695 bits/s. While issues with our free-space link prevented a fully distributed QKD experiment from being run, initial results with distributing entangled photon pairs over a single 1,305 m free-space link have seen entangled photon detection rates between 1,500 cps and 3,000 cps at 10 mW of pump power. Extrapolating from these numbers and our previous experiments with the old source, we expect to see a final secret key rate of ~715 bits/s running the source at the full 50 mW of pump power during a two-free space link experiment. This will represent an order of magnitude increase in our secret key rate over our previous experiments.

Acknowledgements. Support for this work by NSERC, QuantumWorks, CIFAR, CFI, CIPI, ORF, OCE, ERA, and the Bell family fund is gratefully acknowledged.

References

1. Marcikic, I., Lamas-Linares, A., Kurtsiefer, C.: Free-Space Quantum Key Distribution with Entangled Photons. Appl. Phys. Lett. 89, 101122 (2006)
2. Ursin, R., Tiefenbacher, F., Schmitt-Manderbach, T., Weier, H., Scheidl, T., Lindenthal, M., Blauensteiner, B., Jennewein, T., Perdigues, J., Trojek, P., Ömer, B., Fürst, M., Meyenburg, M., Rarity, J., Sodnik, Z., Barbieri, C., Weinfurter, H., Zeilinger, A.: Entanglement-Based Quantum Communication Over 144 km. Nature Physics 3, 481–486 (2007)
3. Erven, C., Couteau, C., Laflamme, R., Weihs, G.: Entangled Quantum Key Distribution Over Two Free-Space Optical Links. Opt. Exp. 16, 16840–16853 (2008)
4. Rarity, J.G., Tapster, P.R., Gorman, P.M., Knight, P.: Ground to Satellite Secure Key Exchange Using Quantum Cryptography. New J. Phys. 4, 82 (2002)
5. Aspelmeyer, M., Jennewein, T., Pfennigbauer, M., Leeb, W., Zeilinger, A.: Long-Distance Quantum Communication With Entangled Photons Using Satellites. IEEE J. of Selected Topics in Quantum Electronics 9, 1541 (2003)
6. Perdigues, J., Furch, B., de Matos, C., Minster, O., Cacciapuoti, L., Pfennigbauer, M., Aspelmeyer, M., Jennewein, T., Ursin, R., Schmitt-Manderbach, T., Baister, G., Rarity, J., Leeb, W., Barbieri, C., Weinfurter, H., Zeilinger, A.: Quantum Communications at ESA - Towards a Space Experiment on the ISS. In: 58th International Astronautical Congress, Hyderabad, India (2007)
7. Peloso, M.P., Gerhardt, I., Ho, C., Lamas-Linares, A., Kurtsiefer, C.: Daylight Operation of a Free Space, Entanglement-Based Quantum Key Distribution System (2008), eprint, quant-ph/0812.1880
8. Bennett, C.H., Brassard, G., Mermin, N.D.: Quantum Cryptography without Bell's Theorem. Phys. Rev. Lett. 68, 557 (1992)
9. Ma, X., Fung, C.H., Lo, H.K.: Quantum Key Distribution With Entangled Photon Sources. Phys. Rev. A 76, 012307 (2007)
10. Beaudry, N.J., Moroder, T., Lütkenhaus, N.: Squashing Models for Optical Measurements in Quantum Communication. Phys. Rev. Lett. 101, 093601 (2008)
11. Tsurumaru, T., Tamaki, K.: Security Proof for QKD Systems with Threshold Detectors (2008), http://arxiv.org/abs/0803.4226
12. Koashi, M., Adachi, Y., Yamamoto, T., Imoto, N.: Security of Entanglement-Based Quantum Key Distribution with Practical Detectors (2008), http://arxiv.org/abs/0804.0891

13. Lütkenhaus, N.: Institute for Quantum Computing, University of Waterloo, 200 University Avenue West, Waterloo, ON, N2L 3G1, Canada (personal communication, 2008)

14. Kwiat, P.G., Mattle, K., Weinfurter, H., Zeilinger, A., Sergienko, A., Shih, Y.: New High-Intensity Source of Polarization-Entangled Photon Pairs. Phys. Rev. Lett. 75, 4337 (1995)

15. Kim, T., Fiorentino, M., Wong, F.N.C.: Phase-Stable Source of Polarization-Entangled Photons using a Polarization Sagnac Interferometer. Phys. Rev. A 73, 012316 (2006)

16. Fedrizzi, A., Herbst, T., Poppe, A., Jennewein, T., Zeilinger, A.: A Wavelength-Tunable Fiber-Coupled Source of Narrowband Entangled Photons. Opt. Exp. 15, 15377 (2007)

17. Brassard, G., Salvail, L.: Secret-Key Reconciliation by Public Discussion. In: Helleseth, T. (ed.) EUROCRYPT 1993. LNCS, vol. 765, pp. 410–423. Springer, Heidelberg (1994)

18. Sugimoto, T., Yamazaki, K.: A Study on Secret Key Reconciliation Protocol "Cascade". IEICE Trans. Fundamentals E83A(10), 1987 (2000)

19. Carter, J.L., Wegman, M.N.: Universal Classes of Hash Functions. Journal of Computer and System Sciences 18, 143 (1979)

Solutions for Redundancy-Free Error Correction in Quantum Channel

Laszlo Bacsardi, Laszlo Gyongyosi, and Sandor Imre

Department of Telecommunications, Budapest University of Technology and Economics,
Magyar tudosok krt. 2., H-1111 Budapest, Hungary
{bacsardi,gyongyosi,imre}@hit.bme.hu

Abstract. All free-space quantum communications require the use of a quantum channel, which transports quantum bits in such a way that the quantum mechanical states of the qubits remain preserved from one end of the channel to the other one. In quantum computing the classical error coding methods could not be used, however we can construct a classical channel with zero redundancy error correction for any unitary channel. In our basically new quantum error correction approach, the classical states are coded into the eigenvectors and unitary transformations. In this paper, we show that with our new algorithm it's possible to create redundancy-free quantum error correction. We also consider the redundancy-free implementation of a unitary error correcting operator. Our protocol achieves the redundancy-free quantum communication using local unitary operations and unitary matrices. These solutions could be useful for the free-space quantum communication.

Keywords: Quantum channel, error correction, redundancy-free.

1 Introduction

Quantum theory takes advantage of quantum mechanical principles such as the *superposition* of states and their *no-cloning* principle. Cryptography based on quantum theory principles is known as *quantum cryptography*. In the past few years, quantum key distribution systems have been undertaken a deep study. Because classical cryptographic methods in wired and *wireless security* have been found to have vulnerabilities, new methods based on *quantum mechanical* principles have been deployed.

The first protocol in *quantum cryptography* was the BB84, which however did not take advantage of the full potential of multiple superposition states. The free-space Quantum Key Distribution (QKD) [1] was first introduced over an optical path of about 30 cm in 1991. Several demonstrations increased the usability of QKD by extending it with line-of-site laser communications systems. In 1998, a research group at Los Alamos National Laboratory, New Mexico, USA developed a free-space QKD over outdoor optical paths for up to 950 m under nighttime conditions [2]. Four years later, in 2002 the researchers of the same laboratory have demonstrated that free-space QKD is possible in daylight or at night [3]. In 2006, the distance of 144 km was reached by an international research group [4]. The actual implementation of quantum cryptography

A. Sergienko, S. Pascazio, and P. Villoresi (Eds.): QuantumCom 2009, LNICST 36, pp. 117–124, 2010.

systems would be invaluable, allowing for the first time the practical possibility of one-time-pad-encrypted, undecipherable communication, which will offer an essentially new degree of security in future communications.

Long distance quantum communication technologies and other quantum devices in the future will far exceed the processing capabilities of current silicon-based devices. In current network technology, in order to spread *quantum cryptography*, interfaces able to manage together the quantum and classical channel must be implemented.

In our point of view, the quantum computing algorithms can be used to affirm our free-space communication in the following four ways: [5]

1. *Open-air communication*: usually "horizontal" telecommunication that happens below 100km height. For channel, the air is used instead of optical cable.
2. *Earth-satellite communications*: it happens through greater heights than the Open-air communication, usually between 300 and 800 km altitude. Signal encoding and decoding is used to produce quantum error correction that allows operation in noisy environment.
3. *Satellite broadcast*: Quantum algorithms can improve the effective bandwidth, thus the brand is better utilized as in traditional cases.
4. *Inter-satellite communication*: the communication between satellites where the channel is the free-space. Any kind of coding and encoding can be used, to increase stability [6].

Despite the fine number of results a lot of work has to be done. The existing experiments usually use one of the easiest key distribution protocols. There is a need to trace some adoptable algorithms and apply them to communication problems between Earth and satellite and also between satellites. For this, a well-described channel model should be set up. Correct parameters to describe the noise of the different types of atmosphere should be found. As the quantum channels show few similarities with the classical ones describing those require more sophisticated approaches.

2   General Quantum Channel

A quantum bit, or a qubit is a quantum system used to store information. As opposed to a bit which can be in one of two states "0" and "1", a qubit can exist in a *continuum* of states. Moreover, we can measure the value of a bit with certainty without affecting its state, while the result of measuring a qubit is non-deterministic and the measurement alters its state. Computer and communication systems using quantum effects have remarkable properties. Quantum algorithms allow efficient factoring of large integers with applications to cryptography.

Using quantum channels, the information carrying quantum system is in interaction with the environment as an undesirable noise. This phenomena is named quantum decoherence [1]. The noise appearing from the entanglement with the environment can be observed in Figure 1.

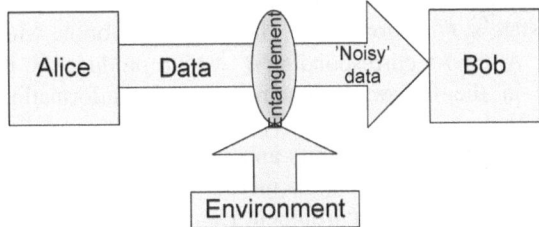

Fig. 1. General model of the quantum channel

For a well functioning communication we need a channel coding to handle the errors appearing in a communication channel. In quantum computing the classical error coding methods could not be used because of the following three reasons [7]:

1. *The errors are continuous.* The errors can results either amplitude or phase decoherence. Moreover both errors have complex coefficients which mean that their codomains are continuous.
2. *Through the No Cloning Theorem* (cloning is allowed only for the classical states e.g., 0 or 1) a simple copy-based redundancy is unadmittable.
3. There are "problems" with the *measurement* of the transmitted states. For the error correction the type of error has to be known but if the quantum bits are measured for determination of the failure then the original bits are lost.

Despite these challenges, several quantum based error correction have been published but they are based on quantum and not classical theorems [8]. In this paper we present a new redundancy-free solutions.

3   Redundancy-Free Channel

The base of our redundancy-free quantum error correction mechanism lies in the fundamental difference between classical and quantum information. In our system the noise of the quantum channel is modeled by a rotation angle. The questions are how to send over a noisy quantum channel certain amount of qubits, to provide error correction. Any physical realization of a quantum channel is likely to be susceptible to errors, because we cannot build perfect physical systems and isolate them from their environments while still maintaining control over the quantum states. We have to use quantum error correction codes to protect quantum information against such errors. The main idea in our redundancy-free theory is the engineering precision, which means that we usually don't need 100 percent perfect solution for an engineering challenge, the 99 percent perfect solution is a good solution. Of course the above described method is only in a rough state, for further use the model further investigations are needed.

The correction of the damaged quantum states is not possible in a classical representation, since the error corretion of qubits is realized by unitary rotations. Our initial assumption is that the channel rotates the qubit with an ω degree, that is consdired to be constant so far. We wish to create a system where error correction is possible. The transmission is considered successful when at the end of the channel the qubit remains in its original state's ε environment. The main question is, whether it is possible to construct such A (and a corresponding B, which produce the inverse of matrix A) transformation in the following scheme, that the information can be processed through the channel

To achieve this we mix the qubits and send them over the channel, as shown in Figure 2. What we expect is that at the measurement, the error for one qubit is distributed among the others in its environment (its neighbors). By being so, the error remains in an ε environment for each qubit.

We use n long qubits so that $2^n = N$, where n is the length of the qubits and N is the size of the space. Let the l th qubit in the sequence sent through the channel be:

$$|\psi_l\rangle = a_l|0\rangle + b_l|1\rangle, \text{ where } |a|^2 + |b|^2 = 1. \tag{1}$$

This case the entire sequence sent through the channel can be described as:

$$|\phi\rangle = \otimes_{i=1}^n |\psi_l\rangle, \tag{2}$$

$$|\phi\rangle = \sum_{i=0}^{2^n-1} \phi_i |i\rangle. \tag{3}$$

One can construct a classical channel with zero redundancy error correction for any unitary channel. Of course the information itself is classical, coded into qubits. This case the channel model is the following: The inputs and outputs are classical bits: $(|0\rangle, |1\rangle)$. Since U is unitary, thus it can be written in the following form:

$$U = \sum_i \lambda_i |u_i\rangle, \tag{4}$$

where λ_i, $|u_i\rangle$ are the eigenvalues and the eigenvectors of matrix U and

$$\lambda_n = e^{j\alpha_n}. \tag{5}$$

Because U is unitary, it acts on each qubit and changes it as

$$|\psi\rangle = U|\psi_k\rangle. \tag{6}$$

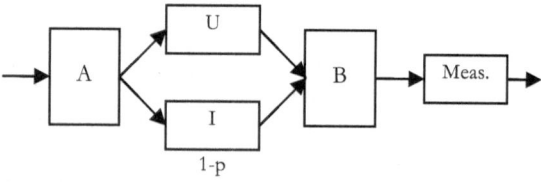

Fig. 2. Our channel model. A transforms the initial qubits into a special form. B has to produce the inverse of matrix A.

Using the eigenvalues from we get the following matrix for U

$$U_k = \begin{bmatrix} e^{j\alpha_{k1}} & 0 \\ 0 & e^{j\alpha_{k2}} \end{bmatrix}. \tag{7}$$

Because of (7) U must like

$$U = \begin{bmatrix} e^{j\pm\alpha_{k1}} & 0 \\ 0 & e^{j\pm\alpha_{k2}} \end{bmatrix}. \tag{8}$$

Now we need to do some assumptions. Let us suppose that we have two. As for the eigenvalues, we have two cases

$$\text{I. } U = \begin{bmatrix} e^{j+\alpha} & 0 \\ 0 & e^{j+\alpha} \end{bmatrix} \otimes \begin{bmatrix} e^{j+\alpha} & 0 \\ 0 & e^{j+\alpha} \end{bmatrix} = \begin{bmatrix} e^{j2\alpha} & 0 \\ 0 & e^{j2\alpha} \end{bmatrix}, \tag{9}$$

$$\text{II. } U = \begin{bmatrix} e^{j+\alpha} & 0 \\ 0 & e^{j+\alpha} \end{bmatrix} \otimes \begin{bmatrix} e^{j-\alpha} & 0 \\ 0 & e^{j-\alpha} \end{bmatrix} = \begin{bmatrix} 1 & 0 \\ 0 & 1 \end{bmatrix} = I. \tag{10}$$

This description lead to a redundancy-free solution because the classical states are coded into the eigenvectors of the U matrix and the eigenvalues can be written in the form shown in (5) in case of a unitary transformation.

With this model one can create redundancy-free error correction. It also works for higher dimensions, not only two. The first simulation results show that with the appropriate selection of the matrix A we can restore one quantum bit sent over the channel without any other (redundant) information.

4 Generalized Redundancy-Free Channel

In this section we consider the redundancy-free implementation of an unitary error correcting operator $\mathcal{R}_\theta$. Our protocol achieves the redundancy-free quantum communication using *local unitary operations* and *unitary matrices*.

4.1 The Redundancy-Free Error Correction

The *error* of the quantum channel is modeled by a unitary transformation, which is denoted by rotation $\mathcal{R}_\theta^\dagger$. In our model the error of the Quantum Channel is an angle $\theta_i \in [0, 2\pi)$ for every classical bit, which prepares the quantum state $|\psi_i\rangle = \cos\theta_i |0\rangle + \sin\theta_i |1\rangle$. In Figure 3. Alice's original qubit is denoted by ψ_A. At the beginning of the communication, Alice sends her quantum state ψ_A on the quantum channel, which changes to $|d\rangle = \mathcal{R}_\theta^\dagger(\psi_A)$ with given probability p. The error of the quantum channel is denoted by $\mathcal{R}_\theta^\dagger$.

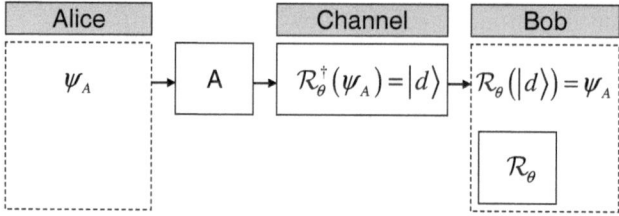

Fig. 3. Generalized channel – initial step

Let the *error-correcting* unitary operation be $\mathcal{R}_\theta \equiv e^{(i\theta\sigma_z/2)}$, for an arbitrary angle $\theta \in [0, 2\pi)$, which corresponds to an arbitrary rotation around the $\hat{z}$-axis of a spin ½ particle. In our error-correcting process when Bob tries to read the sent quantum state, he doesn't know the effect of the quantum channel, thus the required angle state $|\theta\rangle$ nor the complex coefficients a and b.

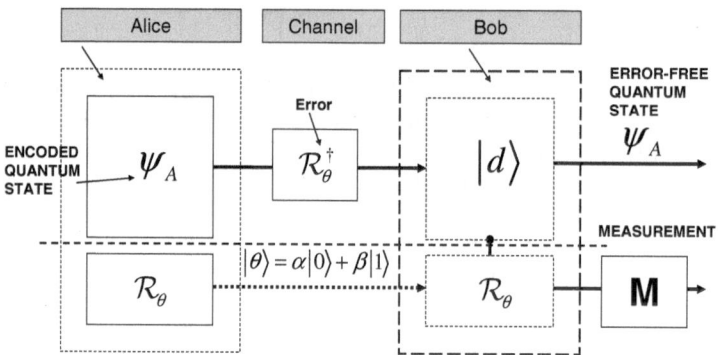

Fig. 4. Our redundancy-free coding mechanism

In Figure 4. we illustrated our redundancy-free coding mechanism, Alice initial state is ψ_A, her correction state is $\mathcal{R}_\theta$. Bob uses a CNOT to correct the error of the quantum channel.

For every quantum state, Bob measures with the projection operator $\mathcal{P} = |0\rangle\langle 0| + |1\rangle\langle 1|$. The result of the linear operator $\mathcal{P} = |0\rangle\langle 0|$ acting on an unknown $|\psi\rangle = \alpha|0\rangle + \beta|1\rangle$ quantum state, projects the state $|\psi_i\rangle$ into state $|0\rangle$ with probability $|\alpha|^2$, while the linear operator $\mathcal{P} = |1\rangle\langle 1|$ projects into state $|1\rangle$ with probability $|\beta|^2$.

4.2 Probabilistic Quantum Error Correction

In our redundancy-free model, in order to read the sent quantum bits correctly, Bob must rotate the i-th data quantum bit by the angle θ_i in the opposite direction of what the *error of the quantum channel rotated*. The error angle $-\theta_0$ of the channel can be corrected using the opposite direction θ_0, which is encoded by the rotation operator $\mathcal{R}(\theta_0)$. The rotation operator $\mathcal{R}(\theta_0)$ in matrix form is:

$$\mathcal{R}(\theta_0) = \begin{pmatrix} \cos\theta_0 & \sin\theta_0 \\ -\sin\theta_0 & \cos\theta_0 \end{pmatrix}. \tag{11}$$

We denoted the sent qubits by $\psi_A = a|0\rangle + b|1\rangle$, and the error of the quantum channel by $\mathcal{R}_{\theta_i}^\dagger$, and on Bob's side, the error-correcting mechanism is realized by a *unitary rotation* $\mathcal{R}(\theta_i)$. Bob has a chance *not greater* than $\varepsilon = \sin^2(\theta_i)$ to correct the sent states, because he doesn't know the original rotation angle θ_i of the quantum channel's error on the *i*-th sent qubit. The rotation operation $\mathcal{R}_\theta$ of the error correcting mechanism can be given by the angle $|\theta\rangle$, since

$$|\theta\rangle = \frac{1}{\sqrt{2}}\left(e^{i\frac{\theta}{2}}|0\rangle + e^{-i\frac{\theta}{2}}|1\rangle\right). \tag{12}$$

The error-correcting method consists of a *control qubit, which* corresponds to the damaged qubit $|d\rangle$, and a *target qubit,* which is equal to the *error-correction* angle state $|\theta\rangle$. To correct state $|d\rangle$ to ψ_A, we use a simple CNOT transformation, thus our state is transformed to

$$|d\rangle \otimes |\theta\rangle \rightarrow \frac{1}{\sqrt{2}}\left(\mathcal{R}_\theta |d\rangle \otimes |0\rangle + \mathcal{R}_\theta^\dagger |d\rangle \otimes |1\rangle\right), \tag{13}$$

and therefore a projective measurement in the $\{|0\rangle, |1\rangle\}$ basis of the correction-state $|\theta\rangle$ will make the damaged qubit $|d\rangle$ collapse either into the desired state $\mathcal{R}_\theta |d\rangle$ or into the wrong state $\mathcal{R}_\theta^\dagger |d\rangle$.

5 Conclusions

Quantum communication specifically provides a method of distributing the secret keys required to provide unconditionally secret communications and its use is guaranteed to reveal the presence of an enemy attempting to compromise the transfer. The redundancy-free channel is not only a solution for wired systems, but could be part of the wireless communication too. This method could be very useful in the long-distance aerial communication, because there would be no need to use redundant error correction codes as nowadays. This way the effective capacity of the satellite link would also be increased.

In this paper we presented a completely new method to correct quantum states. The rotation operations applied in the error correcting mechanism can be implemented with some associated error, which decreases exponentially with the number of quantum states of the error correction state. With redundancy-free solutions we can get over some troubles issued from the atmosphere (in earth-satellite communication) and we can achieve higher bandwidth (effective one) in satellite-communication.

This paper should be also regarded as a starting point for further analysis of the properties and efficiency of our redundancy-free quantum correction system.

References

[1] Imre, S., Ferenc, B.: Quantum Computing and Communications: An Engineering Approach. Wiley, Chichester (2005)

[2] Buttler, W.T., Hughes, R.J., Kwiat, P.G., Lamoreaux, S.K., Luther, G.G., Morgan, G.L., Nordholt, J.E., Peterson, C.G., Simmons, C.M.: Practical free-space quantum key distribution over 1 km, arXiv:quant-ph/9805071

[3] Hughes, R.J., Nordholt, J.E., Derkacs, D., Peterson, C.G.: Practical free-space quantum key distribution over 10 km in daylight and at night. New Journal of Physics 4, 43.1–43.14 (2002)

[4] Schmitt-Manderbach, T., Weier, H., Fürst, M., Ursin, R., Tiefenbacher, F., Scheidl, T., Perdigues, J., Sodnik, Z., Kurtsiefer, C., Rarity, J.G., Zeilinger, A., Weinfurter, H.: Experimental Demonstration of Free-Space Decoy-State Quantum Key Distribution over 144 km. Physical Review Letters PRL 98, 010504 (2007)

[5] Bacsardi, L.: Using Quantum Computing Algorithms in Future Satellite Communication. Acta Astronautica 57(2-8), 224–229 (2005)

[6] Bacsardi, L.: Satellite communication over quantum channel. Acta Astronautica 61(1-6), 151–159 (2007)

[7] Nielsen, M.A., Chuang, I.L.: Quantum Computation and Quantum Information. Cambridge University Press, Cambridge (2000)

[8] Poulin, D.: Stabilizer Formalism for Operator Quantum Error Correction (2005), Quant-ph/0508131

Two-Way Quantum Communication in a Single Optical Fiber with Active Polarization Compensation

G.B. Xavier[1], G. Vilela de Faria[1], T. Ferreira da Silva[1,2], G.P. Temporão[1], and J.P. von der Weid[1]

[1] Center for Telecommunication Studies, Pontifical Catholic University of Rio de Janeiro, R. Marquês de São Vicente, 225, Rio de Janeiro, Brazil
{guix,gian,thiago,temporao,vdweid}@opto.cetuc.puc-rio.br
[2] National Institute of Metrology, Standardization and Industrial Quality, Av. Nossa Sra. das Graças, 50, Duque de Caxias, Brazil

Abstract. We experimentally demonstrate a two-way stable transmission of polarization encoded qubits over 23 km of spooled dispersion-shifted fiber with active polarization control in both directions, while simultaneously exchanging classical data. Two classical reference channels (one containing a telecom 10 Gb/s data stream), wavelength-multiplexed with the quantum signal, are used as feedback. The feasibility of quantum communication is demonstrated in the two opposite directions over 6 hours of continuous operation, as well as a classical error rate better than 1.0×10^{-9}.

Keywords: Quantum communications, Polarization, Optical networks.

1 Introduction

Most quantum communication [1] experiments to date have used a dark optical fiber link for the transmission of qubits between Alice and Bob [2-5]. For a broader insertion of quantum technologies into classical optical telecom environments, it is important to be able to reliably transmit qubits in a fiber populated with telecom traffic. For telecom operators, this is of great interest, since employing a single fiber solely for quantum transmission is not economically attractive.

By far the most widely demonstrated application of quantum information has been quantum key distribution (QKD) [6] employing the BB84 [7] protocol. BB84 is simply a secure way of transmitting the key if the bits are encoded in orthogonal pairs of quantum states which are grouped in two non-orthogonal bases. The quantum channel only needs to be one-way, but a classical communication channel is also needed between Alice and Bob.

From a standard classical optical network point of view, two-way quantum communication makes sense since, in a reconfigurable environment, quantum signals may need to be sent back and forth between Alice, Bob and other eventual network users. In addition to that, the same optical fiber may be shared by many users on both ends, and two-way communication needs to be provisioned in advance by the network operators. Specifically for QKD, there are protocols that require two-way quantum channels [8-11].

A. Sergienko, S. Pascazio, and P. Villoresi (Eds.): QuantumCom 2009, LNICST 36, pp. 125–131, 2010.

In this work we present, to the best of our knowledge, the first experimental demonstration of a two-way quantum communication channel with active continuous polarization control [12,13] in both directions in long-distance optical fibers. This experiment shows that it is possible to perform classical telecom activity and transmit quantum signals simultaneously in an optical fiber with active polarization control. These results can be used to the benefit of telecom operators as well as improving practical implementations of quantum communications.

2 Experimental Theory and Setup

Optical fibers have been extensively used in quantum communication experiments because their properties suit well the transmission of single-photons. In addition to that, they have been extensively optimized by the telecom industry such that one can obtain relatively cheap fibers having low attenuation and chromatic dispersion in the 1550 nm telecom window. They are also widely deployed making it easier to install QKD systems to many different users in the future.

The two most widely used encoding methods of information for quantum communications are polarization [7] and time-bin (phase) [14]. Polarization was the first one used in experimental demonstrations; however, its use in optical fibers was quickly discontinued because of residual birefringence present in the fibers. The process of manufacturing optical fibers is not perfect and therefore a residual birefringence remains present. The problem arises since it depends on environmental factors (temperature and mechanical stresses applied to the fiber), and therefore the polarization state of an optical field at the output of the fiber randomly changes with time. This makes polarization-based quantum communications in optical fibers unfeasible unless active control is employed, which opened the way to phase encoding.

In order to have continuous polarization control enabled, our control system employs two reference channels wavelength-multiplexed with the quantum channel. In order to have optimum control, they need to be located as close as possible to the quantum channel, and the mean differential group delay of the fiber link cannot be much higher than ~ 1 ps [13]. For compatibility purposes and easy availability of components, we place our wavelengths of choice in the ITU-T wavelength grid (International Telecommunication Union - Telecommunication Standardization Sector). We use the 100 GHz grid spacing, meaning that the channels are located 0.8 nm apart in the 1550 nm window. Our quantum channel is located at $\lambda_Q = 1546.12$ nm, with the reference channels located at $\lambda_1 = 1545.32$ and $\lambda_2 = 1546.92$ nm. The experimental setup is shown in Fig. 1.

We have respectively labeled Alice and Bob the transmitter and receiver of the classical information channels, as they only propagate in one direction in our setup. Alice has the classical lasers to transmit the two reference lasers at λ_1 and λ_2 wavelengths. The laser located at λ_1 is a standard telecom distributed feedback (DFB) laser diode operating in continuous wave (CW) mode. The other one is the laser output of a DFB laser inside a bit error rate (BER) test meter operating at 9.953 Gb/s, simulating real telecom traffic.

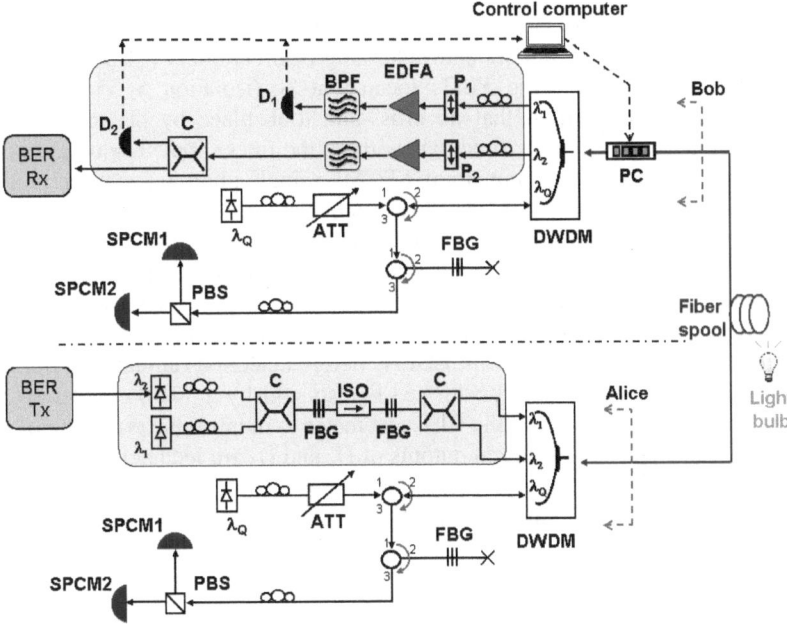

Fig. 1. Experimental setup. The light blue bounded areas at both Alice and Bob's stations represent the control and classical components. The classical channels propagate from Alice to Bob, while the quantum signals propagate simultaneously in both directions. ATT: Optical attenuator; BER Rx and Tx: Bit error rate meter receiver and transmitter respectively; BPF: Band pass filter; D1 and D2: Classical p-i-n photodetectors; DWDM: Dense wavelength division multiplexer; EDFA: Erbium doped fiber amplifier; FBG: Fiber Bragg grating; ISO: Optical isolator; SPCM: Single photon counting module; P1 and P2: Linear polarizers; PC: $LiNbO_3$ polarization controller; PBS: Polarizing beam splitter.

The states of polarization (SOPs) of both channels need to be adjusted with non-orthogonal states using manual polarization controllers before being multiplexed and entering the fiber link [11]. Both channels are combined in a 50/50 coupler, before going through two identical fiber Bragg gratings (FBG) filters, each reflecting 99.9% at $\lambda_Q = 1546.12$ nm. The two FBGs provide ~ 60 dB attenuation at λ_Q, thus reducing the cross-channel amplified spontaneous emission (ASE) noise to ~ 100 dB below the power at the center wavelength of the lasers. One isolator is used between the two FBGs to avoid the creation of a Fabry-Perot cavity. Finally another 50/50 coupler splits the two signals and they are multiplexed in the optical link together with the quantum wavelength λ_Q. The loss the two reference channels suffer when passing through both 50/50 couplers is not of importance since we need to attenuate both channels before inserting them in the optical link to minimize Raman spontaneous scattering noise [15, 16]. The launch powers for both channels after the multiplexer are of -19.8 dBm each.

Alice and Bob are separated by 23 km of dispersion-shifted spooled fiber. The SOPs of the three channels are actively compensated by a $LiNbO_3$ polarization controller (PC) located close to Bob. An incandescent light bulb is used to randomly heat the fiber to simulate temperature changes. Alice and Bob have identical optical hardware to prepare the polarization qubits to be sent and decode the received ones: attenuated lasers at λ_Q, circulators and polarizing beam splitters (PBS), all shown in Fig. 1. Filtering is done using FBGs located at λ_Q (isolation > 35 dB) and the DWDMs, and it was verified that no cross-talk took place by performing photon counting measurements in both directions without the fiber spool. The output ports of the PBS are connected to commercial SPCMs (single-photon counting modules). They operate in Geiger mode [6] at a gate frequency of 100 kHz, with a quantum efficiency of 15% at 1550 nm and a gate width of 2.5 ns. The measured dark count probability is 3.7 x 10^{-5} and 3.2 x 10^{-5} per gate for each SPCMs respectively.

Within Bob's setup, the classical channels are split and pass through linear polarizers P_1 and P_2 oriented at 45° from each other using manual polarization controllers [12, 13]. λ_1 and λ_2 are amplified with EDFAs before detections at classical *pin* photodiodes D_1 and D_2 (300 kHz bandwidth). λ_2 is also split by a 50/50 coupler so that it can also feed the receiver unit of the BER test meter to monitor the error rate of the 10 Gb/s data stream. The two electrical outputs of D_1 and D_2 are fed back into the control computer to close the feedback loop, allowing us to undo any birefringence rotations the fiber may cause for *any* input SOP [13].

3   Experimental Results

A single SOP is sent from each end of the fiber, and the manual polarization controllers before each PBS are adjusted to maximize the counts on one of the SPCMs (the SOP matches the measurement basis). The counts are recorded as a function of time, and the results are plotted in Fig. 2. Initially, the active control system was switched off and the counts recorded by Alice's detectors. We observe that the counts almost immediately begin to drift due to the birefringence variation. The system is then switched on, the states realigned and the counts recorded over a similar time period. The two SPCMs were moved to Bob's side, the counts are recorded and the results from both controlled cases also plotted in Fig. 2. It is clear that the counts in both SPCMs remain stable throughout the experimental runs with the active polarization control switched on. It is also worth mentioning that the BER rate stayed better than 1.0 x 10^{-9} during all measurements, showing that our control system is compatible with simultaneous telecom traffic. The long-term drift observed in the SPCM1 Alice curve is believed to have been caused by a small change in the SOPs before or after the DWDMs. The drop in the count rate when comparing Bob's and Alice's SPCMs is due to ~ 3 dB attenuation caused by the LiNbO$_3$ polarization controller, since in the Alice-Bob path, the controller increases the attenuation of the quantum channel, while in the opposite direction it does not (it is a part of Bob's setup). The launched mean photon number per gate was 1.0 on both sides. It should be noted that our polarization stabilization system is able to compensate for any input polarization state [13]. The visibility can be calculated for the case when the received SOP is compatible with the measurement basis as a function of time. The two controlled cases (co- and counter-propagating) exhibit average visibilities of 0.916 ± 0.025 and 0.931 ± 0.016 respectively, with the deviation from perfect visibility stemming from detector dark counts,

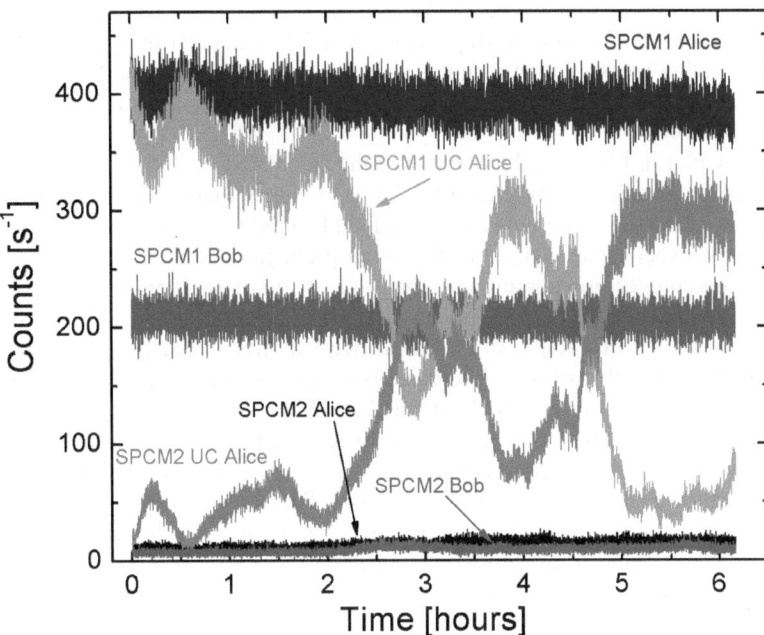

Fig. 2. Experimental results with and without polarization control as a function of time. SPCM1 UC Alice and SPCM2 UC Alice are measurements with the system control turned off.

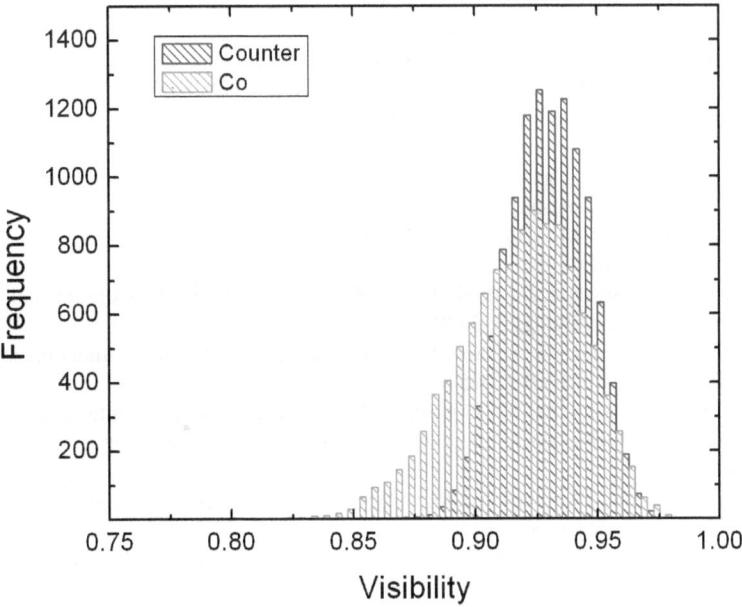

Fig. 3. Distribution of calculated visibilities obtained from the data in figure 2 for co- and counter-propagating directions.

Raman noise, fluctuations added by the polarization stabilizer and imperfect manual alignment of the PBSs with the single photons SOPs. Unsurprisingly the visibility for the uncontrolled measurement wanders randomly across all possible values, showing once again that quantum communication using polarization states without active control is unfeasible.

We now plot in Fig. 3 the histograms of the calculated visibilities from the data of the two controlled directions shown in Fig. 2. It shows that the visibility never goes below 0.8, allowing continuous uninterruptable QKD [6]. As expected from the results in the experiment, visibilities for the co-propagating case are worse when compared to the counter-propagating case because of the extra attenuation in the Alice-Bob direction (due to the LiNbO$_3$ polarization controller).

We have shown that stable fiber optical two-way transmission of polarization encoded qubits is possible in two simultaneous directions in a single long-distance optical fiber, shared with two classical optical channels (one of them containing a 10 Gb/s data stream), used to provide feedback to the active polarization compensation system. Limitations of distance were estimated for our current setup performing simultaneous quantum and classical communications. These results are important for any quantum protocols requiring two-way exchange of qubits, as well as future implementations of hybrid classical-quantum optical networks.

The authors acknowledge financial support from CAPES, FAPERJ and CNPq.

References

1. Gisin, N., Thew, R.: Quantum communication. Nat. Photon. 1, 165 (2007)
2. Takesue, H., et al.: Quantum key distribution over a 40-dB channel loss using superconducting single-photon detectors. Nat. Photon. 1, 343 (2007)
3. Peng, C.-Z., et al.: Experimental Long-Distance Decoy-State Quantum Key Distribution Based on Polarization Encoding. Phys. Rev. Lett. 98, 010505 (2007)
4. Yuan, Z.L., Sharpe, A.W., Shields, A.J.: Unconditionally secure one-way quantum key distribution using decoy pulses. Appl. Phys. Lett. 90, 011118 (2007)
5. Stucki, D., et al.: High speed coherent one-way quantum key distribution prototype (2008), arXiv:0809.5264 [quant-ph]
6. Gisin, N., Ribordy, G., Tittel, W., Zbinden, H.: Quantum cryptography. Rev. of Mod. Phys. 74, 145 (2002)
7. Bennett, C.H., Bessette, F., Brassard, G., Salvail, L., Smolin, J.: Experimental quantum cryptography. J. Cryptology 5, 3 (1992)
8. Lucamarini, M., Mancini, S.: Secure deterministic communication without entanglement. Phys Rev. Lett. 94, 140501 (2005)
9. Cerè, A., Lucamarini, M., Di Giuseppe, G., Tombesi, P.: Experimental test of two-way quantum key distribution in presence of controlled noise. Phys. Rev. Lett. 96, 200501 (2006)
10. Kumar, R., et al.: Two-way quantum key distribution at telecommunication wavelength. Phys. Rev. A 77, 022304 (2008)
11. Boyer, M., Kenigsberg, D., Mor, T.: Quantum key distribution with classical bob. Phys. Rev. Lett. 99, 140501 (2007)

12. Xavier, G.B., Vilela de Faria, G., Temporão, G.P., von der Weid, J.P.: Full polarization control for fiber optical quantum communication systems using polarization encoding. Opt. Express 16, 1867 (2008)
13. Xavier, G.B., et al.: Experimental polarization encoded quantum key distribution over optical fibres with real-time continuous birefringence compensation. New. J. Phys. 11, 045015 (2009)
14. Townsend, P.D., Rarity, J.G., Tapster, P.R.: Enhanced single photon fringe visibility in a 10 km-long prototype quantum cryptography channel. Electron. Lett. 29, 634 (1993)
15. Xavier, G.B., Vilela de Faria, G., Temporão, G.P., von der Weid, J.P.: Scattering Effects on QKD Employing Simultaneous Classical and Quantum Channels in Telecom Optical Fibers in the C-band. In: Ninth international conference on quantum communication, measurement and computing 2009. AIP Conf. Proc., Calgary, vol. 1110, p. 327 (2009)
16. Peters, N.A., et al.: Dense wavelength multiplexing of 1550 nm QKD with strong classical channels in reconfigurable networking environments. New J. Phys. 11, 045012 (2009)

Passive Decoy State Quantum Key Distribution

Marcos Curty[1], Tobias Moroder[2,3], Xiongfeng Ma[2], and Norbert Lütkenhaus[2,3]

[1] Department of Signal Theory and Communications, ETSI Telecomunicación,
University of Vigo, E-36310 Vigo, Spain
mcurty@com.uvigo.es
[2] Institute for Quantum Computing, University of Waterloo,
Waterloo, ON, N2L 3G1, Canada
[3] Quantum Information Theory Group, Institut für Theoretische Physik I,
and Max Planck Institute for the Science of Light, University of Erlangen-Nürnberg,
91058 Erlangen, Germany

Abstract. The use of decoy states enhances the performance of practical quantum key distribution systems significantly by monitoring the quantum channel in a more detailed way. While active modulation of the intensity of the pulses is an effective way of preparing decoy states in principle, in practice passive preparation might be desirable in some scenarios. Known passive methods involve parametric down-conversion. In this paper we show how phase randomized coherent states can be used for the same purpose. Our method involves only linear optics together with a simple threshold photon detector. The performace is comparable to the active decoy methods.

Keywords: Quantum cryptography, quantum key distribution, quantum communication, threshold photon detector.

1 Introduction

Quantum key distribution (QKD) is the first quantum information task to reach the commercial market [1]. It allows two parties (typically called Alice and Bob) to generate a cryptographic key despite the computational and technological power of an eavesdropper (Eve), who interferes with the signals [2]. This secret key is the essential ingredient of the one-time-pad or Vernam cipher, which can provide information-theoretic secure communications.

After the first demonstration of its feasibility [3], several practical implementations of QKD have been realized in recent years [4], [5], [6]. These schemes are typically based on the transmission of weak coherent pulses (WCP); especially since single photon sources are still beyond our present experimental capability. This fact opens an important security loophole. Now, some of the signals contain more than one photon prepared in the same polarization state. In this scenario, Eve is no longer limited by the no-cloning theorem since in these events the signal itself provides her with perfect copies of the signal photon. She can perform, for instance, the so-called *Photon Number Splitting* attack on the multi-photon pulses [7], [8]. This attack provides Eve with full information about the part

A. Sergienko, S. Pascazio, and P. Villoresi (Eds.): QuantumCom 2009, LNICST 36, pp. 132–141, 2010.

of the key generated with the multi-photon signals, without causing any disturbance in the signal polarization. As a result, it turns out that the standard BB84 protocol [9] with WCP can deliver a key generation rate of order $O(\eta^2)$, where η denotes the transmission efficiency of the quantum channel [10,11]. This performance contrasts with the one expected from a QKD scheme using a single photon source, where the key generation rate scales linearly with η.

A significant improvement of the secret key rate can be obtained when the original hardware is slightly modified. In particular, it has been recently shown that decoy state QKD with WCP can basically reach the same performance as single photon sources [12], [13], [14]. In this approach, Alice varies, independently and at random, the mean photon number of each signal state she sends to Bob by employing different intensity settings. This is usually performed by using a variable optical attenuator together with a random number generator. Eve does not know a priori the mean photon number of each signal state sent by Alice. This means that her eavesdropping strategy can only depend on the photon number of these signals, but not on the particular intensity setting used to generate them. From the measurement results corresponding to different intensity settings, the legitimate users can obtain a better estimation of the behavior of the quantum channel. This translates into an enhancement of the resulting secret key rate. This technique has been successfully implemented in several recent experiments [15], [16], [17], which show the practical feasibility of this method.

While active modulation of the intensity of the pulses suffices to perform decoy state QKD in principle, in practice passive preparation might be desirable in some scenarios. For instance, in those setups operating at high transmission rates. Known passive methods rely on the use of a parametric down-conversion source together with a photon detector [18], [19], [20]. In this paper we show that phase randomized WCP can also be used for the same purpose, i.e., one does not need to employ a non-linear optics network preparing entangled states. The main idea is rather simple, although it is counter-intuitive [21]. When two phase randomized coherent states interfere at a beam splitter (BS), the photon number statistics of the outcome signals are classically correlated. This effect contrasts with the one coming from the interference of two pure coherent states at a BS. By measuring one of the two outcome signals, the conditional photon number distribution of the other one varies depending on the result obtained. This measurement can be performed, for instance, with a simple threshold photon detector. In the asymptotic limit of an infinite long experiment, we show that our passive decoy scheme can provide a similar performance to the one achieved with an active source and infinity decoy settings. This idea can also be applied to other practical scenarios with different signals and detectors like, for example, those based on thermal states or even strong coherent pulses in conjunction with a regular photo-detector [22]. In this context, see also [23].

2 Passive Decoy State QKD Setup

The basic setup is illustrated in Fig. 1 (Case A). Suppose two phase randomized WCP,

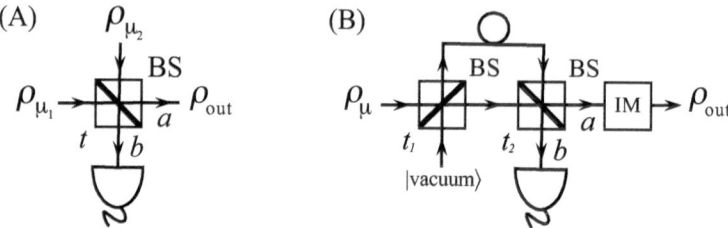

Fig. 1. (A) Basic setup of a passive decoy state QKD scheme: Interference of two phase randomized WCP, ρ_{μ_1} and ρ_{μ_2}, at a beam splitter (BS) of transmittance t. a and b represent the two output modes. (B) Alternative *active* setup with only one weak laser. The delay introduced by one arm of the interferometer is equal to the time difference between two pulses. The intensity modulator (IM) blocks either the even or the odd pulses.

$$\rho_{\mu_1} = e^{-\mu_1} \sum_{n=0}^{\infty} \frac{\mu_1^n}{n!} |n\rangle\langle n| \quad \text{and} \quad \rho_{\mu_2} = e^{-\mu_2} \sum_{n=0}^{\infty} \frac{\mu_2^n}{n!} |n\rangle\langle n| , \qquad (1)$$

interfere at a BS of transmittance t. Here $|n\rangle$ denote Fock states with n photons. Then, it turns out that the photon number statistics of the two outcome signals are classically correlated [21]. To see this, let us first consider the interference of two pure coherent states with fixed phase relationship, $|\sqrt{\mu_1}e^{i\phi_1}\rangle$ and $|\sqrt{\mu_2}e^{i\phi_2}\rangle$, at a BS of transmittance t. The outcome signals are given by

$$\left|\sqrt{\mu_1 t}e^{i\phi_1} + i\sqrt{\mu_2(1-t)}e^{i\phi_2}\right\rangle_a \otimes \left|i\sqrt{\mu_1(1-t)}e^{i\phi_1} + \sqrt{\mu_2 t}e^{i\phi_2}\right\rangle_b . \qquad (2)$$

The joint probability $p_{n,m}$ of having n photons in mode a and m photons in mode b is therefore given by the product of two Poissonian distributions:

$$p_{n,m} = e^{-v/2} \frac{(v\gamma)^n}{n!} \times e^{-v/2} \frac{[v(1-\gamma)]^m}{m!} , \qquad (3)$$

with $v = \mu_1 + \mu_2$, $\gamma = [\mu_1 t + \mu_2(1-t) + \xi\cos\theta]/v$, $\xi = 2\sqrt{\mu_1\mu_2(1-t)t}$ and $\theta = \pi/2 + \phi_2 - \phi_1$. The case of two phase randomized WCP can be solved by just integrating $p_{n,m}$ over all angles θ. We obtain [21]

$$p_{n,m} = \frac{v^{n+m}e^{-v}}{n!m!} \frac{1}{2\pi} \int_0^{2\pi} \gamma^n(1-\gamma)^m d\theta . \qquad (4)$$

By measuring one outcome signal, the conditional photon number statistics of the remaining signal varies depending on the result obtained. For simplicity, from now on we shall consider that this measurement is realized with a simple threshold photon detector. The analysis of other practical scenarios with different signals and detectors can be found in [22]. Such a detector can be characterized by a *Positive Operator Valued Measure* which contains two elements, F_{vac} and F_{click}, given by [24]

$$F_{\text{vac}} = (1-\epsilon) \sum_{n=0}^{\infty} (1-\eta_{\text{d}})^n |n\rangle\langle n| , \qquad (5)$$

and $F_{click} = 1 - F_{vac}$. The parameter η_d denotes the detection efficiency of the detector, and ϵ represents its probability of having a dark count. That is, the outcome of F_{vac} corresponds to no click in the detector, whereas the operator F_{click} gives precisely one detection click, which means at least one photon is detected.

Whenever one ignores the result of the measurement in mode b, the total probability of finding n photons in mode a can be expressed as

$$p_n^t = \sum_{m=0}^{\infty} p_{n,m} = \frac{v^n}{n!} \frac{1}{2\pi} \int_0^{2\pi} \gamma^n e^{-v\gamma} d\theta , \tag{6}$$

which turns out to be a non-Poissonian probability distribution. The joint probability for seeing n photons in mode a and no click in the threshold photon detector has the form

$$p_n^{\bar{c}} = (1 - \epsilon) \sum_{m=0}^{\infty} (1 - \eta_d)^m p_{n,m} = (1 - \epsilon) \frac{v^n e^{-\eta_d v}}{n!} \frac{1}{2\pi} \int_0^{2\pi} \gamma^n e^{-(1-\eta_d)v\gamma} d\theta . \tag{7}$$

If the detector produces a click, the joint probability of finding n photons in mode a is given by

$$p_n^c = p_n^t - p_n^{\bar{c}} . \tag{8}$$

Fig. 2 (Cases A and B) shows the conditional photon number statistics of the outcome signal in mode a depending on the result of the detector (click and no click): $q_n^c \equiv p_n^c/(1 - N)$ and $q_n^{\bar{c}} \equiv p_n^{\bar{c}}/N$, with

$$N \equiv \sum_{n=0}^{\infty} p_n^{\bar{c}} = (1 - \epsilon)e^{-\eta_d[\mu_1(1-t)+\mu_2 t]} I_{0,\eta_d\xi} , \tag{9}$$

and where $I_{q,z}$ represents the modified Bessel function of the first kind [26]. This function is defined as [26]

$$I_{q,z} = \frac{1}{2\pi i} \oint e^{(z/2)(t+1/t)} t^{-q-1} dt . \tag{10}$$

This figure includes as well a comparison between q_n^c and a Poissonian distribution of the same mean photon number (Cases C and D). Both distributions, q_n^c and $q_n^{\bar{c}}$, are also non-Poissonian.

The passive decoy state setup illustrated in Fig. 1 (Case A) requires that Alice uses two pulsed sources of WCP. A similar result can also be obtained with only one source of phase randomized WCP. For instance, Alice could employ the scheme showed as Case B in Fig. 1. This setup has only one laser diode, but includes an intensity modulator (IM) to block either all the even or all the odd outcome pulses. It is, therefore, an active scheme. Still, this setup might be easier to implement than those standard active decoy state schemes where an IM is used to modulate each outcome pulse depending on the result of a random number generator. The main reason to have to block half of the outcome pulses is to

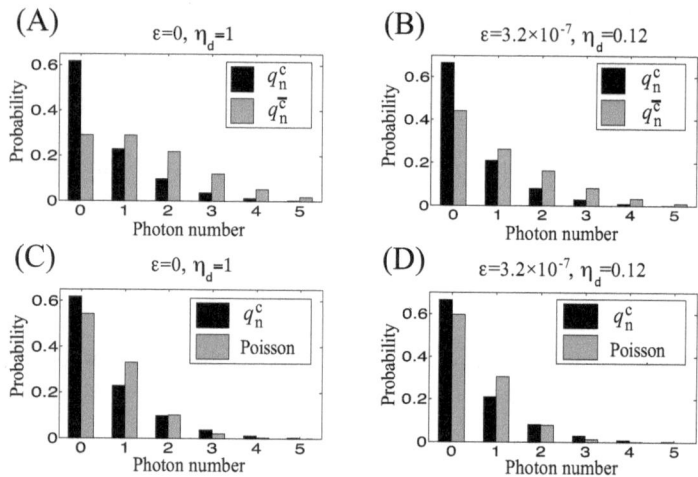

Fig. 2. Conditional photon number distribution in mode a (see Case A in Fig. 1): q_n^c (black) and $q_n^{\bar{c}}$ (grey) for the case $\mu_1 = \mu_2 = 1$ and $t = 1/2$. We consider two situations: (A) A perfect threshold photon detector, *i.e.*, $\epsilon = 0$ and $\eta_d = 1$ [21], and (B) $\epsilon = 3.2 \times 10^{-7}$ and $\eta_d = 0.12$. These last data correspond to the experiment reported in [25]. The Cases C and D represent q_n^c (black) versus a Poissonian distribution of the same mean photon number for the two scenarios described above (perfect and imperfect threshold photon detector).

suppress correlations between them. That is, the action of the IM guarantees that the outcome signals consist of tensor product of classical mixtures of Fock states. Thanks to the one-pulse delay introduced by one arm of the interferometer, together with a proper selection of the transmittance t_1, it can be shown that both cases, A and B in Fig. 1, are equivalent, except from the resulting secret key rate. Specifically, the secret key rate in Case B is half the one that can be obtained in Case A, since half of the pulses are now discarded.

3   Lower Bound on the Secret Key Rate

We consider that Alice and Bob treat no click and click events separately, and they distill secret key from both of them. We use the secret key rate formula provided by [11], [27],

$$R \geq \max\{R^c, 0\} + \max\{R^{\bar{c}}, 0\} , \tag{11}$$

with

$$
\begin{aligned}
R^c &\geq q\{-Q^c f(E^c) H(E^c) + p_1^c Y_1[1 - H(e_1)] + p_0^c Y_0\} \\
&\geq q\{-Q^c f(E^c) H(E^c) + (p_1^c Y_1 + p_0^c Y_0)[1 - H(e_1^u)]\} ,
\end{aligned}
\tag{12}
$$

and similarly for $R^{\bar{c}}$. The parameter q is the efficiency of the protocol ($q = 1/2$ for the standard Bennett-Brassard 1984 protocol [9], and $q \approx 1$ for its efficient version [28]), Q^c is the overall gain of the signals, E^c represents the overall quantum bit error rate (QBER), $f(E^c)$ is the error correction efficiency [typically $f(E^c) \geq 1$ with Shannon limit $f(E^c) = 1$], Y_n denotes the yield of a n-photon signal, $i.e.$, the conditional probability of a detection event on Bob's side given that Alice transmits an n-photon state, e_1 is the single photon error rate, e_1^u represents an upper bound on e_1, and $H(x) = -x \log_2(x) - (1-x) \log_2(1-x)$ is the binary Shannon entropy function.

The quantities Q^c, E^c, $Q^{\bar{c}}$, and $E^{\bar{c}}$ are directly accessible from the experiment. They can be expressed as

$$Q^c = \sum_{n=0}^{\infty} p_n^c Y_n \quad \text{and} \quad Q^c E^c = \sum_{n=0}^{\infty} p_n^c Y_n e_n , \tag{13}$$

and similarly for the case of a no click event. Here e_n denotes the error rate of a n-photon signal ($e_0 = 1/2$ for random background).

To apply the secret key rate formula given by Eq. (11) one needs to estimate a lower bound on the quantity $p_1^c Y_1 + p_0^c Y_0$, together with an upper bound on e_1. For that, we follow the procedure proposed in [29]. This method requires that p_n^t and $p_n^{\bar{c}}$ satisfy certain conditions that we checked numerically. Note, however, that many other estimation techniques are also available, like, for instance, linear programming tools [30]. We find that

$$p_1^c Y_1 + p_0^c Y_0 \geq p_1^c Y_1^l + p_0^c Y_0^u , \tag{14}$$

with

$$Y_1^l \equiv \max \left\{ 0, \frac{p_2^{\bar{c}} Q^t - p_2^t Q^{\bar{c}} - (p_2^{\bar{c}} p_0^t - p_2^t p_0^{\bar{c}}) Y_0^u}{p_2^{\bar{c}} p_1^t - p_2^t p_1^{\bar{c}}} \right\} , \tag{15}$$

where $Q^t = Q^c + Q^{\bar{c}}$, and Y_0^u denotes an upper bound on the background rate Y_0 given by

$$Y_0^u \equiv \min \left\{ \frac{2 E^{\bar{c}} Q^{\bar{c}}}{p_0^{\bar{c}}}, \frac{2 E^t Q^t}{p_0^t} \right\} . \tag{16}$$

The error rate e_1 can be upper bounded as

$$e_1 \leq e_1^u \equiv \min \left\{ \frac{E^{\bar{c}} Q^{\bar{c}} - p_0^{\bar{c}} Y_0^l e_0}{p_1^{\bar{c}} Y_1^l}, \frac{E^c Q^c - p_0^c Y_0^l e_0}{p_1^c Y_1^l}, \frac{p_0^{\bar{c}} E^t Q^t - p_0^t E^{\bar{c}} Q^{\bar{c}}}{(p_0^{\bar{c}} p_1^t - p_0^t p_1^{\bar{c}}) Y_1^l} \right\} , \tag{17}$$

with $Q^t E^t = Q^c E^c + Q^{\bar{c}} E^{\bar{c}}$, and where Y_0^l denotes a lower bound on Y_0 given by

$$Y_0 \geq Y_0^l \equiv \max \left\{ 0, \frac{p_1^t Q^{\bar{c}} - p_1^{\bar{c}} Q^t}{p_1^t p_0^{\bar{c}} - p_1^{\bar{c}} p_0^t} \right\} . \tag{18}$$

The only relevant statistics to evaluate Y_0^l, Y_0^u, Y_1^l, and e_1^u are p_n^t and $p_n^{\bar{c}}$, with $n = 0, 1, 2$. These probabilities can be obtained by solving Eqs. (6)-(7). They are given in the Appendix.

4   Evaluation

For simulation purposes we employ the channel model used in [13], [29]. This model reproduces a normal behavior of a quantum channel, *i.e.*, in the absence of eavesdropping. It allows us to calculate the observed experimental parameters $Q^{\bar{c}}$, $E^{\bar{c}}$, Q^t, and E^t. Our results, however, can also be applied to any other quantum channel, as they only depend on the observed gains and QBERs. In the scenario considered, the yields have the form

$$Y_n = 1 - (1 - Y_0)(1 - \eta_{\text{sys}})^n , \qquad (19)$$

where η_{sys} represents the overall transmittance of the system [13], [29]. That is, η_{sys} includes the transmission efficiency of the quantum channel and that of Bob's detection apparatus. This parameter can be related with a transmission distance l measured in km for the given QKD scheme as $\eta_{\text{sys}} = 10^{-\frac{\alpha l}{10}}$, where α represents the loss coefficient of the optical fiber measured in dB/km. The product $Y_n e_n$ can be expressed as

$$Y_n e_n = Y_0 e_0 + (Y_n - Y_0)e_d , \qquad (20)$$

where e_d is the probability that a photon hits the wrong detector due to the misalignment in the quantum channel and in Bob's detection setup [13], [29]. After substituting these definitions into the gain and QBER formulas we obtain

$$\begin{aligned}
Q^{\bar{c}} &= N - (1 - \epsilon)(1 - Y_0)e^{(\eta_d - \eta_{\text{sys}})\omega - \eta_d v} I_{0,(\eta_d - \eta_{\text{sys}})\xi} , \\
Q^{\bar{c}}E^{\bar{c}} &= (e_0 - e_d)Y_0 N + e_d Q^{\bar{c}} , \\
Q^t &= 1 - (1 - Y_0)e^{-\eta_{\text{sys}}\omega} I_{0,\eta_{\text{sys}}\xi} , \\
Q^t E^t &= (e_0 - e_d)Y_0 + e_d Q^t ,
\end{aligned} \qquad (21)$$

with $\omega = \mu_1 t + \mu_2(1 - t)$.

The resulting lower bound on the secret key rate is illustrated in Fig. 3. The experimental parameters are [25]: $Y_0 = 1.7 \times 10^{-6}$, $e_d = 0.033$, $\alpha = 0.21$ dB/km, and Bob's detection efficiency equal to 0.045. We assume that $q = 1$, $f(E^c) = f(E^{\bar{c}}) = 1.22$, and $t = 1/2$, *i.e.*, we consider a simple $50 : 50$ BS. We study two different situations: (1) $\epsilon = 0$ and $\eta_d = 1$ [21], and (2) $\epsilon = 3.2 \times 10^{-7}$ and $\eta_d = 0.12$ [25]. In both cases the optimal values of the intensities μ_1 and μ_2 are almost constant with the distance. One of them is quite weak (around 10^{-4}), while the other one is around 0.5. Fig. 3 includes as well the case of an active asymptotic decoy state QKD system [13]. The cutoff points where the secret key rate drops down to zero are $l \approx 128$ km (passive setup with two intensity settings) and $l \approx 147$ km (active asymptotic setup). One could reduce this gap further by using a passive scheme with more intensity settings. For instance, one may employ a photon number resolving detector instead of a simple threshold photon detector, or use more threshold detectors in combination with BS [22]. From these results we see that the performance of the passive scheme is comparable to the active one, thus showing the practical interest of the passive setup.

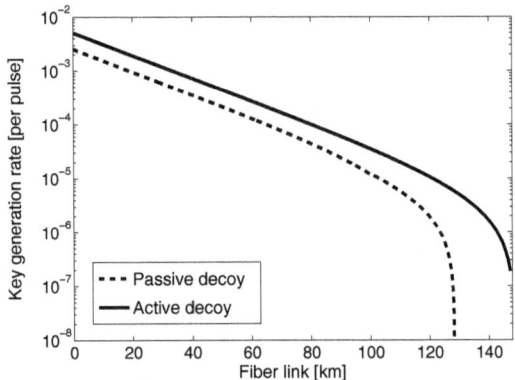

Fig. 3. Lower bound on the secret key rate R given by Eq. (11) in logarithmic scale for the passive decoy state setup given by Fig. 1 (Case A) with two intensity settings. The transmittance of the BS is $t = 1/2$. We consider two possible scenarios: (1) $\epsilon = 0$ and $\eta_d = 1$ [21] (*i.e.*, a perfect threshold photon detector), and (2) $\epsilon = 3.2 \times 10^{-7}$ and $\eta_d = 0.12$ [25]. Both cases provide approximately the same final key rate and they cannot be distinguished with the resolution of this figure (dashed line). The solid line represents a lower bound on R for an active asymptotic decoy state system [13].

5   Conclusion

We have presented a passive decoy state QKD system with phase randomized WCP. This setup uses only linear optical elements and a simple threshold photon detector. It represents an alternative to those active schemes based on the use of a variable optical attenuator together with a random number generator. Moreover, in the asymptotic limit of an infinite long experiment, we have shown that this passive system can provide a similar performance to the one achieved with an active source and infinity decoy settings.

Acknowledgments. The authors wish to thank R. Kaltenbaek, H.-K. Lo, B. Qi, and Y. Zhao for very useful discussions, and in particular M. Koashi for pointing out a reference. M.C. especially thanks the Institute for Quantum Computing (University of Waterloo) for hospitality and support during his stay in this institution. This work was supported by the European Projects SECOQC and QAP, by the NSERC Discovery Grant, Quantum Works, CSEC, and by Xunta de Galicia (Spain, Grant No. INCITE08PXIB322257PR).

References

1. idQuantique, Geneva (Switzerland), www.idquantique.com; MagiQ Technologies, Inc., New York., www.magiqtech.com; and Smartquantum, Lannion (France), www.smartquantum.com

140   M. Curty et al.

2. Scarani, V., Bechmann-Pasquinucci, H., Cerf, N.J., Dušek, M., Lütkenhaus, N., Peev, M.: The Security of Practical Quantum Key Distribution. Rev. Mod. Phys. (accepted for publication, 2009), Preprint quant-ph/0802.4155
3. Bennett, C.H., Bessette, F., Brassard, G., Salvail, L., Smolin, J.: Experimental Quantum Cryptography. Cryptology 5, 3–28 (1992)
4. Marand, C., Townsend, P.D.: Quantum key distribution over distances as long as 30 km. Opt. Lett. 20, 1695–1697 (1995)
5. Muller, A., Zbinden, H., Gisin, N.: Underwater quantum coding. Nature 378, 449–449 (1995)
6. Hughes, R., Morgan, G., Peterson, C.G.: Quantum key distribution over a 48km optical fibre network. J. Mod. Opt. 47, 533–547 (2000)
7. Huttner, B., Imoto, N., Gisin, N., Mor, T.: Quantum Cryptography with Coherent States. Phys. Rev. A 51, 1863–1869 (1995)
8. Brassard, G., Lütkenhaus, N., Mor, T., Sanders, B.C.: Limitations on Practical Quantum Cryptography. Phys. Rev. Lett. 85, 1330–1333 (2000)
9. Bennett, C.H., Brassard, G.: Quantum cryptography: public key distribution and coin tossing. In: IEEE International Conference on Computers, Systems and Signal Processing, pp. 175–179. IEEE Press, New York (1984)
10. Inamori, H., Lütkenhaus, N., Mayers, D.: Unconditional security of practical quantum key distribution. Eur. Phys. J. D 41, 599–627 (2007)
11. Gottesman, D., Lo, H.-K., Lütkenhaus, N., Preskill, J.: Security of quantum key distribution with imperfect devices. Quantum Inf. Comput. 4, 325–360 (2004)
12. Hwang, W.-Y.: Quantum Key Distribution with High Loss: Toward Global Secure Communication. Phys. Rev. Lett. 91, 57901 (2003)
13. Lo, H.-K., Ma, X., Chen, K.: Decoy State Quantum Key Distribution. Phys. Rev. Lett. 94, 230504 (2005)
14. Wang, X.-B.: Beating the Photon-Number-Splitting Attack in Practical Quantum Cryptography. Phys. Rev. Lett. 94, 230503 (2005)
15. Zhao, Y., Qi, B., Ma, X., Lo, H.-K., Qian, L.: Experimental Quantum Key Distribution with Decoy States. Phys. Rev. Lett. 96, 070502 (2006)
16. Rosenberg, D., Harrington, J.W., Rice, P.R., Hiskett, P.A., Peterson, C.G., Hughes, R.J., Lita, A.E., Nam, S.W., Nordholt, J.E.: Long-Distance Decoy-State Quantum Key Distribution in Optical Fiber. Phys. Rev. Lett. 98, 010503 (2007)
17. Schmitt-Manderbach, T., Weier, H., Fürst, M., Ursin, R., Tiefenbacher, F., Scheidl, T., Perdigues, J., Sodnik, Z., Kurtsiefer, C., Rarity, J.G., Zeilinger, A., Weinfurter, H.: Experimental Demonstration of Free-Space Decoy-State Quantum Key Distribution over 144 km. Phys. Rev. Lett. 98, 010504 (2007)
18. Mauerer, W., Silberhorn, C.: Quantum Key Distribution with Passive Decoy State Selection. Phys. Rev. A 75, 050305(R) (2007)
19. Adachi, Y., Yamamoto, T., Koashi, M., Imoto, N.: Simple and Efficient Quantum Key Distribution with Parametric Down-Conversion. Phys. Rev. Lett. 99, 180503 (2007)
20. Ma, X., Lo, H.-K.: Quantum Key Distribution with Triggering Parametric Down-Conversion Sources. New J. Phys. 10, 073018 (2008)
21. Curty, M., Moroder, T., Ma, X., Lütkenhaus, N.: Non-Poissonian statistics from Poissonian light sources with application to passive decoy state quantum key distribution. Accepted for publication in Opt. Lett. (2009)
22. Curty, M., Ma, X., Qi, B., Moroder, T., Lütkenhaus, N.: In preparation (2009)
23. Adachi, Y., Yamamoto, T., Koashi, M., Imoto, N.: Passive decoy-state quantum cryptography with pseudo-single-photon sources. In: 8th Asian Conference on Quantum Information Science (AQIS 2008), Seoul, pp. 25–26 (2008)

24. Rohde, P.P., Ralph, T.C.: Modelling photo-detectors in quantum optics. J. Mod. Opt. 53, 1589–1603 (2006)
25. Gobby, C., Yuan, Z.L., Shields, A.J.: Quantum key distribution over 122 km of standard telecom fiber. Appl. Phys. Lett. 84, 3762–3764 (2004)
26. Arfken, G.: Mathematical Methods for Physicists, 3rd edn. Academic Press, London (1985)
27. Lo, H.-K.: Getting something out of nothing. Quantum Inf. Comput. 5, 413–418 (2005)
28. Lo, H.-K., Chau, H.F., Ardehali, M.: Efficient quantum key distribution scheme and a proof of Its unconditional security. J. Cryptology 18, 133–165 (2005)
29. Ma, X., Qi, B., Zhao, Y., Lo, H.-K.: Practical decoy state for quantum key distribution. Phys. Rev. A 72, 012326 (2005)
30. Bazaraa, M.S., Jarvis, J.J., Sherali, H.D.: Linear Programming and Network Flows, 3rd edn. Wiley, Chichester (2004)

Appendix: Probabilities p_n^t and $p_n^{\bar{c}}$

In this Appendix we provide explicit expressions for the probabilities p_n^t and $p_n^{\bar{c}}$, with $n = 0, 1, 2$. In particular, we have that

$$
\begin{aligned}
p_0^t &= I_{0,\xi} e^{-\omega} , \\
p_1^t &= (\omega I_{0,\xi} - \xi I_{1,\xi}) e^{-\omega} , \\
p_2^t &= \frac{1}{2}[\omega^2 I_{0,\xi} + (1 - 2\omega)\xi I_{1,\xi} + \xi^2 I_{2,\xi}] e^{-\omega} ,
\end{aligned}
\tag{22}
$$

with $\omega = \mu_1 t + \mu_2(1 - t)$. The probabilities $p_n^{\bar{c}}$ have the form

$$
\begin{aligned}
p_0^{\bar{c}} &= \tau I_{0,(1-\eta_{\mathrm{d}})\xi} , \\
p_1^{\bar{c}} &= \tau\left(\omega I_{0,(1-\eta_{\mathrm{d}})\xi} - \xi I_{1,(1-\eta_{\mathrm{d}})\xi}\right) , \\
p_2^{\bar{c}} &= \frac{\tau}{2}\left\{\omega^2 I_{0,(1-\eta_{\mathrm{d}})\xi} + \left[\frac{1}{1-\eta_{\mathrm{d}}} - 2\omega\right]\xi I_{1,(1-\eta_{\mathrm{d}})\xi} + \xi^2 I_{2,(1-\eta_{\mathrm{d}})\xi}\right\} ,
\end{aligned}
\tag{23}
$$

where $\tau = (1 - \epsilon)e^{-[\eta_{\mathrm{d}}\upsilon + (1-\eta_{\mathrm{d}})\omega]}$.

QKD in Standard Optical Telecommunications Networks

D. Lancho[1], J. Martinez[1], D. Elkouss[1], M. Soto[2], and V. Martin[1]

[1] Facultad de Informática, Universidad Politécnica de Madrid,
Campus de Montegancedo, Madrid 28660, Spain
Vicente@fi.upm.es
[2] Depto. Seguridad en Redes y Servicios, Telefónica Investigación y Desarrollo,
Emilio Vargas 6, Madrid 28043, Spain

Abstract. To perform Quantum Key Distribution, the mastering of the extremely weak signals carried by the quantum channel is required. Transporting these signals without disturbance is customarily done by isolating the quantum channel from any noise sources using a dedicated physical channel. However, to really profit from this technology, a full integration with conventional network technologies would be highly desirable. Trying to use single photon signals with others that carry an average power many orders of magnitude bigger while sharing as much infrastructure with a conventional network as possible brings obvious problems. The purpose of the present paper is to report our efforts in researching the limits of the integration of QKD in modern optical networks scenarios. We have built a full metropolitan area network testbed comprising a backbone and an access network. The emphasis is put in using as much as possible the same industrial grade technology that is actually used in already installed networks, in order to understand the throughput, limits and cost of deploying QKD in a real network.

Keywords: Quantum Key Distribution, Passive Optical Networks.

1 Introduction and Testbeds

To date, QKD development has mostly focused in non-shared point-to-point links and in networks made of these links [1]. However, the benefits of integrating QKD with the commercial network infrastructure are clear. With the possibility of a scalable deployment and without requiring a massive initial investment, QKD can reach a broader market. Current networks are evolving towards optical, passive infrastructures and this opens up a window of opportunity for QKD integration, since an all optical path among two points in the network is no longer unfeasible nor extremely expensive. However, the simultaneous propagation of quantum and classical signals over a shared link, presents important problems due to the spilling of photons coming from classical signals that typically have 100 dBm more power.

Research has been done about the use of QKD in optical networks [5,7,9], including some recent studies [2] using the 200 GHz (1.6 nm) ITU DWDM grid

A. Sergienko, S. Pascazio, and P. Villoresi (Eds.): QuantumCom 2009, LNICST 36, pp. 142–149, 2010.

(Dense Wavelength Division Multiplexing). Their aim was to characterize the sources of noise and to demonstrate the use of reconfigurable optical add and drop multiplexers (ROADM) for QKD purposes. Here we go one step beyond in completing a metropolitan area optical network able to establish a transparent path among two of their nodes in order to sustain a QKD link.

Metro networks are logically divided into backbone and access parts. The backbone is specialized in high speed communications. The number of nodes is limited and the cost per node is less of a concern, sharing importance with throughput, reliability, serviceability and upgradeability. Clients are connected to the backbone through an access network. This is a point to multipoint network, with one end connected to the backbone at the carrier company premises, while the other gives service to several clients. A shared link goes from the backbone to some form of splitter that is located in the vicinity of the clients. An exclusive use link connects the splitter to each one of the clients. In the access network, throughput is not as relevant, but cost and maintenance are important issues, since many of them have to be deployed and part of the equipment is either sold or leased to the client.

The new generations of access networks, that are being massively deployed in the current fiber to the home infrastructures, are designed as optical and passive, which opens the possibility of using them for QKD.

The purpose of the present paper is to investigate the limits, trade offs and compatibility issues that the integration of QKD with the existing or newly deployed optical infrastructure could pose. In order to do so, a full metro network comprising backbone and access parts is used.

The backbone testbed is built as a ring composed of three CWDM (Coarse WDM [12]) ROADM nodes, the most common technology in metro and regional networks. Typically, the CWDM ITU grid is implemented in a spectrum of 18 channels separated 20 nm starting at a central wavelength of 1270 nm. The bigger channel spacing makes for an easier filtering of the spurious photons coming from other channels and made the use of amplifiers less common in this kind of networks, since no single EDFA amplifier is able to amplify all of the CWDM channels, thus making the option less attractive from a cost/benefit perspective than in DWDM links.

The access network uses the GPON (Gigabit Passive Optical Network) standard [13], again, the most used in new deployments of fiber to the business. GPON multiplexes three wavelengths over the same fiber to connect the OLT (Optical Line Termination) on the backbone side to the ONT (Optical Network Termination) on the client side. A splitter located at some point in between the OLT and the set of the ONTs —typically much closer to the latter— divides the signal among all the ONTs. The standard defines the 1490 nm wavelength to transport the downstream channel, 1310 nm the upstream channel and 1550 nm for analog video broadcast. The last one is rarely used for its intended purpose and we are using it for the quantum channel. This is just a matter of convenience and other channel could have been chosen.

A brief report of these setups, was given in [3]. It is to be noted that the emphasis is put in what can be readily achieved with currently deployable technology, using as much off the shelf systems as possible and limiting to a minimum the modifications done to what, in essence, is a standard network environment.

2   Results

Data were obtained using id Quantique two way QKD systems, models 3000 and 3100, using BB84. Their maximum admissible loss budget in order to obtain at least a few secure bits per second is around 15 dB. The maximum throughput under ideal conditions (QBER=0), imposed by the detector deadtime, is 100 kbits/sec. In our case, mean photon number was set to simulate a decoy state protocol [17,18] with signal plus one decoy. The optimal mean number for our setup, 0.79, was calculated according to [19]. A full protocol stack, including LDPC (Low Density Parity Check) error correction with 1.05 efficiency [15] and privacy amplification was used.

The backbone results were obtained using three wavelengths. Two wavelengths, 1510 and 1470 nm are used for classical signals (co and counterpropagating, respectively) while 1550 is reserved for the quantum channel. This is not an specially favorable case for QKD, where the quantum channel should have been located at the shorter wavelength. Modules to add and drop the 1550 channel as desired are included at each node. Extra filtering to further isolate the quantum channel was needed and standard DWDM 100 GHz (0.8 nm) filters were used. It is to be noted that, due to the spreading of CWDM and DWDM technology, very narrow and high quality filters are starting to be readily available for commercial use. 50 GHz (0.4 nm) filters are now reasonably common, hence we have extrapolated the data to this case. Losses in this scenario, without the fiber, are 8 dB. A full description will be published elsewhere.

In the backbone experiment, two QKD systems are used to test point to point key growing, key pass-through and key forwarding through the ring while data signals are being simultaneously transmitted at full rate. In Fig. 1 we present an example of a point to point key growing using a pass-through configuration in the middle node. The Quantum Bit Error Rate and key throughput, at different stages of the key distillation process, are shown as a function of the fiber length connecting the first and second ROADM nodes. Only a short, fixed length, fiber was used among the second and third. This is a worst case scenario for QKD since it produces more spurious photons due to Raman Scattering than the situation in which, for the same total distance, the second ROADM node is located somewhere else in between. The QBER value obtained, although high, is well below the 11% threshold above which no secret key rate can be obtained in the usual setting with one way error correction and privacy amplification. A secure key rate around half a kbit/sec. was obtained for distances till 6 Km, although is reduced to 100 bits/sec. for 10 Km, a length larger than the typical for ROADM nodes. It is to be noted that these scenarios are dominated by absorptions in the nodes and not in the fiber. Fiber length is relevant as long as it contributes

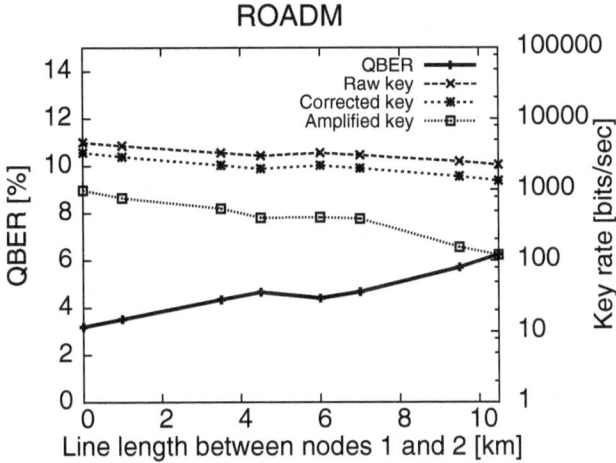

Fig. 1. Results obtained in the backbone testbed. Quantum Bit Error Rate and key rate is presented as a function of the fiber length connecting ROADM nodes 1 and 2, the worst situation for QKD. QBER is depicted on the left axis while key rate in their different distillation phases, till completing privacy amplification, is on the right axis. Data, obtained with 100 GHz (0.8 nm) filters, are extrapolated to 50 GHz (0.4 nm) filters. Final secure key rate approaching half a kbit/sec. can be obtained at 6 Km and even after 10 Km there is a net key throughput of around 100 bits/sec. The change in trend that can be seen in the data before and after 4.5 km is due to the use of different types of fiber (both standard and commercial) that produce different amounts of Raman scattering. Note also that the different segments of fiber are joined through connectors and, for the longer distances, four of them were needed. This means that (in regard to losses but not to Raman scattering) the actual line length is of around 14 km. The average distance among ROADM nodes in real metro network is usually well below 10 km.

to produce Raman scattering. It is important to put this result in perspective: a fully populated DWDM link, with all the 160 channels transmitting at 2.4 Gbit/sec. and secured with a 1 kbit/sec. key, will have to encrypt less than 2^{37} bits/sec. per 256 bit AES key. This is less than the known reasonable security limit of 2^{40} for DES [21], an assumed lower security encryption method. For standard links, key rates as low as a few tens of bits/sec. would suffice to have a much higher security level than what is standard today.

The number of spurious photons entering the quantum channel depends on the total power sent through the fiber, but also on the fiber itself and some other characteristics like the directivity and crosstalk of the components used to build the ROADM, which in our case is a standard model with no special modifications. In a CWDM set up, four wave mixing is not an issue and only Raman scattering is a real threat. Apart from the line length, the only controllable variable is the total launch power. Here it was attenuated (differently for each link length) respect to the standard power used, but well above the detection limit of the

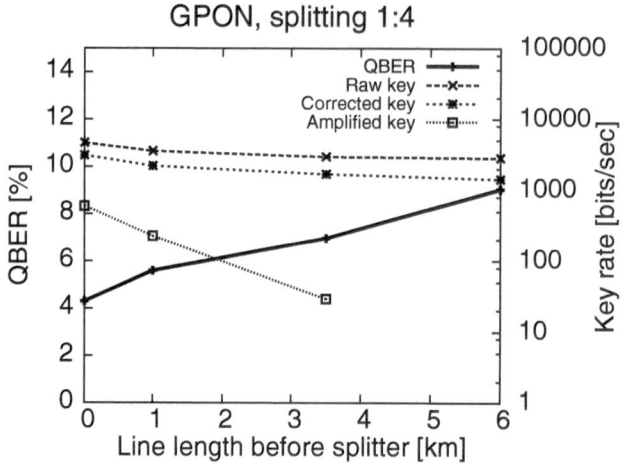

Fig. 2. Results obtained in the access network. Quantum Bit Error Rate and key rate is presented as a function of fiber length connecting the OLT with the splitter, the most populated part of the link and the worst configuration for QKD. Typical OLT-splitter distances can differ a lot with the geographical environment, but are usually on the shorter side for big cities. QBER is depicted on the left axis while key rate, in their different distillation phases, till completing privacy amplification is on the right axis.

transponders, so no increased data error rate resulted in the classical channels working at full speed.

In the access network experiment with GPON, a continuous data flux was established among the OLT (backbone side) and ONT (client side) using the 1490 nm (downstream) and 1310 nm (upstream) channels. Again, QKD used the 1550 nm channel. Due to the more integrated nature of the access equipment (the client part needs to be compact and cheap) there are less modifications that can be readily done. The launch power is fixed and only a small attenuation can be introduced in the OLT and only in the 1490 channel. The filtering used was the same (100 GHz) and the maximum splitting factor possible with our QKD equipment is four. Losses without fiber are 9 dB. The results are shown in Fig. 2. This time, QBER and key rates are presented as a function of the line length connecting the OLT and the splitter, which is the longest segment in access networks and, again, the one that more heavily penalizes the quantum channel. This scenario is more demanding than the ROADM but this is mainly due to the difficulty of attenuating the total power in the line without HW modifications of the OLT/ONT. There is an advantage, however, in that there is no actual need to have the fiber populated with classical signals at the same time that the quantum channel is being used. Because GPON works on a Time Division Multiplexing control, it is easy in theory to assign time slots to the quantum channel. In practice, this is more difficult, since it requires access to reprogram the micro controllers and that would require the involvement of the manufacturer. The QBER obtained is quite high, although still clearly below the threshold. It already starts, for a 0 Km line, at 4% due to the crosstalk in the internal components, mostly coming from the 1310 nm channel that is impossible to attenuate in our setup. This produces a secure key rate around 500 bits/sec. at 0 km that rapidly reduces to 20 bits/sec. at 3.5 km. In our next measurement, at 4.5 km, there is no secret key rate. This would call for a time slot assignment for the QKD channel or a redesign of the OLT/ONT pair to work on a much lower power budget. Although the final key rate was lower than in the backbone, it is still enough to sustain a 256 bits AES key renewal rate faster than it is usual today at short distances. It is to be noted that typical OLT/ONT distances are of the order of 1-2 km in big cities.

3   Conclusion

In this paper we have demonstrated the use of QKD within standard metropolitan optical networks comprising a backbone and access network with all the testbed built around readily available systems and components. The emphasis is put in finding out the limits of QKD integration in real networks rather than in absolute security against any conceivable attack, since finite key effects are not taken into consideration. Some test cases are the worst possible for QKD and they are not the way in which QKD would be reasonably integrated in a network; they are just the most straightforward without resorting to modify the classical communications equipment. They serve the purpose of bounding the performance limits. Some freedom exists when integrating QKD with standard optical networks and a mixture of scenarios is what is to be expected in the real world. For example, in the backbone network, the existence of unused dark fiber for backup purposes is quite common, hence, in the ROADM scenario, bypassing heavily loaded nodes is a real possibility. In GPON, the mentioned time slot assignment for QKD is an obvious solution that is not difficult to implement and would not hit significantly the performance of the access network. A direct connection without shared access network is also a possibility for medium to big companies. In metro scenarios, where limits to the integration are imposed by noise and absorptions in the optical equipment rather than in the fiber, moving the quantum channel to the second window ($\sim$1300 nm) would be advantageous. All this considerations support the view that the integration of QKD in modern optical networks is a real alternative to exclusive use quantum links and trusted-nodes QKD networks.

In the frame of the present testbed and with current QKD equipment, we observe that in the CWDM backbone scenario, the limited absorption budget tolerable by the QKD systems can be as stringent as the Raman scattering from classical channels that, apart from filtering, can be controlled through attenuation and very sensitive transponders without disrupting the classical transmission. The access scenario tested is essentially a worst case (for QKD) GPON in which no modification is done neither to the GPON management nor to the HW itself. In this case, a maximum 1:4 splitting is possible with the current QKD devices. In this scenario, it is possible to greatly increase the key throughput by modifying the device profiles to assign time slots for QKD transmission. Without HW or SW modifications, the GPON scenario is mainly limited by Raman scattering and components crosstalk. In order to obtain the reported QBER and key rates mentioned above, a decoy state protocol is needed. In such a heavily penalized scenarios, it is important not to waste bits of the raw key and more efficient error correction protocols [14] would be a welcome addition [15]. On the other hand, next generation QKD devices like the ones now in the labs, able to withstand more than 30 dB losses, would allow for far higher bit rates and would open the possibility of crossing a full metro network without the need of trusted repeaters.

The authors are indebted to M. Curty for his helpful comments. This work was supported by CDTI, Ministry of Trade and Industry of Spain under project Segur@, CENIT-2007 2004 and UPM 178/Q06 1005-127.

References

1. Poppe, A., Peev, M., Maurhart, O.: Outline of the SECOQC Quantum-Key-Distribution Network. Int. J. Quantum Inf. 6(2), 209–218 (2008)
2. Peters, N.A., et al.: Dense Wavelength Multiplexing of 1550 nm QKD with Strong Classical Channels in Reconfigurable Networking Environments. New. J. of Phys. 11, 045012 (2009)
3. Lancho, D., et al.: Quantum Key Distribution in Commercial Optical Networks. Report to the SECOQC Conference (October 2008), http://www.secoqc.net
4. Bennett, C.H., Brassard, G.: The Dawn of a New Era for Quantum Cryptography: The experimental prototype is working! Sig. Act News 20(4), 78 (1989)
5. Toliver, P., et al.: Experimental Investigation of Quantum Key Distribution Through Transparent Optical Switch Elements. IEEE Photonics Tech. Lett. 15(11), 1669–1671 (2003)
6. Runser, R.J., et al.: Quantum Key Distribution for Reconfigurable Optical Networks. In: 2006 Optical Fiber Communication Conference. Contribution OFL1 (2006)
7. Runser, R.J., et al.: Demonstration of 1.3 μm Quantum Key Distribution Compatibility with 1.5 μm Metropolitan Wavelength Division Multiplexed Systems. In: 2005 Optical Fiber Communication Conference. Contribution OWI2 (2005)
8. Subacius, D., Zavriyev, A., Trifonov, A.: Backscattering Limitation for Fiber-optic Quantum Key Distribution Systems. Appl. Phys. Lett. 86, 011103 (2005)
9. Xia, T.J., et al.: In-Band Quantum Key Distribution (QKD) on Fiber Populated by High-Speed Classical Data Channels. In: 2006 Optical Fiber Communication Conference. Contribution OTuJ7 (2006)
10. Toliver, P., et al.: Demonstration of 1550 nm QKD with ROADM-based DWDM Networking and the Impact of Fiber FWM. In: 2007 Optical Fiber Communication Conference. Contribution CThBB1, vol. 1 (2007)
11. DWDM: ITU-T G.694.1 (06/2002): Spectral grids for WDM applications: DWDM frequency grid, http://www.itu.int/rec/T-REC-G.694.1-200206-I/en
12. CWDM: ITU-T G.694.2 (12/2003): Spectral grids for WDM applications: CWDM wavelength grid, http://www.itu.int/rec/T-REC-G.694.2-200312-I/en
13. GPON: ITU-T G.984.1 (03/2008): Gigabit-capable Passive Optical Networks (GPON): General Characteristics,
http://www.itu.int/rec/T-REC-G.984.1-200803-I/en

14. Brassard, G., Salvail, L.: Secret-key reconciliation by public discussion. In: Helleseth, T. (ed.) EUROCRYPT 1993. LNCS, vol. 765, pp. 410–423. Springer, Heidelberg (1994)
15. Elkouss, D., Martinez, J., Lancho, D., Martin, V.: Rate Compatible Protocol for Information Reconciliation: An application to QKD. Submitted to IEEE Information Theory Workshop 2010 (ITW 2010), Cairo (2010)
16. Bennett, C.H., Bessette, F., Brassard, G., Salvail, L., Smolin, J.: First QKD experiment. J. of Cryptology 5, 3 (1992)
17. Lo, H.-K., Ma, X., Chen, K.: Decoy State Quantum Key Distribution Phys. Rev. Lett. 94, 230504 (2005)
18. Hwang, W.-Y.: Quantum Key Distribution with High Loss: Toward Global Secure Communication. Phys. Rev. Lett. 91(5) (2003)
19. Ma, X., Qi, B., Zhao, Y., Lo, H.-k.: Practical Decoy State for Quantum Key Distribution. Phys. Rev. A 72, 012326 (2005)
20. Wegman, M.N., Carter, J.L.: New hash functions and their use in authentication and set equality. Journal of Computer and System Sciences 22, 265–279 (1981)
21. Van Assche, G.: Quantum Cryptography and Secret-Key Distillation. Cambridge University Press, Cambridge (2006)

Properties of Cascade Switch Superconducting Nanowire Single Photon Detectors

M. Ejrnaes[1], A. Casaburi[1,2], R. Cristiano[1], O. Quaranta[2], S. Marchetti[2],
N. Martucciello[2,3], S. Pagano[1,4], A. Gaggero[5], F. Mattioli[5], and R. Leoni[5]

[1] Istituto di Cibernetica "E. Caianiello" del C.N.R., 80078 Pozzuoli, Italy
m.ejrnaes@cib.na.cnr.it
http://www.cib.na.cnr.it/
[2] Dipartimento di Fisica "E. R. Caianiello", Università di Salerno,
84081 Baronissi, Italy
[3] Laboratorio Regionale SuperMat, CNR-INFM Salerno, 84081 Baronissi, Italy
[4] Dipartimento di Matematica e Informatica, Università di Salerno,
84081 Baronissi, Italy
[5] Istituto di Fotonica e Nanotecnologie del C.N.R., 00156 Roma, Italy

Abstract. Superconducting nanowire single photon detectors have been realized using an innovative photon induced cascade switch of parallel nanowires. We demonstrate that this configuration allows, at the same time, a fast response and a large active area, with the additional advantage of signal pulses with a larger signal to noise ratio. These improvements are obtained maintaining the good quantum efficiency of traditional meandered superconducting detectors. We show that due to the high speed of the parallel nanowire detector special attention is needed to avoid latching, a phenomenon which in the parallel nanowire detector degrades the detector efficiency. We describe how the latching problem can be avoided using a proper nanowire configuration.

Keywords: Single photon detector, detector design, non-equilibrium superconductivity, nanotechnology.

1 Introduction

Superconducting nanowire single photon detectors (SNSPDs) are object of significant research mainly because they offer very good performance at 1550 nm wavelength. Quantum efficiencies as high as 57%, 70 ps timing jitter and low dark count rates have been reported [1,2,3,4]. High efficiency is obtained using 100 nm wide nanowires made from 4 nm thick ultrathin NbN films [5]. To achieve area coverage, many light absorbing nanowires are connected in series using a meander pattern. The serial connection makes the total detector inductance, L_{DET}, sizeable due to the high kinetic inductance of each nanowire. This limits the time constant, $\tau = L_{DET}/Z$, where Z is the impedance of the SNSPD (usually formed by the 50 Ω coaxial readout), of the current return into the SNSPD after a photon detection event. Because the bias current increases

A. Sergienko, S. Pascazio, and P. Villoresi (Eds.): QuantumCom 2009, LNICST 36, pp. 150–157, 2010.

exponentially the SNSPD efficiency the use of a serial connection turns out to be the main limitation of the maximum obtainable count rate of meander SNSPDs [3]. Furthermore, the serial connection makes the maximum count rate of meander SNSPDs inversely proportional to detector area and is around 50 MHz for 10x10 μm^2 SNSPDs, which is significantly below the possibilities offered by NbN material. An open issue is how to obtain high SNSPD maximum count rate with a large active area without degradation of other key performance parameters. Current research is now investigating the use of parallel nanowires, which offer the possibility to sidestep the serial connection and thereby increase the maximum count rate, to fully exploit the speed of NbN. However, the use of parallel nanowires is not trivial: when a photon is absorbed in a nanowire the other parallel nanowires creates a shortcircuit and this makes it difficult to measure the signal pulses.

2   The Cascade Switch SNSPD

One way to enable the use of nanowires connected in parallel for single photon detection is to solve the read out problem by forcing *all* the parallel nanowires to develop normal state regions when a photon is absorbed in one nanowire. We have shown that this can be accomplished by polarizing all the parallel nanowires close to their critical currents [6]. Since SNSPDs usually are biased close to their critical currents, this is not a problem. For convenience we briefly review how the cascade switch works when M nanowires are connected in parallel. When a nanowire becomes resistive due to a photon absorption event, it will deviate its bias current onto the other $M - 1$ parallel nanowires and the load. In order to confine the current within the parallel nanowires at this stage, an inductor is inserted in series with the load. When the other $M - 1$ nanowires are biased close to their critical currents, the extra current will make the total current flowing through these nanowires exceed the critical current of each nanowire. This will drive all the parallel nanowires normal and subsequently they will all deviate their bias current onto the load and a signal will appear. The reduction of current in the parallel nanowires reduces the Joule heating and permit the cooling to recover superconductivity again. At this point, the current will return into the detector which is then ready for another photon. A key point is that the in series inductance can be realized using a serial connection of blocks of parallel light sensitive nanowires. This configuration mimics the serial connection used in the meander SNSPD, with the difference that each element has a M times lower inductance while it covers M time more area. For this reason, the inductance of a SNSPD based on parallel nanowires is reduced by a factor M^2 when compared to a meander SNSPD. Another key point is that the signal pulse of the parallel SNSPD is higher since more current is deviated onto the load. Finally, it is important to remember that the parallel SNSPD does not change the nanowires themselves, it only acts on their interconnection. This should leave the key detector properties, efficiency and dark count rate, unchanged, since they are only related the nanowires themselves. In this work we present measurements of

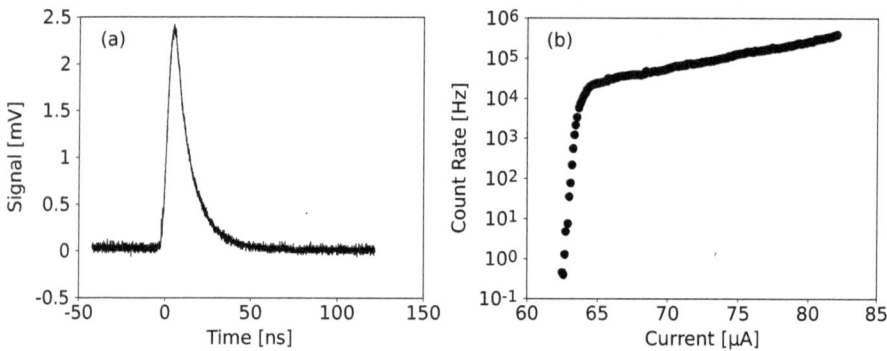

Fig. 1. (a) Parallel SNSPD signal pulse. (b) Measured photon count rate of a parallel SNSPD.

parallel SNSPDs showing that the efficiency at 850 nm is comparable to reported efficiency values of comparable meander SNSPDs. We also present measurement of the bias current dependence of the detector response at 1550 nm that is similar to what has been observed in meander SNSPDs.

In figure 1a we have shown a typical signal pulse from a parallel SNSPD biased in the useful bias range. The signal pulses are characterized by a short risetime and a significantly longer falltime, governed by the inductance of the parallel SNSPD. As seen, the signal to noise ratio (SNR) of the signal pulse is very good due to the large amount of current deviated onto the load. Also shown in figure 1b is the rate of photon induced signal pulses as a function of bias current, clearly demonstrating the presence of a useful bias range where the cascade switch works and solves the read-out problem.

3   Optical Experiments with Parallel SNSPD

Extensive details concerning the fabrication, configuration and measurements of the parallel SNSPDs reported in this work can be found in [7]. In brief, the parallel SNSPDs were made from 9 nm ultra-thin NbN film deposited on MgO substrates and patterned into 100 nm wide nanowires which were configured into parallel SNSPDs with M=4, 8, 12 and 24. In figure 2b we have shown a scanning electron micrograph of one of the parallel SNSPDs. All the devices covered an area of 5x6 μm^2, and the devices with M=8, 12 and 24 incorporated an on-chip in series inductance implemented using a wider NbN strip to ensure sufficient confinement of the bias current during the cascade switch stage. The devices were measured in a continuous flow cryostat with an optical window through which the parallel SNSPDs were illuminated using laser light pulses at 850 nm and 1550 nm. During operation the parallel SNSPDs were current biased through cold RC-filters and read out through a coaxial cable as shown in figure 2a. At room temperatures the signal pulses were amplified with a 1 GHz bandwidth RF amplifier and digitized with either 1 GHZ or 9 GHz bandwidth oscilloscopes.

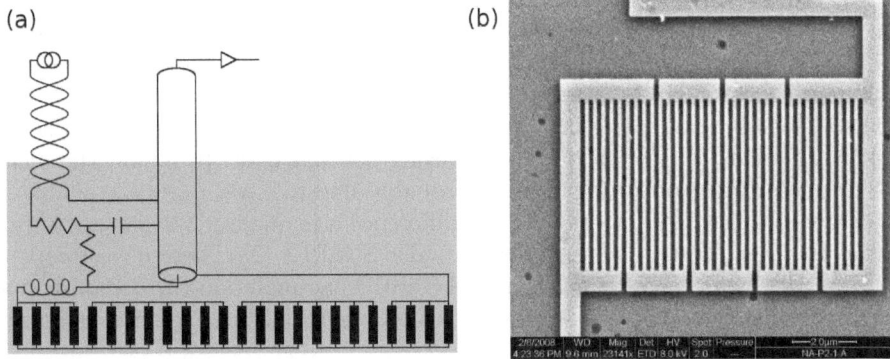

Fig. 2. (a) Diagram of the electrical biasing and readout of the parallel SNSPDs used in the measurements. The grey box indicates the parts located on the cold finger in the cryostat. (b) Scanning electron micrograph of the parallel SNSPD based on a serial connection of blocks each using four parallel nanowires.

In figure 3a we show the signal pulse height for the four different parallel SNSPDs when measured at a bias current of 9.75 μA per nanowire and illuminated with 850 nm laser light pulses. A clear linear trend is found, confirming that all the parallel nanowires are participating in the signal formation mechanism. By attenuating the light pulse intensity, we confirm that the parallel SNSPD exhibits single photon response across the entire useful bias range. In figure 3b we also show the parallel SNSPD quantum efficiency defined as the probability of receiving a signal pulse per photon incident on the detector area for 850 nm light. In the useful bias range, the quantum efficiency shows a characteristic exponential increase also seen in comparable measurements of meander

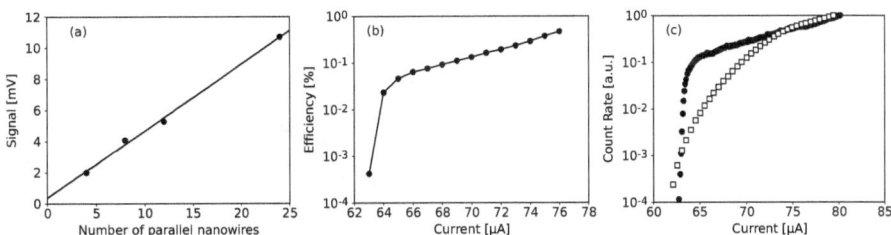

Fig. 3. (a) Measured maximum signal as a function of the number of parallel nanowires used in the parallel SNSPD. The bias current was 9.75 μA/nanowire and the temperature was 5 K. (b) Measured parallel SNSPD Efficiency vs bias current for a parallel SNSPD based on eight parallel nanowires. The measurement was performed at 5 K. The line is a guide for the eye. (c) Normalized parallel SNSPD photon count rate vs bias current for two different wavelengths: 850 nm (filled circles) and 1550 nm (open squares). The SNSPD was based on eight parallel nanowires and was measured at a temperature of 5 K.

SNSPDs [8]. Also the peak quantum efficiency of about 0.5% is in good agreement with previous measurements on meander SNSPDs with about the same thickness, width and at this temperature [5]. Finally we have also measured the parallel SNSPD photo-response when illuminated with 1550 nm laser light pulses (see figure 3c). The bias dependence of the detector count rate is found to be different from what is observed using 850 nm light. In detail, the onset of the detector response is smeared, probably due to lower energy available in 1550 nm photons. The presence of this difference is in qualitative agreement with what has been observed in thinner meander SNSPDs [2]. These measurements sustain the hypothesis that the parallel SNSPD can maintain the demonstrated detector efficiency of meander SNSPDs because it only changes the nanowire interconnections.

4   Parallel SNSPD Latching

The parallel SNSPD based on the cascade switch mechanism increases the maximum count rate roughly a factor M^2, where M is the number of parallel nanowires, and has been demonstrated working with up to at least 24 parallel nanowires [7]. Such a high speed gain easily triggers the new problem of SNSPD latching, which is the next limit of the maximum count rate in parallel SNSPDs. Latching occurs when the parallel SNSPD does not return into the superconducting state after a detection event, effectively stopping the detector from responding to subsequent photons, even if the bias current is lower than the critical current. For this reason, latching can be observed as an apparent reduction in the critical current (see figure 4a) of the detector which is detrimental because it inhibits detector operation in the bias range where it is most sensible to photons (see figure 4b).

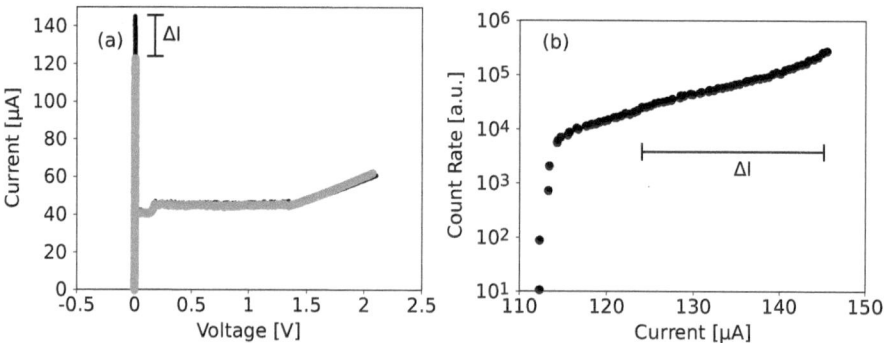

Fig. 4. (a) Current-voltage characteristics of the same parallel SNSPD measured with an extra in series inductance of 470 nH (small black circles) and 100 nH (large grey circles) demonstrating the difference in maximum bias current ΔI. (b) Measured photon count rate at 470 nH illustrating the missing bias region, ΔI.

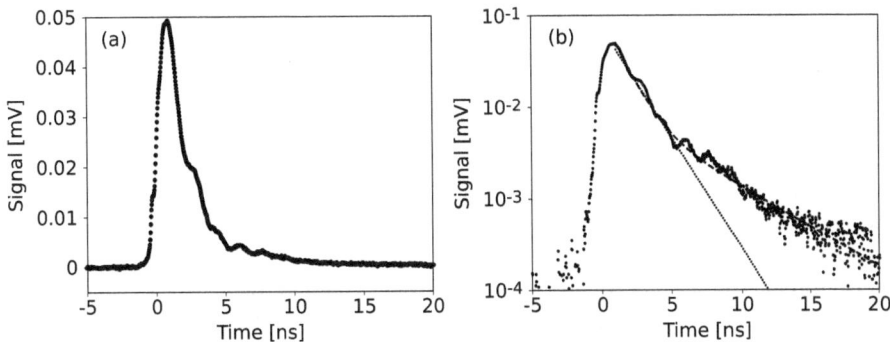

Fig. 5. (a) Averaged signal pulse of a parallel SNSPD based on four parallel nanowires (points). (b) Same as (a) but with a logarithmic signal amplitude scale. The lines are fit to the signal decay using a single exponential decay (dotted line) and a double exponential decay (dashed line).

The phenomenon has been observed in meander SNSPDs when the load impedance was increased or when the detector inductance was lowered [9]. Parallel SNSPDs are particularly susceptible to latching, probably due to the fact that operation is faster while the energy dissipation in the detector is higher [10]. Experimentally, this manifests itself as a need of large detector inductance values as increased M is used, to achieve proper operation [7]. By realizing the inductance with light sensitive nanowires one can achieve fast operation of large area detectors. Latching in parallel SNSPDs can be avoided, for a fixed load of 50 Ω, by properly choosing the number M according to the needed detector area. Smaller detectors must use smaller M whereas larger detectors can use more parallel nanowires without triggering the latching phenomenon. It is also likely that larger M can be used if the load resistance is reduced. According to [9] this could also speed up the SNSPD because the inductance can be reduced more than the load thereby lowering the time constant that limits the return of the current into the detector. This possibility comes at the cost of reduced signal pulse amplitude and therefore also reduces the signal to noise ratio of the detector signal.

In figure 5a we have shown a signal pulse from a parallel SNSPD based on four parallel nanowires measured with only 47 nH in series, i.e. going towards fast (~100 MHz) maximum count rate operation. In this condition we observe no latching and for this reason the efficiency should not be degraded. As can be seen the device is effectively ready for another photon after 10 ns. Interestingly, the high SNR have also allowed us to observe, for the first time, that the signal pulse decay is not well described by a single exponential decay in this condition (see figure 5b). This is contrary to what we have observed when slower operation is performed, where the decay is in fact well described by a single exponential. We can fit the data rather well with a bi-exponential decay. The time constant of the fast decay ($\tau = 1.2$ ns) can be identified with the electrical time constant whereas the origin of the second time constant ($\tau = 5$ ns) is not identified. The presence of this second exponential decay may impose a new limitation to overcome before very fast (> 1 GHz) maximum count rate of parallel SNSPDs can be achieved.

In conclusion, we have presented a new SNSPD configuration based on the use of parallel nanowires which can achieve significant gains in maximum count rate for large area SNSPDs. Measurements of the parallel SNSPD quantum efficiency and the change in detector efficiency with photon wavelength sustains that the parallel SNSPD maintains the good efficiency of meander SNSPDs. The gain in maximum count rate is however limited by new problem of SNSPD latching. We have discussed how latching can be avoided in parallel SNSPDs by increasing the detector size. Finally we have reported for the first time, that the signal pulse decay is not well described by a single exponential decay when fast operation is performed. In our measurements, the signal decay seems bi-exponential with a second fairly long time constant of 5 ns. This may have to be overcome if very fast maximum count rate is to be achieved with parallel SNSPDs.

This work was carried out in the framework of the EU project SINPHONIA NMP4-CT-2005-016433.

References

1. Goltsman, G.N., Okunev, O., Chulkova, G., Lipatov, A., Semenov, A., Smirnov, K., Voronov, B., Dzardanov, A., Williams, C., Sobolewski, R.: Picosecond superconducting single-photon optical detector. Appl. Phys. Lett. 79, 705 (2001)
2. Korneev, A., Kouminov, P., Matvienko, V., Chulkova, G., Smirnov, K., Voronov, B., Goltsman, G.N., Currie, M., Lo, W., Wilsher, K., Zhang, J., Sysz, W., Pearlman, A., Verevkin, A., Sobolewski, R.: Sensitivity and gigahertz counting performance of NbN superconducting single-photon detectors. Appl. Phys. Lett. 84, 5338 (2004)
3. Kerman, A.J., Dauler, E.A., Keicher, W.E., Yang, J.K.W., Berggren, K.K., Goltsman, G., Voronov, B.: Kinetic-inductance-limited reset time of superconducting nanowire photon counters. Appl. Phys. Lett. 88, 111116-1-3 (2006)
4. Rosfjord, K.M., Yang, J.K.W., Dauler, E.A., Kerman, A.J., Anant, V., Voronov, B.M., Goltsman, G.N., Berggren, K.K.: Nanowire Single-photon detector with an integrated optical cavity and anti-reflection coating. Opt. Express 74, 527 (2006)
5. Lipatov, A., Okunev, O., Smirnov, K., Chulkova, G., Korneev, A., Kouminov, P., Goltsman, G., Zhang, J., Slysz, W., Verevkin, A., Sobolewski, R.: An ultrafast NbN hot-electron single-photon detector for electronic applications. Supercond. Sci. Technol. 15, 1689–1692 (2002)
6. Ejrnaes, M., Cristiano, R., Quaranta, O., Pagano, S., Gaggero, A., Mattioli, F., Leoni, R., Voronov, B., Goltsman, G.: A cascade switching superconducting single photon detector. Appl. Phys. Lett. 91, 262509-1-3 (2007)
7. Ejrnaes, M., Casaburi, A., Quaranta, O., Marchetti, S., Gaggero, A., Mattioli, F., Leoni, R., Pagano, S., Crisiano, R.: Characterization of parallel superconducting nanowire single photon detectors. Supercond. Sci. Technol. 22, 055006-1-7 (2009)

8. Verevkin, A., Zhang, J., Sobolewski, R., Lipatov, A., Okunev, O., Chulkova, G., Korneev, A., Smirnov, K., Goltsman, G.N., Semenov, A.: Detection efficiency of large-active-area NbN single-photon superconducting detectors in the ultraviolet to near-infrared range. Appl. Phys. Lett. 80, 4687–4689 (2002)
9. Kerman, A.J., Yang, J.K.W., Molnar, R.J., Dauler, E.A., Berggren, K.K.: Electrothermal feedback in superconducting nanowire single-photon detectors. Phys. Rev. B 79, 100509-1–4 (2009)
10. Ejrnaes, M., Casaburi, A., Cristiano, R., Quaranta, O., Marchetti, S., Pagano, S.: Maximum count rate of large area superconducting single photon detectors. J. Mod. Optics 56, 390–394 (2009)

Nano-Optical Studies of Superconducting Nanowire Single Photon Detectors

John A.O'Connor[1], Paul A. Dalgarno[1], Michael G. Tanner[1], Richard J. Warburton[1], Robert H. Hadfield[1], Burm Baek[2], Sae Woo Nam[2], Shigehito Miki[3], Zhen Wang[3], and Masahide Sasaki[3]

[1] Heriot Watt University, Edinburgh, United Kingdom
[2] National Institute of Standards and Technology (NIST), Boulder, Colorado, USA
[3] National Institute of Information and Communications Technology (NICT), Japan
jao10@hw.ac.uk

Abstract. Single-photon detectors based upon superconducting nanowires offer single-photon sensitivity from the visible well into the infrared, encompassing the crucial telecommunication wavelengths. In addition they benefit from low dark counts (Hz) and low timing jitter (sub 100 ps). These detectors have recently been employed in photon-counting applications such as quantum cryptography and characterization of quantum emitters. Key challenges in the development of this emerging single-photon detector technology are improving device uniformity and improving optical coupling efficiency. We report on photoresponse mapping of superconducting nanowire single-photon detectors using nano-optical techniques. The device is mounted on a high resolution translation stage and illuminated with a focal spot significantly smaller than the device area (20 μm x 20 μm). Using a confocal microscope configuration, we achieve a diffraction limited spot size of 800 nm at $\lambda = 950$ nm. We have previously shown that uniform (high detection efficiency) and constricted (low efficiency) nanowire devices give contrasting photoresponse signatures – a broad plateau response across the whole device and a single point response respectively. Recent work on purposely-designed low fill-factor devices (100 nm linewidth, 1 mm period) has allowed us to resolve individual 100 nm width wires.

Keywords: Superconducting nanowire single photon detectors, SSPD, SNSPD.

1 Introduction

Advances in optics and nanotechnology allow the interaction between light and matter to be studied on ever decreasing length scales which may hold the key to gaining insights into fundamental physics and developing new technologies. The use of nano-optical techniques [namely confocal microscopy in conjunction with high refractive index solid immersion lens (SIL)] allows us to focus light with sub-micrometre precision onto nanostructured superconducting wires and to map their photoresponse[1].

Superconducting nanowire single-photon detectors (SNSPDs) based on superconducting nanowires hold great promise as a new type of high-speed, high sensitivity single-photon detector, with a spectral range from the visible well into the infrared. The basic device concept was pioneered by Gol'tsman *et al.*[2]. A 100 nm wide wire is patterned by electron beam lithography and reactive ion etching in an ultrathin

A. Sergienko, S. Pascazio, and P. Villoresi (Eds.): QuantumCom 2009, LNICST 36, pp. 158–166, 2010.

(~4 nm thick) NbN superconducting film. The superconducting wire (operated in the temperature range 1.5–5 K) is biased close to its critical current I_C. An incident visible or infrared photon perturbs the current distribution, triggering a fast voltage pulse with picosecond rise time. The largest area nanowire SSPDs now available[3] consist of a meander wire (100 nm linewidth, 200 nm period) covering an area up to $20 \times 20 \mu m^2$, compatible with a single mode telecommunications fibre. SSPDs have been successfully employed in quantum information processing applications, spanning quantum cryptography[4,5], quantum emitter characterization[6-10], and quantum logic circuits[11].

2 Photoresponse Mapping

Figure 1 shows the mapping setup. This apparatus provides a platform for excellent optical and spatial resolution, by combining a confocal microscope (using aspheric

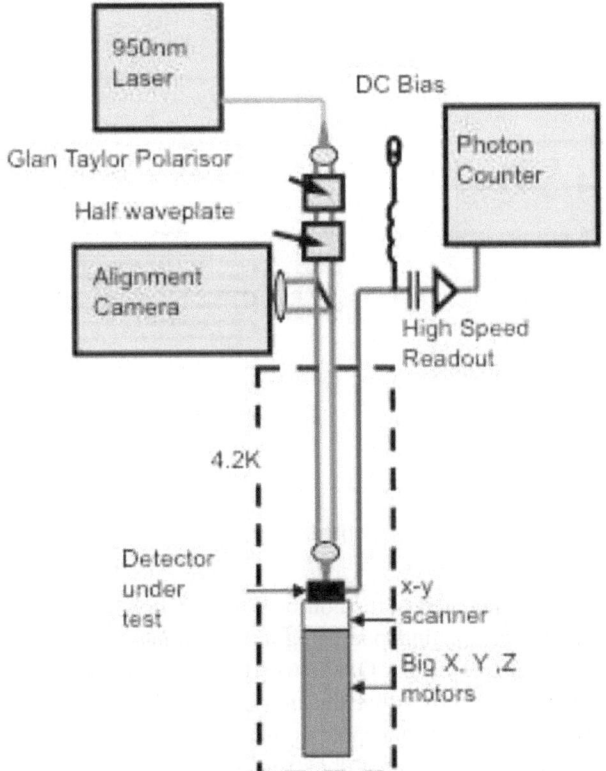

Fig. 1. Photoresponse mapping setup. This configuration affords high resolution scanning over a $30 \times 30 \mu m2$ area. The confocal microscope has a short working distance and high NA. The sample is mounted on low temperature piezoelectric linear positioners. The device is electrically connected for biasing and readout of photodetection events. A Glan Taylor polarizer is used to define the state of linear polarisation and the half waveplate is used to rotate the linear polarisation state (0-360°).

lenses) with XYZ piezoelectric linear positioners at 4.2K. The camera aids optical alignment. The sample space is filled with He vapour and is immersed in liquid He.

The resolution limit by Sparrow's criterion[12] is

$$FWHM = \frac{0.52\lambda}{NA} \tag{1}$$

Where FWHM is the full width at half maximum of the focal spot, λ is the wavelength, NA is the numerical aperture of the objective lens. The resolution can be boosted further (by a factor of the refractive index n) via use of a hemispherical solid immersion lens[1,13].

In this study, the devices tested were both 20 x 20 μm^2 area[3], but of differing fill factors: 10% (100 nm linewidth, 1 μm period) and 50% (100 nm linewidth, 200 nm period).

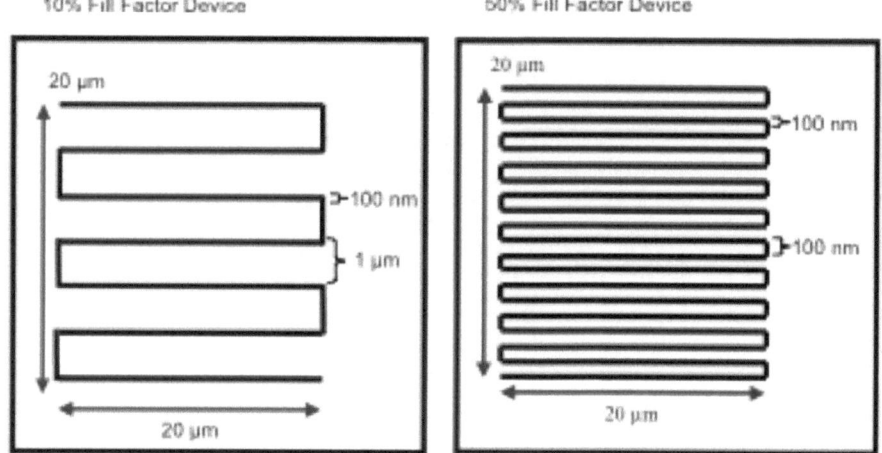

Fig. 2. Schematic of 10% and 50% fill factor devices

The 10% device was mounted in the confocal microscope. Using an objective lens with NA=0.68, the device was mapped using a continuous wave (cw) laser at λ=950 nm and a pulsed laser at λ=975 nm. Qualitatively the same response was seen at both wavelengths. Figure 3(A) shows that the individual wires in the device have been resolved. The one thing that is noticeable is that there is a missing line in the device that is electrically connected but is not photosensitive to the same extent as the rest of the device; this is due to that section of the wire having a lower current density than across the rest of the device. Possibly this is due to this wire being wider or thicker than the rest of the device.

The 50% device was mounted and mapped in the same way as the 10% device. Figure 3(B) shows a reasonably uniform response indicated by the plateau, which

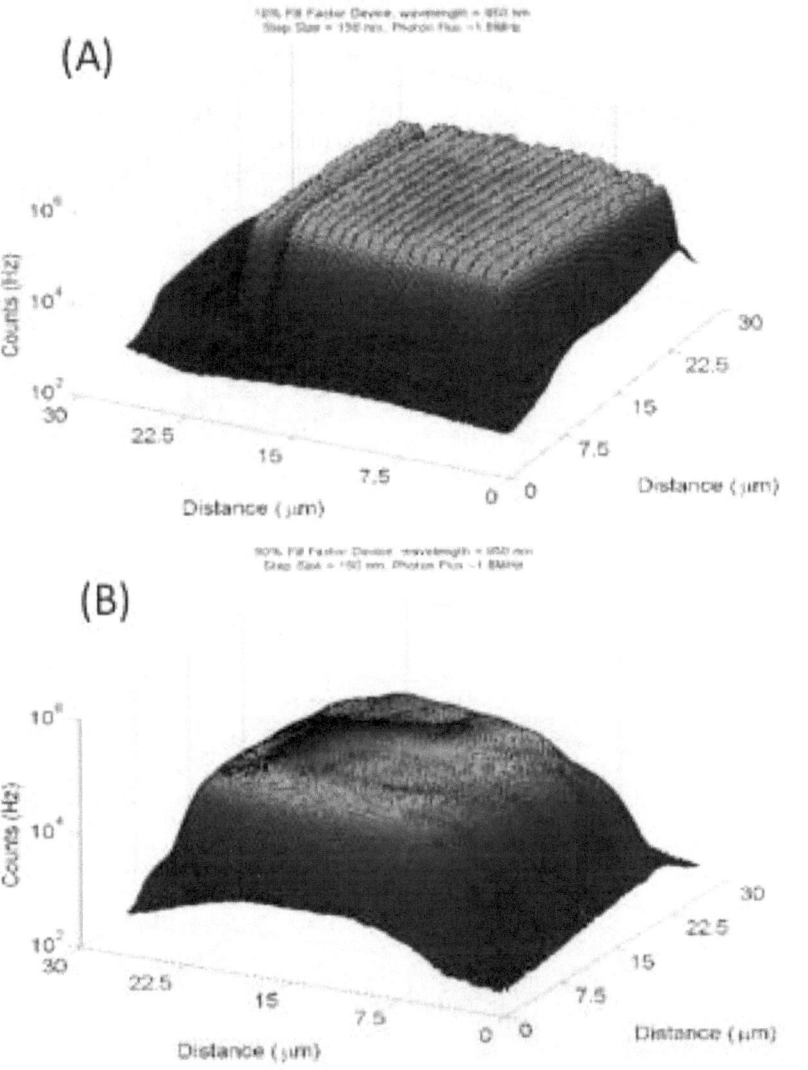

Fig. 3. Photoresponse maps, photon flux ~2MHz, device area 20 μm x 20 μm, λ=950 nm, FWHM spot size 800 nm. (A) 10% device (100 nm linewidth, 1μm period) with wires distinguishable. (B) 50% device (100 nm linewidth, 200 nm period) with uniform photoresponse.

varies by no more than a factor of 2. Again at both wavelengths qualitatively the same response was seen. For these measurements the background count rate was ~200Hz and the photon flux was ~2MHz.

Kinetic inductance is observed to increase as the critical current is approached[17] due to the increased current density. This increase is less in a device with constrictions as the current density in the majority of the device is still far from the critical current density. Through fitting with a suitable model the limiting constriction in the device can be characterised, in this C = 0.654 and C = 0.376 for the 10% and 50% fill factor devices respectively, where C is the ratio of the constriction cross sectional area to the unconstructed superconducting cross section of the wire. As the value for the

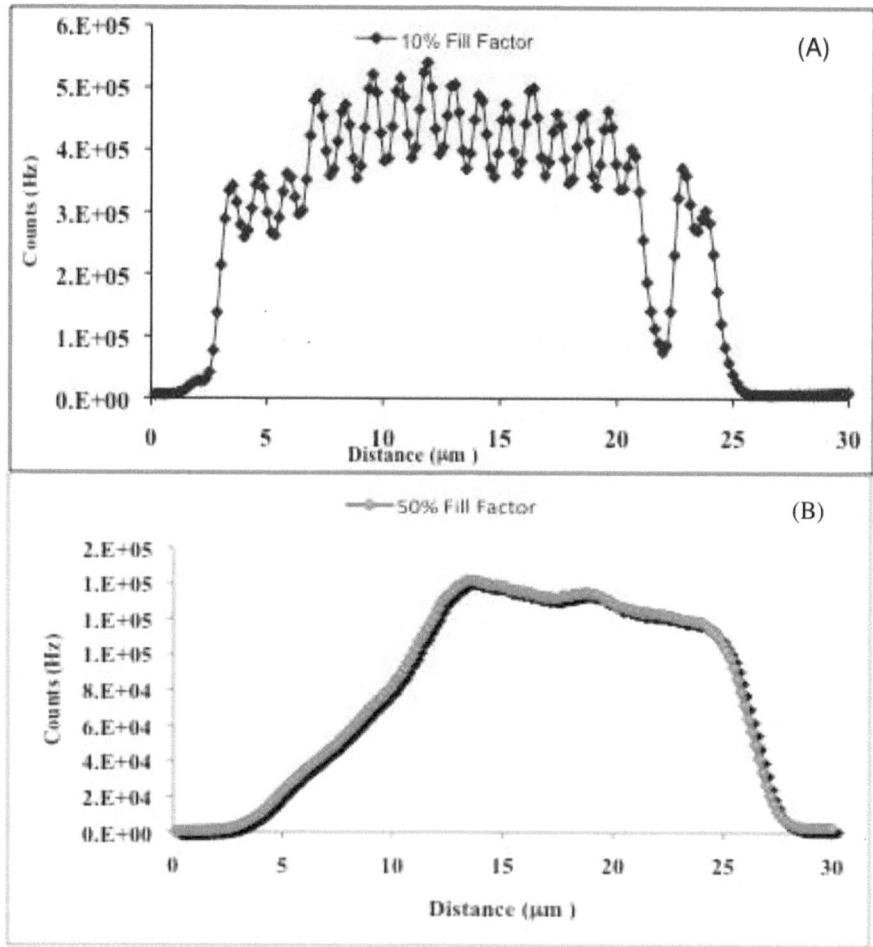

Fig. 4. (A) 10% Fill Factor device line scan perpendicular to the wires of the device. Missing line (still electrically connected) is visible. (B) 50% Fill Factor device line scan perpendicular to the wires. One edge is well defined and the other is less so depicted by the smaller gradient slope.

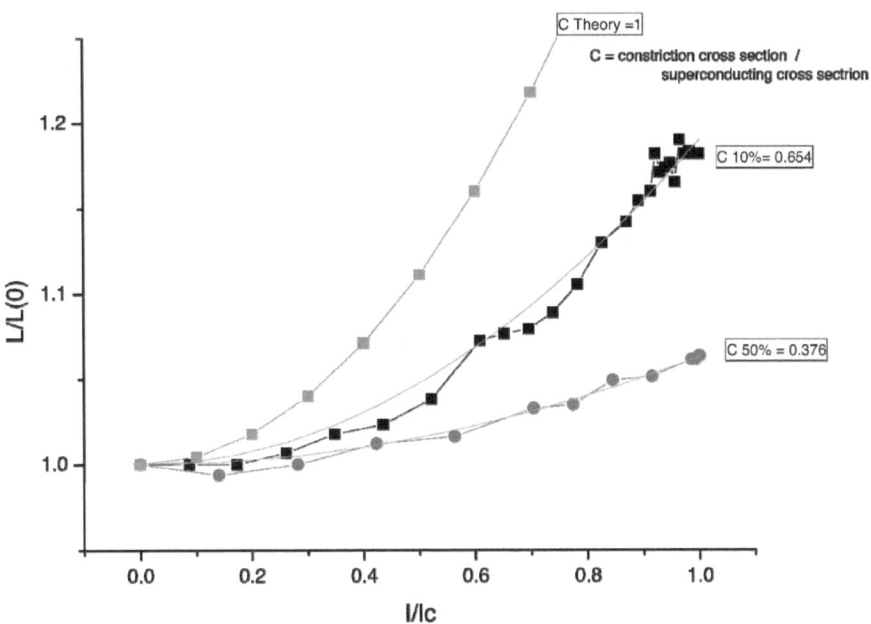

Fig. 5. Inductance trend approaching critical current (black squares for 10% fill factor and red circles for 50% fill factor devices) with fit from theory (green lines) and reference fit of unconstructed device (magenta squares).

10% device is close to 1 the whole length of the nanowire is able to approach the critical current achieving the high uniformity observed. The high value of C again suggests good fabrication was achieved. While conversely the lower value of C for the 50% device indicates a possible constriction in the nanowire suggesting the uniformity of the fabrication is not a good as the 10% device.

3 Polarisation Dependence

The linear polarisation state of the light that is incident on the device can affect the performance and efficiency of the devices. Polarisation effects of the devices have been shown and reported before[14-16]. We measured this effect in on both 10% and 50% fill factor devices.

The results show that polarisation dependence of both devices gives a maximum count rate with the light polarised parallel to the wires and a minimum count rate with the light polarised perpendicular to the wires. This agrees with theoretical models of absorption that have been carried out and reported by Anant et al[16].

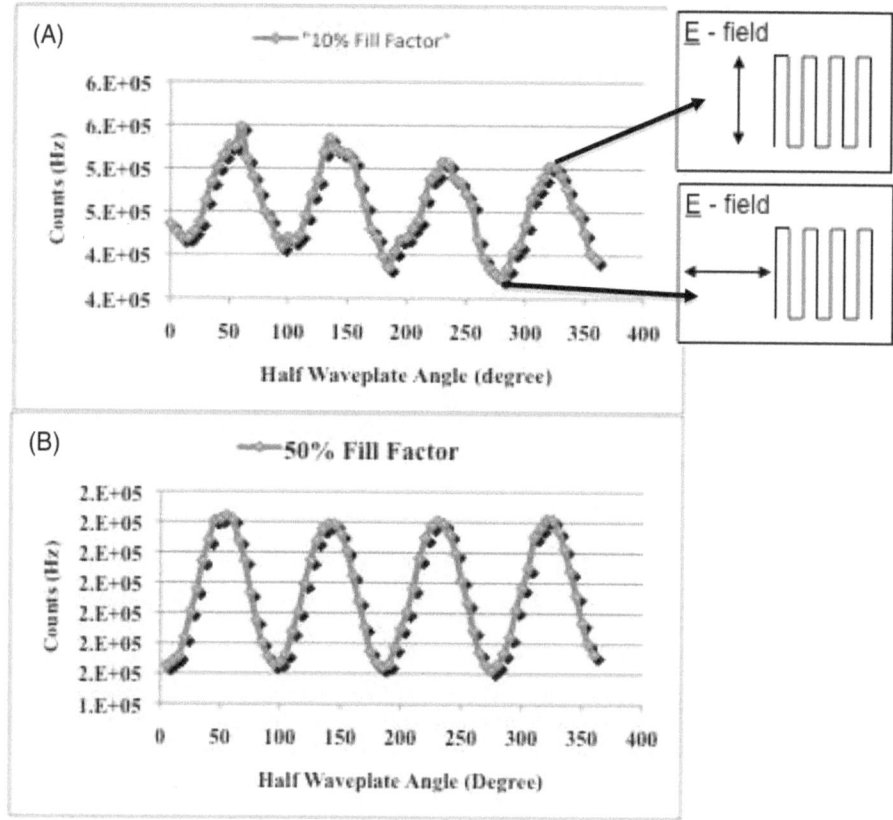

Fig. 6. Results of polarisation dependence for (A) 10% fill factor device and, (B) 50% device illuminated by 950nm laser light. Where 50° (and repeats every 90°) aligns the polarisation of the light to be parallel to the wires of the device, and 95° (and repeating every 90°) is aligned perpendicular to the wires.

4 Conclusion and Outlook

We have successfully carried out photoresponse mapping studies with submicrometer precision using a confocal microscope configuration. In recent experiments we have made complete photoresponse maps of 10% and 50% fill factor devices and are able to resolve individual wires in the former case. Both the 10% and 50% fill factor devices show polarisation dependences which varies in extent with the wavelength of the incident light. We plan to extend these studies to investigate local variations in timing jitter. Furthermore this enhanced optical coupling will allow us to study and explore novel device designs (e.g. nanoantennae) which may prove to be easier than the meander design to fabricate with higher quantum efficiencies, allowing scaling up to multi-pixel devices.

Acknowledgements

JAO, PAD, RJW, RHH thank the UK EPSRC for support. RHH is also supported by a Royal Society of London University Research Fellowship.

References

1. Hadfield, R.H., Dalgarno, P.A., O'Connor, J.A., Ramsay, E., Warburton, R.J., Gansen, E.J., Baek, B., Stevens, M.J., Mirin, R.P., Nam, S.: Submicrometer photoresponse mapping of nanowire superconducting single-photon detectors. Appl. Phys. Lett. 91, 241108 (2007)
2. Gol'tsman, G.N., Okunev, O., Chulkova, G., Lipatov, A., Semenov, A., Smirnov, K., Voronov, B., Dzardanov, A., Williams, C., Sobolewski, R.: Picosecond superconducting single-photon optical detector. Appl. Phys. Lett. 79, 705 (2001)
3. Miki, S., Fujiwara, M., Sasaki, M., Baek, B., Miller, A.J., Hadfield, R.H., Nam, S., Wang, Z.: Large sensitive-area NbN nanowire superconducting single-photon detectors fabricated on single-crystal MgO substrates. Appl. Phys. Lett. 92, 061116 (2008)
4. Hadfield, R.H., Habif, J.L., Schlafer, J., Schwall, R.E., Nam, S.: Quantum key distribution at 1550 nm with twin superconducting single-photon detectors. Appl. Phys. Lett. 89, 241129 (2006)
5. Takesue, H., Nam, S., Zhang, Q., Hadfield, R.H., Honjo, T., Tamaki, K., Yamamoto, Y.: Quantum key distribution over a 40-dB channel loss using superconducting single-photon detectors. Nat. Photonics 1, 343 (2007)
6. Hadfield, R.H., Stevens, M.J., Gruber, S.S., Miller, A.J., Schwall, R.E., Mirin, R.P., Nam, S.: Single photon source characterization with a superconducting single photon detector. Opt. Express 13, 10846 (2005)
7. Stevens, M.J., Hadfield, R.H., Schwall, R.E., Nam, S., Mirin, R.P., Gupta, J.A.: Fast lifetime measurements of infrared emitters using a low-jitter superconducting single-photon detector. Appl. Phys. Lett. 89, 031109 (2006)
8. Liang, C., Lee, K.F., Medic, M., Kumar, P., Hadfield, R.H., Kumar, P.: Characterization of fiber-generated entangled photon pairs with superconducting single-photon detectors. Opt. Express 15, 1322 (2007)
9. Hadfield, R.H., Stevens, M.J., Nam, S., Mirin, R.P.: Single-photon source characterization with twin infrared-sensitive superconducting single-photo detectors. J. Appl. Phys. 101, 103104 (2007)
10. Zinoni, C., Alloing, B., Li, L.H., Marsili, F., Fiore, A., Lungi, L., Gerardino, A., Vakhtominm, Y.B., Smirnov, K.V., Gol'tsman, G.N.: Single-photon experiments at telecommunication wavelengths using nanowire superconducting detectors. Appl. Phys. Lett. 91, 031106 (2007)
11. Chen, J., Altepeter, J.B., Medic, M., Lee, K.F., Gokden, B., Hadfield, R.H., Nam, S., Kumar, P.: Demonstration of a quantum controlled-NOT gate in the telecommunications band. PRL 100, 133603 (2008)
12. Sparrow, C.M.: On spectroscopic resolving power. Astrophys. J. 44, 76 (1916)
13. Serrels, K.A., Ramsay, E., Dalgarno, P.A., Gerardot, B., O'Connor, J.A., Hadfield, R.H., Warburton, R.J., Reid, D.T.: Solid immersion lens applications for nanophotonic devices. J. Nanophotonics 2, 021854 (2008)

14. Driessen, E.F.C., Braakman, F.R., Reiger, E.M., Dorenbos, S.N., Zwiller, V., de Dood, M.J.A.: Polarization measurements on NbN SNSPDs 480 nm – 1550 nm. Eur. Phys. J. Appl. Phys. 47, 10701 (2009)
15. Dorenbos, S.N., Reiger, E.M., Akopian, N., Perinetti, U., Zwiller, V., Zijlstra, T., Klapwijk, T.M.: Superconducting single photon detectors with minimized polarization deendence. APL 93, 161102 (2008)
16. Anant, V., Kerman, A.J., Dauler, E.A., Yang, J.K.W., Rosfjord, K.M., Berggren, K.K.: Optical properties of superconducting nanowire single-photon detectors. Opt. Express 16, 10750 (2008)
17. Kerman, A.J., Dauler, E.A., Keicher, W.E., Yang, J.K.W., Berggren, K.K., Gol'tsman, G.N., Voronov, B.: Kinetic-inductance-limited reset time of superconducting nanowire photon counters. Appl. Phys. Lett. 88, 111116 (2006)

Examples of Quantum Dynamics in Optomechanical Systems

Max Ludwig, Georg Heinrich, and Florian Marquardt

Arnold-Sommerfeld Center for Theoretical Physics, Center for NanoScience
and Department of Physics, Ludwig-Maximilians Universität München,
Munich, Germany

Abstract. Optomechanical systems exploit the interaction between the
optical radiation field and mechanical resonators in a laser-driven cavity.
In the past few years, these systems have been the focus of considerable experimental and theoretical attention, yielding promising successes,
particularly in using optomechanical cooling to reduce the thermal occupation of the resonators. This offers the prospect of observing quantum
dynamics involving the motion of macroscopic mechanical objects. We
review two features: First, the nonlinear self-induced mechanical oscillations induced by a strong laser drive can exhibit interesting quantum
behaviour at low temperatures. Second, a mechanically driven membrane inside an optical cavity can 'shuttle photons' around, and this
system exhibits intricate dynamical interference effects (Landau-Zener-
Stueckelberg oscillations).

Keywords: Optomechanics, quantum electrodynamics, quantum
physics, photons, nanomechanics, optical cavity.

1 Introduction

The interaction of light with matter is one of the fundamental topics in physics. A
new chapter has been opened by the recent combination of nano- and micromechanical systems with optical cavities. The light circulating inside the cavity
exerts a radiation pressure force on a moveable mirror (attached to the mechanical resonator), thus inducing mechanical motion. Conversely, any motion results
in a change of the cavity length, displacing the optical resonance frequency, and
thus influencing the circulating intensity, i.e. the force. In this way, an interesting
coupled dynamics of the light field and the mechanical vibration is produced.
Among other effects, the light can be used to cool the mechanical motion (in
analogy to laser cooling of atoms). This has been exploited to cool mechanical
resonators to near the ground state of the mechanical motion of one of their vibrational eigenmodes (for recent experimental examples see [1,2]). Once ground
state cooling is achieved, a wealth of quantum effects may be observed, creating
nonclassical states in optomechanical systems. We refer the reader to our recent
review [3] for a non-technical introduction and comprehensive references to the
literature.

A. Sergienko, S. Pascazio, and P. Villoresi (Eds.): QuantumCom 2009, LNICST 36, pp. 167–179, 2010.
© Institute for Computer Sciences, Social-Informatics and Telecommunications Engineering 2010

In the present contribution, we discuss and review our results concerning two topics connected to the dynamics of optomechanical systems.

First, we discuss the nonlinear self-induced mechanical oscillations induced by a strong laser drive that can exhibit interesting quantum behaviour at low temperatures. When the laser beam is blue detuned with respect to the cavity, the back-action of the light onto the mechanical motion tends to diminish the mechanical damping rate. This can lead to an instability which makes the system settle into a pattern of periodic oscillations. The power dissipated by intrinsic mechanical friction is then offset by the power fed into the system via the radiation. We have analyzed theoretically the behaviour in this regime for a system cooled down to low temperatures, into the quantum regime of mechanical motion. Using a master equation description, we have found the quantum state of the oscillator, taking full account of the effects of quantum fluctuations.

Second, a mechanically driven membrane inside an optical cavity can 'shuttle photons' around, and this system exhibits intricate dynamical interference effects. When a partially transparent membrane is placed between two fixed end-mirrors, it essentially divides the cavity in two halves. Photons then can tunnel through the membrane from one side to the other. A photon being in the left or the right side of the cavity realizes a simple two-state system. When applying a mechanical drive to the membrane, one can shuttle photons from the left half to the right half, which shows up in the amount of light transmitted through the whole setup. When the drive is too fast for this regular shuttle process, one can see interesting interference phenomena. Essentially, this setup thus allows to observe the physics of driven two-level systems in an optomechanical setting. This includes, in particular, interference effects at stronger drive, known as Landau-Zener-Stueckelberg oscillations.

2   Quantum Nonlinear Dynamics in Optomechanical Systems

The delayed light force in an optomechanical system changes the mechanical damping rate. The additional contribution Γ_{opt} can become negative, decreasing the total damping rate Γ_{full}. When Γ_{full} becomes negative as well, the system settles into a nonlinear dynamical regime of self-sustained oscillations [4,5]. In this section, we will describe the quantum behaviour in this regime. An extensive previous analysis has been published in [5]. We will illustrate the main features by some additional numerical examples not to be found in [5].

2.1   The Model

In the following we will focus on comparing the quantum dynamics of the coupled cavity-oscillator system to the results of the analytic solution of the classical equations of motion. In doing so we will discuss the effect of quantum fluctuations and identify a quantum parameter that allows to keep track of the quantum-to-classical transition. In particular we will present results of a fully quantum mechanical approach based on a master equation in Lindblad form, and results of a semi-classical approach based on Langevin equations where quantum effects are mimicked by fluctuating noise terms.

We consider the following Hamiltonian to describe a generic optomechanical system consisting of a driven optical cavity whose resonance frequency depends on the position of a mechanical element:

$$\hat{H} = \hbar\left(-\Delta + g_M(\hat{b} + \hat{b}^\dagger)\right)\hat{a}^\dagger\hat{a} + \hbar\omega_M\hat{b}^\dagger\hat{b} + \qquad (1)$$
$$\hbar\alpha_L(\hat{a} + \hat{a}^\dagger) + \hat{H}_\kappa + \hat{H}_{\Gamma_M} = \hat{H}_0 + \hat{H}_\kappa + \hat{H}_{\Gamma_M},$$

which is written in the rotating frame of the driving laser field whose frequency is denoted by ω_L and whose amplitude is set by α_L. The laser is detuned by $\Delta = \omega_L - \omega_{cav}$ with respect to the optical cavity mode which is described by photon annihilation and creation operators $\hat{a}$ and $\hat{a}^\dagger$. The mechanical oscillator is characterized by its frequency ω_M and mass m_M, and its displacement is given by $\hat{x}_M = x_{ZPF}(\hat{b} + \hat{b}^\dagger)$, with a mechanical zero-point amplitude of $x_{ZPF} = \sqrt{\hbar/(2m_M\omega_M)}$. The optomechanical coupling, i.e. the coupling between the optical field and the mechanical displacement, is given by the parameter g_M. In the simplest case, with a movable, fully reflecting mirror at one end of an optical cavity of length L, we have $g_M = -\omega_{cav}x_{ZPF}/L$, and thus $g_M(\hat{b} + \hat{b}^\dagger) = -\omega_{cav}\hat{x}_M/L$. The radiation pressure force corresponding to this coupling term is given by $\hat{F}_{rad} = -\hbar g_M\hat{a}^\dagger\hat{a}/x_{ZPF} = \hbar\omega_{cav}\hat{a}^\dagger\hat{a}/L$. The decay of the cavity photons and the mechanical damping of the cantilever are captured by $\hat{H}_\kappa$ with a cavity ring down rate κ, and by $\hat{H}_{\Gamma_M}$ with a damping rate Γ_M, respectively.

2.2 Classical Dynamics

The classical equations of motion following from the Hamiltonian (1) can be rescaled (see ref. [5]) to

$$\frac{d\tilde{\alpha}}{d\tilde{t}} = \left[i\left(\frac{\Delta}{\omega_M} + \tilde{x}\right) - \frac{1}{2}\frac{\kappa}{\omega_M}\right]\tilde{\alpha} + \frac{1}{2}$$
$$\frac{d^2\tilde{x}}{d\tilde{t}^2} = -\tilde{x} + \mathcal{P}|\tilde{\alpha}|^2 - \frac{\Gamma_M}{\omega_M}\frac{d\tilde{x}}{d\tilde{t}}, \qquad (2)$$

where $\tilde{t} = \omega_M t$, $\tilde{\alpha} = i\alpha\omega_M/(2\alpha_L)$, $\tilde{x} = gx/(\omega_M x_{ZPF})$ and the field operator $\hat{a}$ is replaced by the classical amplitude α and the position operator $\hat{x}$ by the classical quantity x. Note that the system's dynamics is governed by four dimensionless parameters only, the driving strength $\mathcal{P} = 8|\alpha_L|^2 g_M^2/\omega_M^4$, the detuning Δ/ω_M, the cavity decay κ/ω_M and the mechanical dissipation Γ_M/ω_M, and that it is purely deterministic as both photon shot noise and mechanical fluctuations are ignored.

Above a certain threshold of driving power the mechanical resonator will perform self-sustained oscillations of the form $x(t) \approx \bar{x} + A\cos(\omega_M t)$ in steady state, as has been seen experimentally [6,7,8]. Hence the oscillator's dynamics is fully determined once the parameters $\{\bar{x}, A\}$ are found as a function of

$\{\mathcal{P}, \Delta/\omega_M, \kappa/\omega_M, \Gamma_M/\omega_M\}$. This task, as presented in detail in [4,5], is accomplished by demanding that, firstly, the total force on the mechanical resonator has to vanish on average, and, secondly, the power transferred to the mechanical resonator from the light field has to balance the friction loss, i.e. $P_{rad}/P_{fric} \equiv 1$.

Hence after eliminating $\bar{x}$ from the force balance condition, the attractors for the oscillation amplitude A or equivalently for the oscillation energy $E_{M,cl} = \frac{1}{2}m\omega_M^2 A^2$, can be determined implicitly from the power balance condition as demonstrated in Figure 1(a): The plot shows the ratio P_{rad}/P_{fric} as a function of detuning Δ/ω_M and oscillation energy $E_{M,cl}$. The contour lines correspond to $P_{rad}/P_{fric} \equiv 1$ for different values of the mechanical damping rate and represent solutions of the oscillation energy (and thus A) as a function of Δ/ω_M for given values of $\mathcal{P} = 0.12$, $\kappa/\omega_M = 0.3$ and $\Gamma_M/\omega_M = 0.074$ (black), 0.071 (blue), 0.019 (white). Note that the oscillation energy E_M is scaled by a characteristic energy scale of the system, $E_0 = \frac{1}{2}m\omega_M^2 x_{FWHM}^2$, where x_{FWHM} denotes the full width at half maximum (FWHM) of the cavity translated into a length scale ($x_{FWHM} = \kappa L/\omega_{cav}$, where L is the cavity's length).

For $\kappa < \omega_M$ the attractor diagram shows pronounced features at the sidebands, where the detuning Δ equals an integer multiple of the mechanical frequency ω_M. However, due to the large driving power $\mathcal{P}$, the resonance peaks are strongly tilted, shifted towards lower detunings and show discontinuities. At the second sideband the strongly damped solution (black contour) does not feature self-oscillations. After a small reduction of the damping rate, however, the attractor at the second sideband appears, i.e. oscillations at finite amplitude (blue line) are possible solutions of the system. Because $A = 0$ remains to be a stable solution at the second sideband, the system already shows bistable behaviour (and, more generally, multistable behaviour for other parameter sets).

2.3   The Quantum Parameter

In contrast to the classical equations of motion, the set of parameters, $\mathcal{P}, \Delta/\omega_M$, $\kappa/\omega_M, \Gamma_M/\omega_M$ does not suffice for a quantum mechanical treatment, where $\hbar$ enters as well. The quantum mechanical nature of the system can be described by a "quantum parameter" ζ, which compares the magnitude of the cantilever's zero-point fluctuations, x_{ZPF}, with the cavity linewidth x_{FWHM}:

$$\text{quantum parameter}: \zeta = \frac{x_{ZPF}}{x_{FWHM}} = \frac{g_M}{\kappa}. \tag{3}$$

Obviously, this turns out to be equal to the ratio of the optomechanical coupling frequency g_M and the cavity decay rate κ, such that values larger than one indicate a strong-coupling regime. The quantum parameter ζ vanishes in the classical limit $\hbar \to 0$, as the zero-point fluctuations x_{ZPF} of the cantilever go to zero. The magnitude of ζ determines the effect of quantum fluctuations, i.e. the shot noise of the cavity and the mechanical zero-point fluctuations on the dynamics of the coupled cavity-cantilever system.

We note that there is an alternative way to introduce the quantum parameter (3) by comparing the mechanical zero-point momentum fluctuations to the momentum a single intracavity photon transfers to the cantilever. This approach

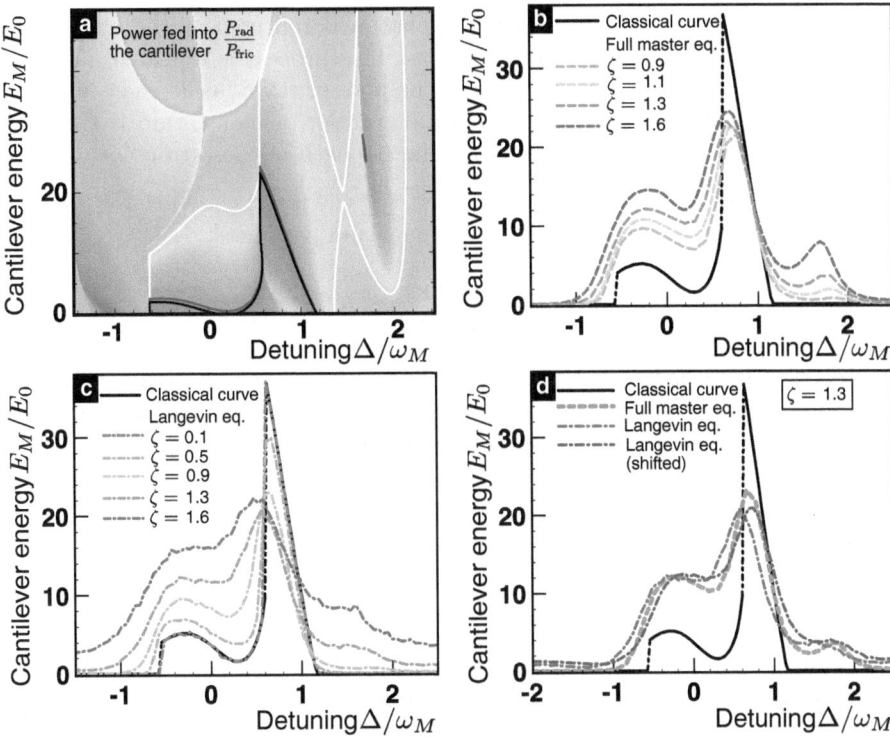

Fig. 1. Self-induced optomechanical oscillations in the quantum regime. Plots of cantilever oscillation energy vs. detuning for various parameters. Results from the Langevin equation (which mimics quantum noise) are compared to the purely deterministic classical solution and to the results from the full quantum master equation. The parameters are $\kappa/\omega_M = 0.3$, $\Gamma_M = 0.074$, $\mathcal{P} = 0.12$. (a) The classical solution for the oscillation energy is given by the black contour line in the attractor diagram, while the blue (gray) and white contours depict the solutions for lower damping rates, $\Gamma_M = 0.071$ and $\Gamma_M = 0.019$ respectively. Self-oscillations at the second sideband occur already for slightly modified parameters (see the blue contour line). (b) In the solutions of the full quantum master equation (dashed curves), we observe a shift of the resonances towards lower detunings and a smooth behaviour, in contrast to the sharp structures and discontinuities of the classical result (black curve). For high values of the quantum parameter ζ [see main text], the quantum curves show features that occur in the classical solution for lower damping rates only: The peak at the second sideband appears and the peaks of the first and second sideband merge. (c) The results of the Langevin equation recover the classical curves for the case of very weak quantum fluctuations, i.e. for $\zeta = 0.1$. They also reproduce the main features of the curves that come from the master equation. However, this approach fails to match the results of the master equation for large values of ζ and outside the region of instability. (d) Replacing the radiation pressure term of equation (7) by $\frac{\hbar g}{m x_{ZPF}}(|\alpha|^2 - \frac{1}{2})$ shifts the semi-classical curve towards lower detunings, but does not lead to a better agreement with the curve from the quantum master equation.

has been presented in [9] where the mechanical element was composed of an ensemble of ultracold atoms. When the photon is reflected at the mechanical oscillator, it transfers a momentum of $2\hbar k$. This process is repeated after one cavity round-trip time $\frac{2L}{c}$ for as long as the photon stays inside the cavity, i.e. for a span of time given by κ^{-1}. The total transfer of momentum is therefore given by $p_{\text{phot}} = \hbar k \frac{c}{L} \kappa^{-1} = \hbar\omega_{\text{cav}}/\kappa L$. The strength of the zero-point momentum fluctuations is given by $p_{\text{ZPF}} = \sqrt{\frac{\hbar m_M \omega_M}{2}} = \frac{\hbar}{2x_{\text{ZPF}}}$. Taking the ratio of these two quantities leads directly to the quantum parameter:

$$\frac{p_{\text{phot}}}{p_{\text{ZPF}}} = \frac{2x_{\text{ZPF}}}{\kappa L/\omega_{\text{cav}}} = 2\zeta. \tag{4}$$

We see that for a large quantum parameter a single phonon of the cantilever causes a detectable shift of the cavity resonance. In turn, a single photon causes the cantilever to change its momentum noticeably.

2.4   Quantum Dynamics: Master Equation

We will now investigate the quantum dynamics in the regime of self-induced oscillations. We choose to focus on the extreme quantum regime, where the bath temperature is set to zero and the quantum parameter is made to achieve values around one. This demonstrates the effects of quantum fluctuations most clearly. Although this limit cannot be reached (at present) using micro- or nanomechanical resonators (where bulk temperatures are high and the quantum parameter is less than 10^{-3}), it might be achievable in cold atom systems or other variants.

For a fully quantum mechanical treatment we aim at finding the steady state solution of the reduced density matrix $\hat{\rho}$ for the mechanical oscillator and the optical mode of the cavity. The time evolution of $\hat{\rho}$ is given by

$$\frac{d}{dt}\hat{\rho} = \frac{[\hat{H}_0, \hat{\rho}]}{i\hbar} + \Gamma_M \, \mathcal{D}[\hat{b}] + \kappa \, \mathcal{D}[\hat{a}], \tag{5}$$

where $\mathcal{D}[\hat{A}] = \hat{A}\hat{\rho}\hat{A}^\dagger - \frac{1}{2}\hat{A}^\dagger\hat{A}\hat{\rho} - \frac{1}{2}\hat{\rho}\hat{A}^\dagger\hat{A}$ denotes the standard Lindblad operator and we assumed the mechanical bath to be at zero temperature.

For the numerical evaluation, we rewrite equation 5 as $d\hat{\rho}/dt = \mathcal{L}\hat{\rho}$, with a Liouvillian super-operator $\mathcal{L}$. We then interpret the density matrix as a vector with a finite length in a truncated Fock-state basis of the coupled system. The time evolution of $\hat{\rho}$ is then governed by the matrix $\mathcal{L}$ and the density matrix at long times (in steady state) is given by the eigenvector of $\mathcal{L}$ with eigenvalue 0. The numerical calculation of this eigenvector is much more efficient than a simulation of the full time evolution. Since we are dealing with large sparse matrices, it is convenient to employ an Arnoldi method that finds a few eigenvalues and eigenvectors of $\mathcal{L}$ by iterative projection. For Hermitean matrices, the Arnoldi method is also known as the Lanczos algorithm.

For a comparison with the classical solution, the mechanical energy is evaluated from the system's steady state density matrix as $E_{M,\text{qm}} = \hbar\omega_M \langle \hat{n}_M \rangle =$

$\hbar\omega_M\mathrm{Tr}(\hat{n}_M\hat{\rho})$. Note that we exclude the zero-point energy and define the occupation number operator as $\hat{n}_M = \hat{b}^\dagger\hat{b} + \frac{1}{4x_{\mathrm{ZPF}}^2}(\langle\hat{x}_M\rangle^2 - 2\hat{x}_M\langle\hat{x}_M\rangle)$ in order to exclude the effects of a static displacement of the mechanical resonator, i.e. $\langle\hat{x}_M\rangle \neq 0$. Only in this case the expectation value of $\hat{n}_M$ corresponds to the oscillation energy. The scale of the mechanical energy is again set by the classical quantity $E_0 = \frac{1}{2}m\omega_M^2 x_{\mathrm{FWHM}}^2$ such that $E_M/E_0 = 4\zeta^2\langle\hat{n}_M\rangle$.

In figure 1(b) we demonstrate the influence of the quantum parameter $\zeta = x_{\mathrm{ZPF}}/x_{\mathrm{FWHM}}$ determining the crossover from the quantum regime towards classical behaviour. In particular we compare the oscillation energy obtained from the classical solution to the results coming from the master equation approach. The classical (black) curve depicts the oscillation energy as a function of the detuning parameter and corresponds to the black contour in the attractor diagram. It features self-oscillations around the resonance ($\Delta = 0$) and the first sideband ($\Delta/\omega_M = 1$) and discontinuities at the left slopes (dash-dotted black lines). Results of the solution of the quantum master equation are shown for four different values of the quantum parameter $\zeta = 0.9, 1.1, 1.3, 1.6$. When compared to the classical curve the peaks appear gradually broadened, reduced in height, and shifted to lower detuning for increasing values of the quantum parameter ζ.

As expected, the discrepancy between the quantum mechanical and the classical result reduces with diminishing quantum parameter ζ. In this parameter regime, however, the full convergence towards the classical curve cannot be illustrated as the numerical costs become very large for small ζ, where the occupation numbers become large. The curves from the quantum master equation for large ζ seem to show features observed in the classical solution for a lower mechanical damping rate (or higher driving strength). The resonance at the second sideband emerges with increasing ζ and the gap between the first and the second sideband disappears (see also the blue and white contours in the attractor diagram).

2.5 Langevin Equations

We complement our analysis by comparing the results of the quantum master equation to numerical simulations of semiclassical Langevin equations that try to mimic the effects of quantum noise. To imitate both the zero-point fluctuations of the mechanical oscillator and the shot-noise inside the cavity, white noise terms, denoted by α_{in} and ξ, are added to the classical equations of motion:

$$\dot{\alpha} = [i(\Delta + g\frac{x}{x_{\mathrm{ZPF}}}) - \frac{\kappa}{2}]\alpha - i\alpha_L + \sqrt{\kappa/2}\,\alpha_{in} \qquad (6)$$

$$\ddot{x} = -\omega_M^2 x + \frac{\hbar g}{m x_{\mathrm{ZPF}}}|\alpha|^2 - \Gamma_M\dot{x} + \sqrt{\hbar\omega_M\Gamma/m}\,\xi, \qquad (7)$$

where $\langle\alpha_{in}\rangle = \langle\xi\rangle = 0$ and $\langle\alpha_{in}(t)\alpha_{in}^*(t')\rangle = \langle\xi(t)\xi(t')\rangle = \delta(t - t')$. The coefficients in front of the noise terms are chosen such that in the absence of optomechanical coupling we obtain the zero-point fluctuations, i.e. $\langle|\alpha|^2\rangle = 0.5$ away from optical resonance, and $\frac{m\omega_M^2}{2}\langle x^2\rangle = \frac{\hbar\omega_M}{4}$.

The results of a numerical simulation of the Langevin equations (6) and (7) are presented in figures 1(c) and (d) where again the oscillation energy (not including the mechanical zero-point energy) is plotted as a function of the detuning parameter. For negligible quantum fluctuations, i.e. for a small quantum parameter of $\zeta = 0.1$, the semiclassical curve recovers the classical solution of the attractor diagram as required for consistency. Increasing the quantum parameter allows to observe the complete transition into the quantum regime. The curves from the Langevin approach show the main features of the corresponding curves from the master equation approach (see figure 1(b)): The broadening of the peaks, the shift towards lower detunings and the emergence of the second sideband with increasing quantum parameter ζ.

Still, the Langevin approach can mimick the results from the master equation only partially. The approximation gets worse when dealing with low photon numbers and very large values of the quantum parameter ζ. In particular, the oscillation energy of the cantilever is overestimated by the semi-classical approach in the regions away from or in between the resonances. This is because the Langevin equation introduces artificial fluctuations of the radiation pressure force in the vacuum state. Indeed, $|\alpha|^2$ has a finite variance even in the ground state of the photon field, in contrast to $\hat{a}^\dagger \hat{a}$. To give a few numbers on the occupation numbers of the cavity for the parameters of figures 1(c) and (d) and $\zeta = 1.3$, we record that the photon number at $\Delta/\omega_M = -1$, $\Delta/\omega_M = 1.5$ and $\Delta/\omega_M = 2$ has dropped to values below 0.1 from a maximal value of 4.4 at the resonance. The semi-classical approach overestimates the quantum fluctuations, as becomes more and more apparent for large values of the quantum parameter. We observe that for $\zeta = 1.6$ the semi-classical curve of 1(c) deviates strongly from its fully quantum mechanical counterpart of figure 1(b) over the whole range of the detuning parameter.

Another inconsistency of the Langevin approach is the fact that the zero-point occupation of the cavity field does not lead to radiation pressure on the cantilever. To eliminate this effect, one might therefore try and replace the radiation pressure term of equation (7) by $\frac{\hbar g}{m x_{\text{ZPF}}}(|\alpha|^2 - \frac{1}{2})$. The resulting curve in figure 1(d) is shifted towards higher detunings, but does not improve the comparison to the result from the quantum master equation. The manipulation of the radiation pressure term shows up in an easily visible artefact: the corresponding curve even shows an increase in the cantilever energy on the cooling side $(\Delta/\omega_M \lesssim -1)$, where the actual cavity occupation should drop down to zero.

In summary, quantum fluctuations may significantly alter the features of self-induced oscillations in optomechanical systems. A semi-classical Langevin approach provides a useful tool to describe the quantum-classical crossover, complementing the master equation simulation of the full quantum dynamics.

3   Dynamical Interference in the "Photon Shuttle"

The standard approach to optomechanical systems is to record the transmission or reflection and thereby influence and detect the mechanical motion in steady

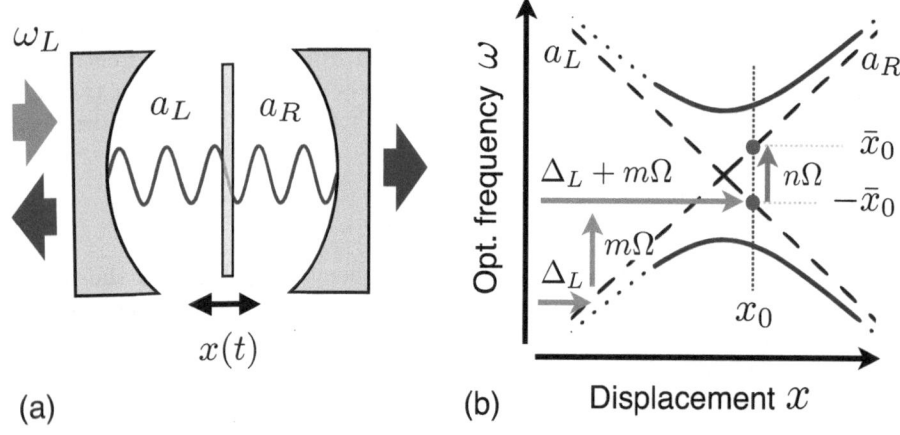

Fig. 2. The "photon shuttle". (a) Setup: a membrane placed inside a cavity couples two modes a_L, a_R. The left hand side is excited by a laser of frequency ω_L. The transmission to the right is recorded. (b) Resonance frequency as function of displacement. The membrane's motion linearly changes the bare mode frequencies (dashed). Due to the optical coupling (transmission through the membrane) there is an avoided crossing (dotted). The membrane is periodically driven around the mean position x_0. To see transmission, two multiphonon transitions are involved. First, excitation of the left cavity mode, Eq. (8). Second, transferring photons from the left to the right one, Eq. (9).

3.2 Physical Picture: Multiphonon Transitions

The observation of transmission through the cavity will be determined by two subsequent processes. First of all, the laser drive has to excite the cavity to insert photons into the left half. Secondly, the internal dynamics must be able to transfer these photons into the right mode such that transmission can be observed. In general, both processes are inelastic and therefore require energy to be transferred between the light field and the oscillating membrane. In the following, we restrict to the case of strong mechanical drive $\bar{A} \gg \Omega, g$ (see [10] for more details and other regimes).

Concerning the excitation process, the left mode is optically driven by the laser at frequency Δ_L. Due to the additional mechanical drive of the membrane, the left mode's frequency, $-\bar{x}(t) = -\bar{A}\cos(\Omega t) - \bar{x}_0$, is oscillating around its average value $-\bar{x}_0$. The resonance pattern of such a single periodically modulated optical mode, apart from the obvious resonance at $\Delta_L = -\bar{x}_0$, comprises mechanical sidebands due to multiphonon transitions. Thus, the condition to insert photons into the left mode of the cavity reads

$$\Delta_L + m\Omega = -\bar{x}_0, \qquad (8)$$

see Fig. 2b. Here, $m\Omega$ is an adequate multiple of Ω. The width of the individual resonances is determined by the cavity decay rate κ. Note that due to the strong

state. Going beyond this standard scheme, we want to ask what happens w.
the mechanical element is driven (periodically). It turns out that the most
teresting features can be obtained when turning from the standard setup tc
slightly different one, where a vibrating membrane is placed between two fix
end mirrors. The features reviewed in the following are discussed in more deta
in a recent work of ours [10].

3.1   Model

We consider an optomechanical system with a membrane placed in the middle
between two fixed mirrors, separating the cavity into two parts, Fig. 2a. This
is different from the standard setup with a moveable end-mirror, but such a
system was recently realized experimentally in [11]. The partially transparent
membrane couples two modes a_L and a_R of the left and the right cavity half,
respectively. The displacement $x(t)$ of the membrane changes the bare modes'
frequency ω_0. Because of the coupling there is an avoided level crossing in the
resonance frequency of the cavity, see Fig.2b. The system Hamiltonian reads

$$H_{sys} = \omega_0 \hbar \left(1 - \frac{x(t)}{l}\right) a_L^\dagger a_L + \omega_0 \hbar \left(1 + \frac{x(t)}{l}\right) a_R^\dagger a_R + g\hbar \left(a_L^\dagger a_R + a_R^\dagger a_L\right).$$

Here, $a_L^\dagger a_L$ and $a_R^\dagger a_R$ are the number operators for the left and right mode,
respectively, l is the length of both cavity halves and g represents the optical
coupling frequency. We assume that the membrane is driven with a mechanical
frequency Ω and at an amplitude A around a mean position x_0, so the displacement reads:

$$x(t) = A\cos(\Omega t) + x_0.$$

Such a system may be termed a "photon shuttle": Photons will be transferred
from the left half of the cavity to the right half by the mechanical driving. Here we
will study in particular the regime where the timescale of photon exchange, governed by the optical coupling frequency g, becomes comparable to the timescale
of the mechanical motion ($g \simeq \Omega$). In this case the system's dynamics can no
longer be treated as quasi-static in terms of the hyperbola branches depicted
in Fig. 2b (dotted). Instead, non-equilibrium photon dynamics driven by mechanical motion must be taken into account. We note that recently the optical
coupling g has been significantly reduced [12] and is tunable down to 200 kHz
at present. Thus, for typical mechanical drive in the range of 1 MHz, the regime
$g \simeq \Omega$ is experimentally accessible.

To investigate the system, one may drive the left hand side of the cavity with
a laser at frequency ω_L and examine the transmission $T(t)$ through the cavity
at the right hand side. The coupling to external modes is treated in Markov
approximation by a constant decay rate κ. Terms for dissipation and for the
laser drive must be added to the system Hamiltonian. We consider the laser drive
relative to the bare optical mode frequency ω_0 and define the detuning $\Delta_L = \omega_L - \omega_0$. To simplify the notation we express all length in terms of frequency:
$\bar{A} = (\omega_0/l)A$, $\bar{x}_0 = (\omega_0/l)x_0$.

drive the effective optical coupling between modes is very small and thus does not contribute.

The dynamics of the second process transferring photons from the left into the right mode turns out to be analogous to the one of a two state system, periodically driven through an avoided-crossing, where Landau-Zener (LZ) transitions [13,14] might occur. A LZ transition splits the photon state into a coherent superposition, the two amplitudes gather different phases and interfere the next time the system transverses the avoided crossing. For atomic systems such oscillations in the final state population, depending on the parameters of the drive, are known as Stueckelberg oscillations [15]. The same process might as well be depicted as a second multiphonon transition from the left mode with average frequency $-\bar{x}_0$ to the right one at $+\bar{x}_0$, see Fig. 2b. Thus, the second resonance condition becomes apparent, namely

$$n\Omega = 2\bar{x}_0. \tag{9}$$

3.3   Behaviour of the Optical Transmission in the Driven System

Rigorously, the transmission $T(t) = \kappa \langle a_R^\dagger(t) a_R(t) \rangle / (b^{in})^2$ is calculated using input-output theory, where b^{in} is the amplitude of the laser drive, and can be expressed as [10]

$$T(t) = \kappa^2 \left| \int_{-\infty}^{t} G(t,t') e^{-i\Delta_L t' - (\kappa/2)(t-t')} dt' \right|^2,$$

where the phase factor includes laser drive and cavity decay, while the Green's function $G(t,t')$ describes the amplitude for a photon to enter the cavity from the left at time t' and to be found in the right mode later at time t. The latter can be factorized $G(t,t') = \tilde{a}_R(t,t') e^{-i(\bar{A}/\Omega)\sin(\Omega t')}$, such that each factor describes one of the two processes discussed above. For the excitation, $e^{-i(\bar{A}/\Omega)\sin(\Omega t')} = \sum_m J_m(\bar{A}/\Omega) e^{-im\Omega t'}$ describes possible multiphonon transitions $m\Omega$, see (8), whose individual strength is determined by a Bessel function $J_m(\bar{A}/\Omega)$. Similarly, the strength of the second multiphonon transition $n\Omega$, originating from the internal dynamics described by $a_R(t,t')$, that is necessary to transfer photons from left to right, turns out to be determined by $J_n(2\bar{A}/\Omega)$. This second process and its dependence on $\bar{A}$ involves the physics of Stueckelberg oscillations as mentioned above.

In Fig. 3 we present numerical results for the time-averaged transmission ($\bar{A} \gg \Omega, g$) as a function of mean position x_0 and laser detuning Δ_L. This illustrates the two resonance conditions as well as the modulation by $|J_m(\bar{A}/\Omega)|$ and $|J_n(2\bar{A}/\Omega)|$. For $\Delta_L = 0$, the conditions (8) and (9) are met for $\bar{x}_0$ being a multiple of Ω. However, for the parameters used, $|J_{\pm 1}(\bar{A}/\Omega)|$ is near a maximum while $|J_0(\bar{A}/\Omega)|$ is close to a minimum. Thus, the transmission for $\bar{x}_0 = 0$ is significantly smaller than for $\bar{x}_0 = \pm\Omega$. The periodicity of the Bessel function becomes clear from its asymptotic form for large arguments, $J_k(y) \simeq \sqrt{2/\pi y} \cos(y - k\pi/2 - \pi/4)$. For $m = -1$,

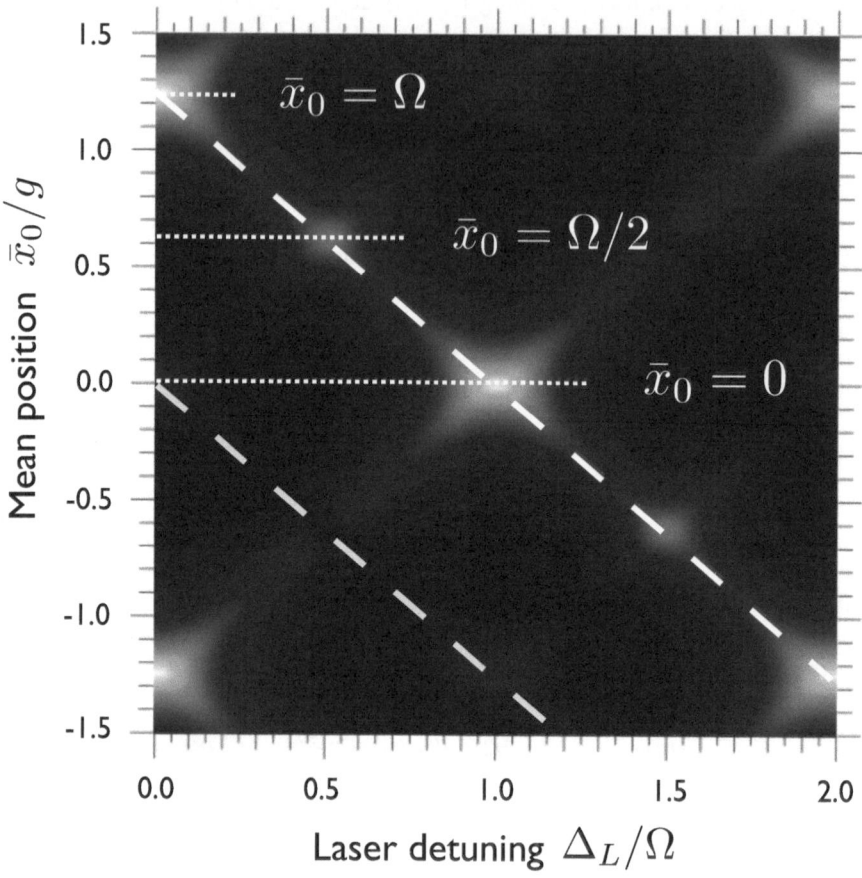

Fig. 3. Time-averaged transmission through an optical cavity with an oscillating partially transparent membrane in the middle. Transmission plotted as function of mean membrane position x_0 and laser detuning Δ_L. Further parameters are $\Omega/2\pi = 0.2g$, $\bar{A} = 54g$. Transmission can be observed if both resonance conditions (8) and (9) are met. Nevertheless, it is modulated by the product of two Bessel functions $\left| J_m(\bar{A}/\Omega) \right|$ and $\left| J_n(2\bar{A}/\Omega) \right|$ with appropriate indices according to (8) and (9). Dashed: $\bar{x}_0 = -\Delta_L - m\Omega$, see Eq. (8), with $m = -1$ (white) and $m = 0$ (yellow).

Eq. (8) is fulfilled along the white, dashed line. If we increase Δ_L we tune out of resonance and the transmission reduces. For $\Delta_L = \Omega/2$, (8) and (9) can be met for $\bar{x}_0$ being an *odd* multiple of $\Omega/2$, and so on. If we take into account the modulation by $\left| J_m(\bar{A}/\Omega) \right|$ and $\left| J_n(2\bar{A}/\Omega) \right|$ with the appropriate indices, we understand all the features of Fig. 3.

Even at low drive (small A), interesting features may be observed. In that regime, mechanically induced Rabi oscillations show up via an Autler-Townes splitting in the transmitted light. For more details, we refer the reader to [10].

Acknowledgments

We thank Björn Kubala and Jack Harris for collaborating on the quantum non-linear dynamics and the photon shuttle, respectively. This work has been supported by the German Science Foundation (DFG) via NIM, SFB 631 and the Emmy-Noether program, as well as by the DIP and GIF programs.

References

1. Gröblacher, S., Hertzberg, J.B., Vanner, M.R., Cole, G.D., Gigan, S., Schwab, K.C., Aspelmeyer, M.: Demonstration of an ultracold micro-optomechanical oscillator in a cryogenic cavity. Nat. Phys. 5(7), 485–488 (2009)
2. Schliesser, A., Arcizet, O., Riviere, R., Anetsberger, G., Kippenberg, T.J.: Resolved-sideband cooling and position measurement of a micromechanical oscillator close to the Heisenberg uncertainty limit. Nat. Phys. 5(7), 509 (2009)
3. Marquardt, F., Girvin, S.M.: Optomechanics. Physics 2, 40 (2009)
4. Marquardt, F., Harris, J.G.E., Girvin, S.M.: Dynamical multistability induced by radiation pressure in high-finesse micromechanical optical cavities. Phys. Rev. Lett. 96, 103901 (2006)
5. Ludwig, M., Kubala, B., Marquardt, F.: The optomechanical instability in the quantum regime. New Journal of Physics 10, 095013 (2008)
6. Höhberger, C., Karrai, K.: Self-oscillation of micromechanical resonators. In: Nanotechnology 2004, Proceedings of the 4th IEEE conference on nanotechnology, p. 419 (2004)
7. Kippenberg, T.J., Rokhsari, H., Carmon, T., Scherer, A., Vahala, K.J.: Analysis of radiation-pressure induced mechanical oscillation of an optical microcavity. Phys. Rev. Lett. 95, 033901 (2005)
8. Metzger, C., Ludwig, M., Neuenhahn, C., Ortlieb, A., Favero, I., Karrai, K., Marquardt, F.: Self-induced oscillations in an optomechanical system driven by bolometric backaction. Phys. Rev. Lett. 101, 133903 (2008)
9. Murch, K.W., Moore, K.L., Gupta, S., Stamper-Kurn, D.M.: Observation of quantum-measurement backaction with an ultracold atomic gas. Nat. Phys. 4(7), 561–564 (2008)
10. Heinrich, G., Harris, J.G.E., Marquardt, F.: The photon shuttle: Landau-Zener-Stueckelberg dynamics in an optomechanical system (2009), arXiv:0909.2164
11. Thompson, J.D., Zwickl, B.M., Jayich, A.M., Marquardt, F., Girvin, S.M., Harris, J.G.E.: Strong dispersive coupling of a high-finesse cavity to a micromechanical membrane. Nature 452(7183), 72–75 (2008)
12. Sankey, J.C., Jayich, A.M., Zwickl, B.M., Yang, C., Harris, J.G.E.: Improved "position squared" readout using degenerate cavity modes. In: Cote, R., Gould, P.L., Rozman, M. (eds.) Proceedings of the XXI International Conference on Atomic Physics. World Scientific, Singapore (2008)
13. Zener, C.: Non-adiabatic crossing of energy levels. Proc. R. Soc. London A 137(Non-Adiabatic Crossing of Energy Levels), 696 (1932)
14. Landau, L.D.: On the theory of transfer of energy at collisions ii. Phys. Z. USSR 2, 46–51 (1932)
15. Stückelberg, E.C.G.: Theorie der unelastischen Stösse zwischen Atomen. Helv. Phys. Acta 5, 369–422 (1932)

A 2D Electron Gas for Studies on Tunneling Dynamics and Charge Storage in Self-assembled Quantum Dots

Bastian Marquardt[1], Hicham Moujib[1], Axel Lorke[1], Dirk Reuter[2], Andreas D. Wieck[2], and Martin Geller[1]

[1] Experimental Physics and CeNIDE, University of Duisburg-Essen, 47058 Duisburg, Germany
bastian.marquardt@uni-due.de
[2] Chair of Solid-State Physics, Ruhr University Bochum, 44780 Bochum, Germany

Abstract. The carrier tunneling dynamics of self-assembled InAs quantum dots (QD) is studied using time-resolved conductance measurements of a nearby two-dimensional electron gas (2DEG). The coupling strength (tunneling time) between the QDs and the 2DEG is adjusted by different thicknesses of the spacer layers. We demonstrate a strong influence of charged QDs on the conductance on the 2DEG, even for very weak coupling, where standard C-V spectroscopy is unsuitable to investigate the electronic structure of these QDs.

Keywords: III-V semiconductors, indium compounds, self-assembly, semiconductor quantum dots, tunnelling, two-dimensional electron gas.

For the last 15 years, self-assembled quantum dots have attracted much attention because of their special electronical behavior as *artificial atoms* [1]. These systems are of great interest, not only for studying the fundamental properties of the 0D-electron system, but also because of possible device application like flash memories, single electron devices or single photon sources [2,3,4]. To realize such possible applications the carrier charging and emission characteristics are of importance. Previous work has shown that the capacitance-voltage (C-V) spectrocopy is a valuable tool to study electronic structures und charge carrier dynamics of coupled low-dimensional electron systems [5,6]. However, these capacitance measurements have their experimental limitations in both time and spatial resolution [7] and studying single QDs with long retention times is almost impossible using C-V spectroscopy. We present here a new method that enables us to detect the tunneling dynamics with high spatial and time resolution.

The mentioned limited time resolution of the frequency-dependent capacitance measurement can clearly be identified in Fig. 1. Three samples with different coupling strenghts between the QDs and the 2DEG were investigated using C-V spectroscopy. The three samples consist of an inverted HEMT structure with embedded self-assembled InAs QDs [5]. The QD layer of the strongly-coupled sample #1 is separated from the 2DEG by a 25 nm thick GaAs-tunneling barrier

A. Sergienko, S. Pascazio, and P. Villoresi (Eds.): QuantumCom 2009, LNICST 36, pp. 180–188, 2010.

from the 2DEG [as schematically shown in the inset of Fig. 1(a)]. The tunneling barriers of sample #2 and #3 consist of a 10 nm and 20 nm $Al_{0.34}Ga_{0.64}As$, respectively, and 20 nm GaAs [see the insets of Fig. 1(b) and (c)]. This results in charge tunneling times (i. e. electron tunneling between QDs and 2DEG), which are orders of magnitudes longer than those of sample #1. At low temperature (all measurements were performed at 4.2 K) sample #1 can be well characterized by C-V spectroscopy [Fig. 1(a)]. The observed maxima in the capacity can be directly linked to the individual electron states of the QDs [5,6,7,8]. Figure 1(a) shows the C-V spectrum of a rather strongly-coupled QD/2DEG-system. The average tunneling time between the QD states can be estimated to about 100 μs in frequency-dependent C-V measurements [6]. The applied high frequency (f= 10 kHz) leads to a smooth curve (i. e. high signal-to-noise-ratio), hence, every individual QD-state can be resolved.

Figure 1(b) depicts the capacitance versus the gate bias of a weakly-coupled electron system. Using frequency-dependent C-V spectroscopy [6], the tunneling time of the first s-state is determined to be $\tau_{s1} \approx$ 6 ms and the tunneling time of the p-states to be $\tau_p \approx$ 1.4 ms. The very weak coupling between the 2DEG and the QDs requires low-frequency modulation (f= 23 Hz), which makes it difficult to obtain high-quality C-V spectra [see Fig. 1(b)]. However, a comparsion with C-V studies of sample #1 [see Fig. 1(a)] allows us to identify the double-peak structure around -0.5 V and the broad feature between -0.2 V and 0.4 V with charging of the s and p shell [8], respectively. Accordingly, at a gate bias smaller than the charging voltage of the first s-state ($V_{g,s1} \approx$ -0.6 V) the QDs

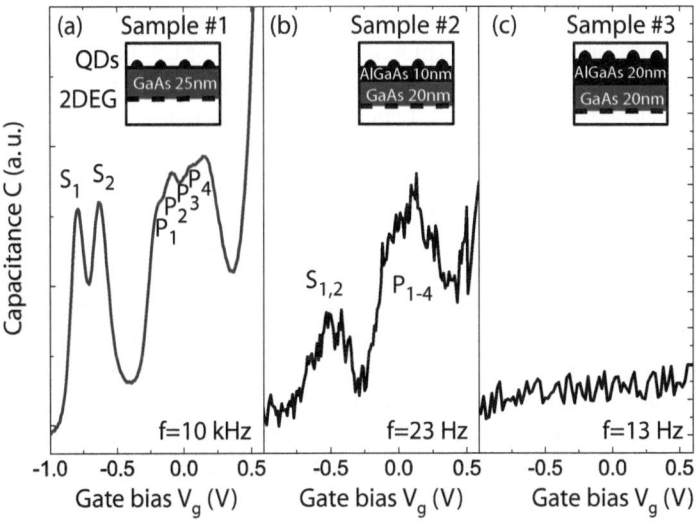

Fig. 1. C-V spectra of three samples (#1, #2 and #3) with different coupling strengths between the QDs and a 2DEG, adjusted by different thicknesses and compositions of the AlGaAs/GaAs tunneling barrier

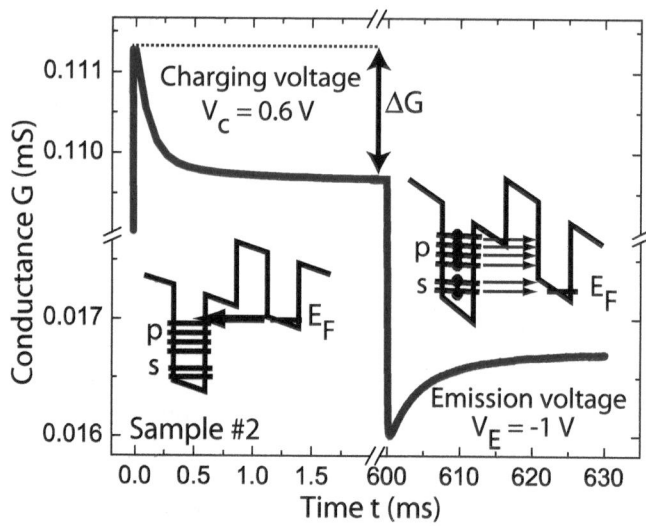

Fig. 2. Charging and emission transient of sample #2

are empty, and they are fully occupied (6 electrons per dot) at a gate bias larger than 0.4 V.

In C-V studies performed on sample #3 [see Fig. 1(c)], no charging peaks could be observed even for frequencies down to 1 Hz because of the height and thickness of the tunneling barrier. Corresponding to the exponential dependence of the tunneling process on the thickness and height of the barrier a much longer retention time is expected and C-V measurements can not be used to characterize the internal electronic structure of such very weakly-coupled QD/2DEG system as for sample #3. Hence, a different *frequency-independent* measurement tool has to be used to investigate this structure as described in the following.

We use a time-resolved measurement technique, where the charging state of the QDs is altered by a gate voltage, while the conductance of the 2DEG is measured in a two-terminal geometry. Using different charging and emission voltages, applied to the gate contact, allows us to observe the electron tunneling between the 2DEG and the QDs time-resolved. Figure 2 shows the conductance of sample #2 as a function of time when the gate bias is changed abruptly.

For instance, the operation starts with a 600 ms long QD-charging pulse ($V_c = 0.6$ V) applied to the top gate of the macroscopic electron channel. In this case, the Fermi-level E_F is energetically above the highest p-state, tunneling occurs from the 2DEG to the QD states, and hence the QD states are filled with electrons by the 2DEG [schematically depicted in the left inset in Fig. 2]. The charging of the QDs depletes the nearby 2DEG which results in decrease of the conductance. This can be understood either as screening of the gate potential by the QD charges or (equivalently) as QD-charge-induced image charges in the 2DEG. At t= 600 ms, an emission bias of $V_E = -1$ V is applied, such that the

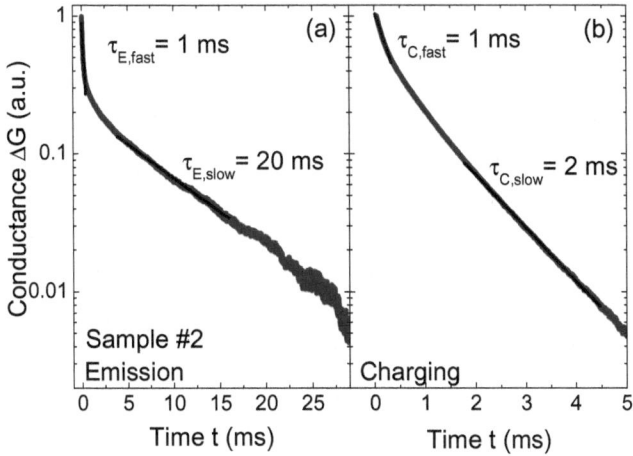

Fig. 3. Charging and emission transient of sample #2 in a semi-logarithmic plot

Fermi-level E_F is now below the s-states [depicted in the right inset of Fig. 2] and tunneling from the QD states to the 2DEG takes place.

To quantitatively evaluate the transient times, Fig. 3 shows the emission and charging transients of Fig. 2 on a semi-logarithmic scale. The emission transient [Fig. 3(a)] shows a multi-exponential decay with time-constants between $\tau_{E,fast}$ = 1 ms and $\tau_{E,slow}$ = 20 ms. Because tunneling is fast for high-energy states and slow for low-energy states, we attribute the escape rate of $\tau_{E,fast}$ = 1 ms to tunneling out of the p-states. The tunneling times of the s-states can be determined to be $\tau_{E,slow}$ = 20 ms. This is in acceptable agreement with the frequency-dependent C-V measurements mentioned above with $\tau_p \approx$ 1.4 ms and $\tau_{s1} \approx$ 6 ms, if the difficulties of estimating multi-exponential decays are considered [9].

The charging process [Fig. 3(b)] also reflects a multi-exponential transient. Surprisingly, however, only time-constants τ_c between 1 ms and 2 ms are observed (see corresponding linear fits in red). This discrepancy can be understood as a result of non-equilibrium tunneling processes as depicted by the insets in Fig. 2. During the emission process, the s-electrons have to penetrate a relatively high tunneling barrier (lowest arrows in Fig. 2, right). During the charging process, on the other hand, because of the large positive bias, the electrons can be all injected into high-lying states with short tunneling times. The subsequent relaxation processes (p→s) are known to be of the order of picoseconds for electrons in self-assembled QDs [10].

To compare the time-resolved measurements with standard CV data [see Fig. 1(b)] we have used in the past the *charge-selective* method [11] and evaluated the transients to obtain a high-quality charging spectrum of the dots [12]. A similar operation, which is based on time-dependent conductance measeurments

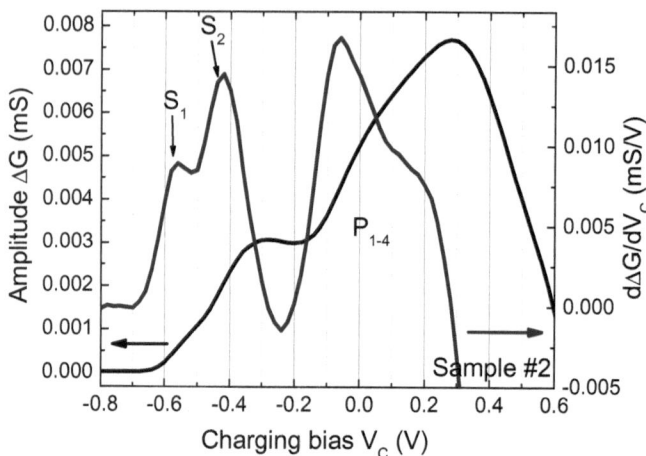

Fig. 4. Conductance amplitude ΔG (black line) and its derivative (blue)

is described in the following. The operation cycle starts with a QD depletion pulse (depletion bias V_{depl}= -1 V) and a following charging pulse (charging bias V_C= 0.6 V). The depletion bias is constant, whereas the charging bias is scanned from 0.6 V (fully occupied dots) to -1 V (empty dots).

Figure 4 depicts the charging transient amplitude ΔG [see Fig. 2] and its derivative (blue curve) as a function of the applied charging bias. The conductance amplitude ΔG of the charging transient depends on the number n_{QD} of the involved QDs and the average number N of occupied QD states. By changing the charging bias V_C, the number of occupied states $N(V_C)$ can be tuned. From the C-V data [see Fig. 1(b)] the charging voltage of the QD states is known. Between -0.8 V and -0.7 V no QD states will be charged, hence, no transient (amplitude) can be measured. When the ground state s_1 becomes occupied at about V_c= -0.6 V the transient amplitude increases. A small change in the curve can be observed if the second s-state is charged with electrons at -0.5 V. This change in the conductance amplitude due to the s-state charging can be well resolved in the first derivative [blue curve in Fig. 4]. The amplitude ΔG saturates between -0.3 V and -0.2 V as the number of involved QD states N is now constant (N=2) and start to raise again while the p-states are charged with electrons. The maximum ΔG is observed when all QD states are filled (V_c= 0.3 V). For a charging bias V_c>0.3 V the amplitude decreases again as no further QD states can be charged. The derivative of the amplitude offers a much better resolved charging spectrum (with two s-states maxima and a broad peak of the p-states) than the C-V curve in Fig. 1(b).

As mentioned above, capacitance studies can not be used to probe the sample #3, because of the very long tunneling times of the charge carriers. Charging transients, however, can be recorded on time scales well above 10^4 s. Figure 5 shows the time-resolved charging and emission transients of the 2DEG of sample #3 on a semi-logarithmic scale and indeed we are able to observe conductance

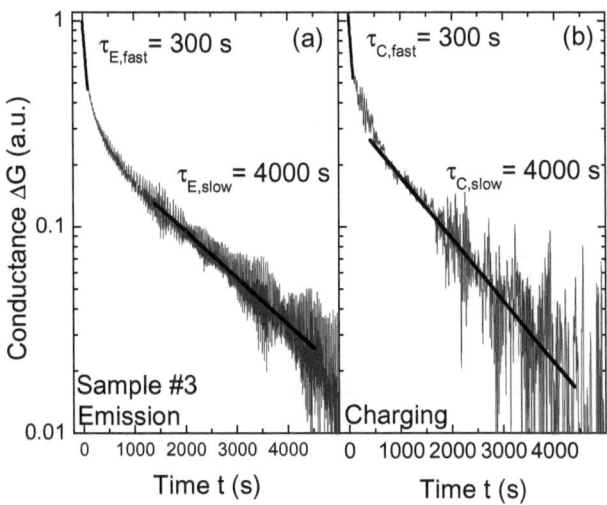

Fig. 5. Charging and emission transient of sample #3 in a semi-logarithmic plot

transients which are six orders of magnitudes slower than the transients of sample #2 [see Fig. 3].

In Fig. 5 two different time constants of the charging and the emission transients of sample #3 can be observed. The emission and the charging transients are very similar. The fast decay has a time constant of about 300 s, whereas the slow decay has a time constant of 4000 s. We attribute the fast conductance decay to tunneling into and out of the InAs wetting layer and the slow decay to tunneling into and out of the highest p-level. The main difference to measurements of sample #2 [see Fig. 3] is the fact that the tunneling times of the charging process do not differ from the emission process. One possible explanation is that the emission transient in Fig. 5 are not only be given by tunneling but also by thermal excitation of the carriers. At 4.2 K (kT≈0.4 meV) thermal excitation of the order of 50 meV is extremely weak, which explains why it is not observed for sample #2. For sample #3 however, the extremely long retention times may allow electrons from the s-state to escape by a combined process of thermal excitation into the p-state and successive tunneling to the 2DEG. Further measurements and calculations are needed to test this model.

The present devices with very weakly coupled low-dimensional electron systems together with real-time conductance measurements enable us to decouple the applied gate bias from QD charge occupation.

The fact that non-equilibrium states can be prepared in the present sample makes them promising for charge storage as discussed in the following. Figure 6 shows the transfer characteristics of the sample, i. e. the conductance of the 2DEG versus the gate bias for three different scan times. The measurement cycle starts with a 200 ms long discharging pulse (V_{depl}= -1 V), which depletes

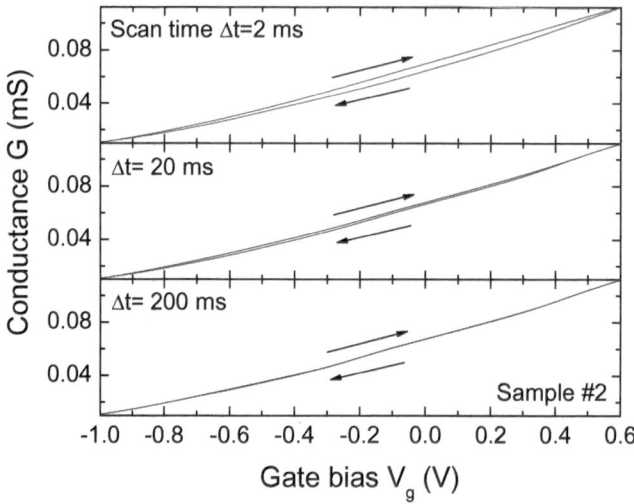

Fig. 6. Hysteresis measurements of sample #2 with different scan times Δt

the InAs QDs as discussed above. Fast bias sweeps (Δt= 2 ms, 20 ms and 200 ms) from the depletion voltage upward to the filling voltage (V_{fill}= 0.6 V) follow. As a consequence, during the fastest sweep [$\Delta t \geq \tau$, see Fig. 6(a)] the QDs remain empty during the entire upward sweep and the 2DEG remains unaffected by the (empty) states of the QDs. Next, during a 200 ms long charging period at a gate bias V_{fill} of 0.6 V, the QDs become completely charged. The reduction in the charge carrier density lowers the conductance of the 2DEG, resulting in the observed hysteresis. The hysteresis decreases by increasing the scan time and, hence, vanishes for sweep times longer than 200 ms, the longest charge carrier storage time in the QD ensemble [see Fig. 6(c)]. The measured hysteresis opening $\Delta G/G$ in Fig. 6(a) for a gate bias of 0 V is about 10 %. In comparison, the relative change in 2D carrier density between fully charged dots (6 electrons per dot) and empty dots amounts to $\Delta n/n \approx 7$ % at V_g= 0 V. Using an approximately constant mobility [5] leads to $\Delta G/G \approx 7$ %, in good agreement with the measured value. Thus, we are able to switch between two different QD charge occupation levels (completely full and empty QDs) for the same applied gate bias. This further supports the conclusion that the observed hysteresis is indeed given by the different QD charging states.

Using Gauss' law to model the three layer system (gate, dot layer, 2DEG) [5] it can easily be shown that for every electron transferred into the QD layer roughly one electron will be depleted from the 2DEG. Equivalently this can be understood either as screening of the gate potential by the QD charges or as QD-charge-induced image charges in the 2DEG. The reduction of the charge carrier density lowers the conductance of the 2DEG, resulting in the observed hysteresis [15].

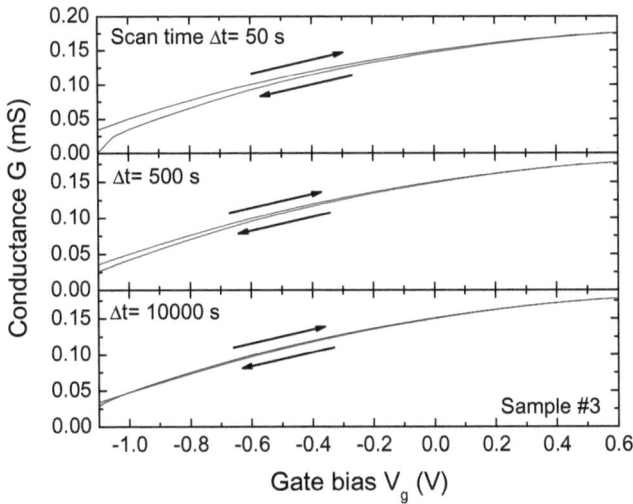

Fig. 7. Hysteresis measurements of sample #3 with different scan times Δt

Figure 7 shows hysteresis measurements of sample #3 for three different scan times Δt. Similar to hysteresis meaurements of sample #2 [see Fig. 6], the hysteresis opening becomes smaller if the scan time Δt is raised. For scan times slower than 10 000 s the hysteresis opening completely vanishes. This is in good agreement with the observed time constants of the transients [see Fig. 5]. This confirms the assumption that the charge storage in the QDs causes the measured hysteresis also for sample #3. Furthermore, because of the low 2DEG charge carrier density of sample #3 ($n_{2DEG} = 3 \times 10^{11} \text{cm}^{-2}$), the hysteresis opening can be enhanced to about 30 % which is of central importance for using a 2DEG as a sensitive read-out in future QD-based memories [16].

In conclusion, we introduced a novel technique which enables to extend the experimental range regarding both tunneling dynamics and number of probed QDs. We have shown that the conductance of the 2DEG can be used as an efficient detector to study the charge tunneling dynamics of the nearby self-assembled QD-layer. The observed signal ΔG is independent on the coupling strength of the heterostructure. Therefore, the technique can be used as a characterization tool for the investigation of very weakly-coupled low-dimensional electron systems with long retention times.

Acknowledgments

The authors gratefully acknowledge financial support under Grant No. 01BM461 and 01BM451 by the German Bundesministerium für Bildung und Forschung (BMBF) and by the DFG in the framework of the NanoSci-E+ project QD2D of the European Comission.

References

1. Bimberg, D. (ed.): Semiconductor Nanostructures. Springer, Berlin (2008); contributions therein
2. Nowozin, T., Marent, A., Geller, M., Bimberg, D., Akcay, N., Öncan, N.: Appl. Phys. Lett. 94, 042108 (2009)
3. Gustavsson, S., Leturcq, R., Simovic, B., Schleser, R., Ihn, T., Studerus, P., Ensslin, K., Studerus, D., Driscoll, D.C., Gossard, A.C.: Phys. Rev. Lett. 96, 076605 (2006)
4. Yuan, Z., Kardynal, B.E., Stevenson, R.M., Shields, A.J., Lobo, C.J., Cooper, K., Beattie, N.S., Ritchie, D.A., Pepper, M.: Science 295, 102 (2002)
5. Russ, M., Meier, C., Lorke, A., Reuter, D., Wieck, A.D.: Phys. Rev. B 73, 115334 (2006)
6. Luyken, R.J., Lorke, A., Govorov, A.O., Kotthaus, J.P., Medeiros-Ribeiro, G., Pedroff, P.M.: Appl. Phys. Lett. 74, 2486 (1999)
7. Miller, B.T., Hansen, W., Manus, S., Luyken, R.J., Lorke, A., Kotthaus, J.P.: Phys. Rev. B. 56, 6764 (1997)
8. Drexler, H., Leonard, D., Hansen, W., Kotthaus, J.P., Pedroff, P.M.: Phys. Rev. Lett. 73, 2252 (1994)
9. Istratov, A.A., Vyvenko, O.F.: Rev. Sci. Instrum. 70, 1233 (1999)
10. Müller, T., Schrey, F.F., Strasser, G., Unterrainer, K.: Appl. Phys. Lett. 83, 3572 (2003)
11. Geller, M., Kapteyn, C., Mller-Kirsch, L., Heitz, R., Bimberg, D.: Appl. Phys. Lett. 82, 2706 (2003)
12. Marquardt, B., Geller, M., Lorke, A., Reuter, D., Wieck, A.D.: Appl. Phys. Lett. 95, 22113 (2009)
13. Russ, M., Meier, C., Marquardt, B., Lorke, A., Reuter, D., Wieck, A.D.: Phase Transitions 79, 765 (2006)
14. Kapteyn, C.M.A., Heinrichsdorff, F., Stier, O., Heitz, R., Grundmann, M., Zakharov, N.D., Bimberg, D., Werner, P.: Phys. Rev. B. 60, 14265 (1999)
15. Note that the influence of Coulomb scattering on the mobility due to the negatively charged dots is negligible, as shown in Ref. [5, 13]
16. Marent, A., Geller, M., Schliwa, A., Feise, D., Pötschke, K., Bimberg, D., Akcay, N., Öncan, N.: Appl. Phys. Lett. 91, 242109 (2007)

Entanglement Purification with Hybrid Systems

S. Mancini[1], G. Morigi[2], S. Rebic[3], and D. Vitali[1]

[1] Physics Department, University of Camerino, I-62032 Camerino, Italy
[2] Grup d'Òptica, Departament de Física, Universitat Autònoma de Barcelona,
E-08193 Bellaterra (Barcelona), Spain
[3] Centre for Quantum Computing Technology, Macquarie University,
Sydney, NSW 2109, Australia

Abstract. We propose a scheme for continuous variable entanglement purification using the interaction of two-mode squeezed beams with cold trapped atomic ensembles. In the limit of very large number of atoms, a collective atomic ensemble excitation becomes a continuous quantity. Measurement of populations in the two lower levels is of non-Gaussian nature that is necessary for the entanglement purification. We foresee the use of this scheme for implementation of a continuous variable quantum repeater.

Keywords: Entanglement manipulation and characterization, quantum description of light-matter interaction, quantum state engineering and measurements.

1 Introduction

The property of entanglement lies at the heart of quantum information processing (quantum communication, teleportation, cryptography) and quantum mechanics in general. Continuous variable (CV) entangled states are very attractive in this regard due to the relative easiness in their generation and manipulation, as compared to their discrete counterpart [1]. However, the degree of entanglement between two distant sites of a quantum network usually decreases exponentially with the length of a connecting channel, calling for implementation of some entanglement purification procedure [2]. This involves purifying mixed entangled states, such as two-mode squeezed states, after their two halves have been distributed through noisy channels. It is well-known that a procedure involving Gaussian local operations and classical communications does not lead to the enhancement of the initial entanglement [3], so non-Gaussian local operations must necessarily make part of a protocol [4].

Here, a scheme for the implementation of CV entanglement purification is proposed, using two-mode squeezed beams and two independent cold, three-level atomic ensembles in a Λ configuration (see Fig. 1). Input is provided by two-mode squeezed light with arbitrary, but less than perfect, squeezing and each mode impinges on one atomic ensemble. Squeezed beams excite one of the two Λ transitions, with the other being excited by another travelling wave field.

A. Sergienko, S. Pascazio, and P. Villoresi (Eds.): QuantumCom 2009, LNICST 36, pp. 189–199, 2010.
© Institute for Computer Sciences, Social-Informatics and Telecommunications Engineering 2010

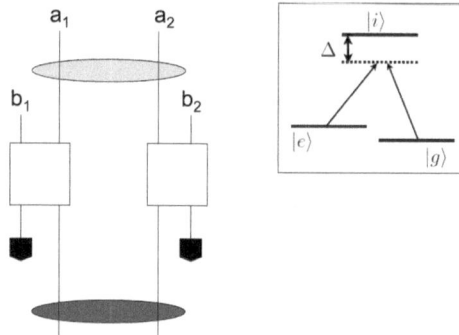

Fig. 1. Proposed setup: Two entangled light fields a_1 and a_2 are impinging on two independent but identical cold atomic ensembles describable by bosonic fields b_1 and b_2. After local interactions, the b fields are subjected to measurement while the a fields should emerge more entangled. The box shows the Λ configuration of atomic transitions. One of the two transitions is driven by the a fields and the other by classical fields.

In the limit of very large number of atoms (continuous limit), a collective ensemble excitation becomes a CV quantity (a bosonic mode) and this collective CV quantity interacts with the optical modes realizing the first part of a purification protocol. The second important part is a non-Gaussian measurement, implemented through the measurement of population of the atomic ground state. It is shown that such a protocol increases two-mode squeezing (entanglement) in the output modes with respect to that of the input ones.

2   The Model

Consider two radiation modes described by field operators a_1 and a_2 (see Fig. 1). Each of the fields drives the $|e\rangle \to |i\rangle$ transition of the ensemble of three level atoms in Λ configuration. The two atomic ensembles are considered to be independent but identical, so the couplings of each field with the atoms are taken to be equal $g_1 = g_2 = g$. Transitions $|g\rangle \to |i\rangle$ are driven by classical fields with Rabi frequencies $\Omega_{1,2}$. The coupling of the fields to the atoms is assumed to be off resonant with a large detuning Δ. The Hamiltonian of the total system can be written as ($\hbar = 1$ throughout the paper)

$$H = \Delta \sum_{l=1}^{2} \sum_{k=1}^{N_l} \sigma_{ii}^{k,l} + i \sum_{l=1}^{2} \sum_{k=1}^{N_l} \left(g^* \sigma_{ie}^{k,l} a_l - g \sigma_{ei}^{k,l} a_l^\dagger \right)$$
$$+ i \sum_{l=1}^{2} \sum_{k=1}^{N_l} \left(\Omega_l \sigma_{ig}^{k,l} - \Omega_l^* \sigma_{gi}^{k,l} \right), \tag{1}$$

where index l labels the radiation mode and the atomic ensemble it impinges on and $\sigma_{nm}^{k,l}$ denotes atomic flip operator for the kth atom (out of total of N_l atoms per ensemble) in the lth ensemble.

In the case of large detuning Δ, the intermediate levels $|i\rangle$ can be adiabatically eliminated and spontaneous emission neglected. Under the collective coupling condition, the interaction shown in Fig. 1 is then described by the following Hamiltonian in the rotating frame [5]

$$H' = i \sum_{l=1}^{2} \frac{\sqrt{N_l}}{\Delta} \left(g^* \Omega_l S_{ge}^{(l)} a_l^\dagger - g \Omega_l^* S_{eg}^{(l)} a_l \right), \qquad (2)$$

where $S_{nm}^{(l)} = \sum_{k=1}^{N_l} \sigma_{nm}^{k,l}/\sqrt{N_l}$ denotes the collective atomic operators for the respective ensembles. In this expression, ac-Stark shifts have been neglected as they can be compensated in practice by refining the laser frequency.

In the Holstein-Primakoff representation [6], the collective atomic operators (that are angular momentum operators) can be associated with bosonic creation and annihilation operators via the relation $S_{ge}^{(l)} = (N_l - b_l^\dagger b_l)^{1/2} b_l$, where b_l denotes the bosonic annihilation operator satisfying the canonical commutation relations $[b_l, b_k^\dagger] = \delta_{lk}$. In the low excitation limit, the number of atoms transferred to state $|e\rangle$ is small compared to the total number of atoms. The collective atomic operators can then be further approximated by $S_{ge}^{(l)} \approx \sqrt{N_l} \, b_l$. The Hamiltonian (2) then reduces to the following form

$$H_{BS} = i \sum_{l=1}^{2} \mu_l (a_l^\dagger b_l - a_l b_l^\dagger), \qquad (3)$$

with $\mu_l = |g||\Omega_l| N_l / \Delta$ (it is possible to choose the phase of the impinging modes and of the driving lasers so that $\arg(g) = \arg(\Omega_l)$). The collective atomic ensemble excitation in the continuous limit of very large number of atoms becomes a CV quantity. The effective coupling coefficients μ_l are in general different because it is extremely difficult to prepare two large atomic ensembles with exactly the same numbers of atoms. However, one can obtain $\mu_1 = \mu_2$ by e.g. adjusting the coupling fields Ω_l accordingly. Hence, in the following it will be assumed that $\mu_1 = \mu_2 = \mu$.

The effective Hamiltonian of Eq. (3) has the familiar form reminiscent of the beam splitter interaction [7]. The coupling constant μ can be thought to be the rotation angle for the two modes a_l and b_l.

In the starting Hamiltonian (1), we could have exchanged the atomic transitions coupled to the fields. That is, we could have considered each of the quantum field driving the $|g\rangle \rightarrow |i\rangle$ transition, whilst classical fields driving transitions $|e\rangle \rightarrow |i\rangle$. The total Hamiltonian in such a case would read

$$H = \Delta \sum_{l=1}^{2} \sum_{k=1}^{N_l} \sigma_{ii}^{k,l} + i \sum_{l=1}^{2} \sum_{k=1}^{N_l} \left(g^* \sigma_{ig}^{k,l} a_l - g \sigma_{gi}^{k,l} a_l^\dagger \right)$$

$$+ i \sum_{l=1}^{2} \sum_{k=1}^{N_l} \left(\Omega_l \sigma_{ie}^{k,l} - \Omega_l^* \sigma_{ei}^{k,l} \right), \qquad (4)$$

Then, by repeating the previous steps we could have arrived to the following effective Hamiltonian

$$H_{PA} = i \sum_{l=1}^{2} \nu_l (a_l^\dagger b_l^\dagger - a_l b_l), \tag{5}$$

with $\nu_l = |g||\Omega_l|N_l/\Delta$ (and choosing again phases so that $\arg(g) = \arg(\Omega_l)$). Again we could assume, without loss of generality, $\nu_1 = \nu_2 = \nu$.

The effective Hamiltonian of Eq. (5) has the familiar form reminiscent of the nondegenerate parametric amplifier [7]. The coupling constant ν can be thought to be proportional to the amplitude of the pump and second-order susceptibility.

3   The Dynamics

We are now going to study the dynamics of the system induced by the Hamiltonian H_{BS}, while that of Hamiltonian H_{PA} will be discussed later on. The Liouville equation $\dot{\rho}(t) = -i[H_{BS}, \rho(t)]$ for the state of the four modes a_1, a_2, b_1, b_2 can be transformed into a partial differential equation for the normally ordered characteristic function

$$\chi(\boldsymbol{\eta}, t) = \mathrm{Tr} \left\{ \rho(t) \prod_i e^{\eta_i \sigma_i^\dagger} \prod_i e^{\eta_i^* \sigma_i} \right\}, \tag{6}$$

with $\boldsymbol{\sigma} = (a_1, a_2, b_1, b_2)$ the vector of bosonic operators and $\boldsymbol{\eta} = (\alpha_1, \alpha_2, \beta_1, \beta_2)$ the vector of corresponding classical complex variables.

The equation of motion for the characteristic function becomes [7]

$$\frac{\partial \chi(\boldsymbol{\eta}, t)}{\partial t} = \mu \left(\alpha_1 \frac{\partial}{\partial \beta_1} + \alpha_1^* \frac{\partial}{\partial \beta_1^*} + \alpha_2 \frac{\partial}{\partial \beta_2} + \alpha_2^* \frac{\partial}{\partial \beta_2^*} \right.$$
$$\left. - \beta_1 \frac{\partial}{\partial \alpha_1} - \beta_1^* \frac{\partial}{\partial \alpha_1^*} - \beta_2 \frac{\partial}{\partial \alpha_2} - \beta_2^* \frac{\partial}{\partial \alpha_2^*} \right) \chi(\boldsymbol{\eta}, t). \tag{7}$$

Let us consider as initial condition a two-mode squeezed thermal state for a_1, a_2 and vacuum state for b_1, b_2, (which means that all the atoms in the ensemble are in the ground state $|g\rangle$), that is

$$\chi(\boldsymbol{\eta}, 0) = \exp \left\{ -\left(\frac{\lambda^2}{1-\lambda^2} + n_T \right) (|\alpha_1|^2 + |\alpha_2|^2) \right\} \exp \left\{ \frac{\lambda}{1-\lambda^2} (\alpha_1 \alpha_2 + \alpha_1^* \alpha_2^*) \right\}, \tag{8}$$

with n_T number of thermal photons and λ the two-mode squeezing parameter. Notice that for $n_T = 0$ the state for the modes a_1, a_2 is a two-mode squeezed state $\sqrt{1-\lambda^2} \sum_n \lambda^n |n\rangle_{a_1} |n\rangle_{a_2}$.

Equation (7) can be solved by assuming a Gaussian solution of the form

$$\chi(\boldsymbol{\eta}, t) = \exp \left[-A(|\alpha_1|^2 + |\alpha_2|^2) - B(|\beta_1|^2 + |\beta_2|^2) \right.$$
$$+ A_{12} (\alpha_1 \alpha_2 + \alpha_1^* \alpha_2^*) + B_{12} (\beta_1 \beta_2 + \beta_1^* \beta_2^*)$$
$$+ E (\alpha_1 \beta_1^* + \alpha_1^* \beta_1 + \alpha_2 \beta_2^* + \alpha_2^* \beta_2)$$
$$\left. + C (\alpha_1 \beta_2 + \alpha_1^* \beta_2^* + \alpha_2 \beta_1 + \alpha_2^* \beta_1^*) \right], \tag{9}$$

and transforming it to a set of linear differential equation for the time dependent coefficients $A, B, E, A_{12}, B_{12}, C$. Then, one can determine these coefficients as

$$A(t) = \left(\lambda^2 + n_T - \lambda^2 n_T\right) \frac{\cos^2 \mu t}{1 - \lambda^2}, \tag{10a}$$

$$B(t) = \left(\lambda^2 + n_T - \lambda^2 n_T\right) \frac{\sin^2 \mu t}{1 - \lambda^2}, \tag{10b}$$

$$E(t) = \frac{1}{2} \left(\lambda^2 + n_T - \lambda^2 n_T\right) \frac{\sin(2\mu t)}{1 - \lambda^2}, \tag{10c}$$

$$A_{12}(t) = \lambda \frac{\cos^2 \mu t}{1 - \lambda^2}, \tag{10d}$$

$$B_{12}(t) = \lambda \frac{\sin^2 \mu t}{1 - \lambda^2}, \tag{10e}$$

$$C(t) = \frac{-\lambda}{2} \frac{\sin(2\mu t)}{1 - \lambda^2}. \tag{10f}$$

4   Measurement

The physical scheme based on the interaction with the collective atomic ensemble is now supplemented by the simplest example of non-Gaussian measurement. This is the Fock basis measurement on modes b_1, b_2 that checks for the presence or absence of excitations in each mode. It corresponds to a measurement of populations in the two lower levels of the atomic systems (see Fig. 1). For each ensemble, laser pulse is applied on state $|e\rangle$ in order to observe quantum jumps. If the jump is not observed then the state $|e\rangle$ is not populated, i.e. all the population resides in $|g\rangle$. This situation corresponds to the projection $|0\rangle_b \langle 0|$. Otherwise, the measurement results in the projection $\mathbb{I}_b - |0\rangle_b \langle 0|$. The described procedure can be formalised by introducing the positive operator valued measure (POVM)

$$\{E_{x_1=0}, \ E_{x_1=1}\} \equiv \{|0\rangle_{b_1} \langle 0|, \ \mathbb{I}_{b_1} - |0\rangle_{b_1} \langle 0|\}, \tag{11a}$$

$$\{E_{x_2=0}, \ E_{x_2=1}\} \equiv \{|0\rangle_{b_2} \langle 0|, \ \mathbb{I}_{b_2} - |0\rangle_{b_2} \langle 0|\}. \tag{11b}$$

Defining $\boldsymbol{x} = (x_1, \ x_2)$, the conditioned state will be

$$\rho_{\boldsymbol{x}} = \frac{1}{p_{\boldsymbol{x}}} \{E_{\boldsymbol{x}} \rho E_{\boldsymbol{x}}^\dagger\} \tag{12}$$

with probablility

$$p_{\boldsymbol{x}} = \mathrm{Tr}\{E_{\boldsymbol{x}} \rho E_{\boldsymbol{x}}^\dagger\}, \tag{13}$$

meaning that the measurement procedure results (in average) in a mixture $\sum_{\boldsymbol{x}} p_{\boldsymbol{x}} \rho_{\boldsymbol{x}} = \sum_{\boldsymbol{x}} E_{\boldsymbol{x}} \rho E_{\boldsymbol{x}}^\dagger$.

The measurement procedure can be expressed in the language of characteristic function. First notice that the density operator for modes a_1, a_2, b_1, b_2 can be written in terms of the normally ordered characteristic function as

$$\rho = \int \frac{d^2\alpha_1}{\pi} \frac{d^2\alpha_2}{\pi} \frac{d^2\beta_1}{\pi} \frac{d^2\beta_2}{\pi} \chi(\alpha_1, \alpha_2, \beta_1, \beta_2) D^\dagger(\alpha_1) D^\dagger(\alpha_2) D^\dagger(\beta_1) D^\dagger(\beta_2)$$

$$\times e^{(-|\alpha_1|^2 - |\alpha_2|^2 - |\beta_1|^2 - |\beta_2|^2)/2}, \tag{14}$$

where D denotes the displacement operator [7].

Then, let us consider the following case

$$
\begin{aligned}
\rho_{01} &= \frac{1}{p_{01}} \mathrm{Tr}_{b_1 b_2} \left\{ \rho E_{01}^\dagger \right\} \\
&= \frac{1}{p_{01}} \int \frac{d^2\alpha_1}{\pi} \frac{d^2\alpha_1}{\pi} \frac{d^2\beta_1}{\pi} \frac{d^2\beta_2}{\pi} \chi(\alpha_1, \alpha_2, \beta_1, \beta_2) \mathrm{Tr}_{b_1} \left\{ D^\dagger(\beta_1) |0\rangle_{b_1} \langle 0| \right\} \\
&\quad \times \mathrm{Tr}_{b_2} \left\{ D^\dagger(\beta_2) \left[I_{b_2} - |0\rangle_{b_2} \langle 0| \right] \right\} D^\dagger(\alpha_1) D^\dagger(\alpha_2) e^{(-|\alpha_1|^2 - |\alpha_2|^2 - |\beta_1|^2 - |\beta_2|^2)/2}, \\
&= \int \frac{d^2\alpha_1}{\pi} \frac{d^2\alpha_1}{\pi} \left\{ \frac{1}{p_{01}} \int \frac{d^2\beta_1}{\pi} \frac{d^2\beta_2}{\pi} \chi(\alpha_1, \alpha_2, \beta_1, \beta_2) \right. \\
&\quad \left. \times \left[\langle \beta_1 | 0\rangle \left(\pi \delta^2(\beta_2) - \langle \beta_2 | 0\rangle \right) \right] e^{(-|\beta_1|^2 - |\beta_2|^2)/2} \right\} \\
&\quad \times D^\dagger(\alpha_1) D^\dagger(\alpha_2) e^{(-|\alpha_1|^2 - |\alpha_2|^2)/2}.
\end{aligned}
\tag{15}
$$

Now the quantity inside the curly brackets can be considered as the normally ordered characteristic function corresponding to the (normalized) state ρ_{01}. It can be rewritten as

$$
\begin{aligned}
\chi_{01}(\alpha_1, \alpha_2) &= \frac{1}{p_{01}} \int \frac{d^2\beta_1}{\pi} \frac{d^2\beta_2}{\pi} \chi(\alpha_1, \alpha_2, \beta_1, \beta_2) \\
&\quad \times \left[\langle \beta_1 | 0\rangle \left(\pi \delta^2(\beta_2) - \langle \beta_2 | 0\rangle \right) \right] e^{(-|\beta_1|^2 - |\beta_2|^2)/2} \\
&= \frac{1}{p_{01}} \int \frac{d^2\beta_1}{\pi} \chi(\alpha_1, \alpha_2, \beta_1, \beta_2 = 0) e^{-|\beta_1|^2} \\
&\quad - \frac{1}{p_{01}} \int \frac{d^2\beta_1}{\pi} \frac{d^2\beta_2}{\pi} \chi(\alpha_1, \alpha_2, \beta_1, \beta_2) e^{-|\beta_1|^2 - |\beta_2|^2}
\end{aligned}
\tag{16a}
$$

The value of the probability p_{01} comes from the normalization condition

$$
\chi_{01}(\alpha_1 = 0, \alpha_2 = 0) = 1.
\tag{17}
$$

Analogously we obtain for the other conditional states

$$
\begin{aligned}
\chi_{10}(\alpha_1, \alpha_2) &= \frac{1}{p_{10}} \int \frac{d^2\beta_2}{\pi} \chi(\alpha_1, \alpha_2, \beta_1 = 0, \beta_2) e^{-|\beta_2|^2} \\
&\quad - \frac{1}{p_{10}} \int \frac{d^2\beta_1}{\pi} \frac{d^2\beta_2}{\pi} \chi(\alpha_1, \alpha_2, \beta_1, \beta_2) e^{-|\beta_1|^2 - |\beta_2|^2}
\end{aligned}
\tag{18a}
$$

$$
\begin{aligned}
\chi_{11}(\alpha_1, \alpha_2) &= \frac{1}{p_{11}} \chi(\alpha_1, \alpha_2, \beta_1 = 0, \beta_2 = 0) \\
&\quad - \frac{1}{p_{11}} \int \frac{d^2\beta_1}{\pi} \chi(\alpha_1, \alpha_2, \beta_1, \beta_2 = 0) e^{-|\beta_1|^2} \\
&\quad - \frac{1}{p_{11}} \int \frac{d^2\beta_2}{\pi} \chi(\alpha_1, \alpha_2, \beta_1 = 0, \beta_2) e^{-|\beta_2|^2} \\
&\quad + \frac{1}{p_{11}} \int \frac{d^2\beta_1}{\pi} \frac{d^2\beta_2}{\pi} \chi(\alpha_1, \alpha_2, \beta_1, \beta_2) e^{-|\beta_1|^2 - |\beta_2|^2}
\end{aligned}
\tag{18b}
$$

$$
\chi_{00}(\alpha_1, \alpha_2) = \frac{1}{p_{00}} \int \frac{d^2\beta_1}{\pi} \frac{d^2\beta_2}{\pi} \chi(\alpha_1, \alpha_2, \beta_1, \beta_2) e^{-|\beta_1|^2 - |\beta_2|^2}
\tag{18c}
$$

The above expressions can be calculated from the integral

$$I(\alpha_1, \alpha_2; u, v) = \int d(u\beta_1)d(v\beta_2)\chi(\alpha_1, \alpha_2, \beta_1, \beta_2)e^{-|\beta_1|^2 - |\beta_2|^2}, \qquad (19)$$

where the measures are defined as follows

$$d(u\beta_1) = \begin{cases} d^2\beta_1 & u = 1 \\ \pi\delta^2(\beta_1)d^2\beta_1 & u = 0 \end{cases}, \qquad (20a)$$

$$d(v\beta_2) = \begin{cases} d^2\beta_2 & v = 1 \\ \pi\delta^2(\beta_2)d^2\beta_2 & v = 0 \end{cases}. \qquad (20b)$$

In summary we can write

$$\chi_{0,0}(\alpha_1, \alpha_2) = \frac{1}{p_{00}}I(\alpha_1, \alpha_2; 1, 1), \qquad (21a)$$

$$\chi_{0,1}(\alpha_1, \alpha_2) = \frac{1}{p_{01}}\left[I(\alpha_1, \alpha_2; 1, 0) - I(\alpha_1, \alpha_2; 1, 1)\right], \qquad (21b)$$

$$\chi_{1,0}(\alpha_1, \alpha_2) = \frac{1}{p_{10}}\left[I(\alpha_1, \alpha_2; 0, 1) - I(\alpha_1, \alpha_2; 1, 1)\right], \qquad (21c)$$

$$\chi_{1,1}(\alpha_1, \alpha_2) = \frac{1}{p_{11}}\left[I(\alpha_1, \alpha_2; 0, 0) - I(\alpha_1, \alpha_2; 1, 0)\right.$$
$$\left. -I(\alpha_1, \alpha_2; 0, 1) + I(\alpha_1, \alpha_2, 1, 1)\right]. \qquad (21d)$$

The value of the probability $p_{x_1 x_2}$ comes from the normalization condition

$$\chi_{x_1 x_2}(\alpha_1 = 0, \alpha_2 = 0) = 1. \qquad (22)$$

The integrals (19) can be easily evaluated and they result:

$$I(\alpha_1, \alpha_2; 0, 0) = \exp\left[-A\left(|\alpha_1|^2 + |\alpha_2|^2\right) + A_{12}\left(\alpha_1\alpha_2 + \alpha_1^*\alpha_2^*\right)\right], \qquad (23a)$$

$$I(\alpha_1, \alpha_2; 0, 1) = \frac{1}{B+1}\exp\left[-\left(A - \frac{C'^2}{B+1}\right)|\alpha_1|^2 - \left(A - \frac{E^2}{B+1}\right)|\alpha_2|^2 \right.$$
$$\left. + \left(A_{12} + \frac{C'E}{B+1}\right)(\alpha_1\alpha_2 + \alpha_1^*\alpha_2^*)\right], \qquad (23b)$$

$$I(\alpha_1, \alpha_2; 1, 0) = \frac{1}{B+1}\exp\left[-\left(A - \frac{E^2}{B+1}\right)|\alpha_1|^2 - \left(A - \frac{C^2}{B+1}\right)|\alpha_2|^2 \right.$$
$$\left. + \left(A_{12} + \frac{CE}{B+1}\right)(\alpha_1\alpha_2 + \alpha_1^*\alpha_2^*)\right],$$

$$I(\alpha_1, \alpha_2; 1, 1) = \frac{1}{(B+1)^2 - B_{12}^2}\exp\left[-\left(A - \frac{(B+1)(E^2 + C^2) + 2B_{12}CE}{(B+1)^2 - B_{12}^2}\right)\right.$$
$$\times\left(|\alpha_1|^2 + |\alpha_2|^2\right)$$
$$+ \left(A_{12} + \frac{2(B+1)CE + B_{12}(C^2 + E^2)}{(B+1)^2 - B_{12}^2}\right)$$
$$\left. \times(\alpha_1\alpha_2 + \alpha_1^*\alpha_2^*)\right], \qquad (23c)$$

Using Eqs.(21), (22) and (23) we also derive

$$p_{0,0} = I(0,0;1,1) = \frac{1}{(B+1)^2 - B_{12}^2}, \tag{24a}$$

$$p_{0,1} = I(0,0;1,0) - I(0,0;1,1) = \frac{1}{B+1} - \frac{1}{(B+1)^2 - B_{12}^2}, \tag{24b}$$

$$p_{1,0} = I(0,0;0,1) - I(0,0;1,1) = \frac{1}{B+1} - \frac{1}{(B+1)^2 - B_{12}^2}, \tag{24c}$$

$$p_{1,1} = I(0,0;0,0) - I(0,0;1,0) - I(0,0;0,1) + I(0,0;1,1)$$
$$= 1 - \frac{2}{B+1} + \frac{1}{(B+1)^2 - B_{12}^2}. \tag{24d}$$

5 Entanglement Purification

Entanglement enhancement at the output is expected for conditioned non-Gaussian states (corresponding to non vacuum measurement outcomes). However, besides being non-Gaussian states these are also non pure state and their entanglement quantification is rather cumbersome. We adopt as an operational measure the teleportation fidelity [8]. That is, the conditional output state is assumed to be initially shared by Alice and Bob for the standard CV teleportation protocol [9]. Then, the average fidelity F for teleporting a coherent state is calculated. Actually, we get from Ref.[9], using the faltung theorem for Fourier transforms,

$$F = \int \frac{d^2\xi}{\pi} |\Phi^{in}(\xi)|^2 [\Phi^{ch}(\xi^*,\xi)]^* \tag{25}$$

where Φ stands for symmetrically ordered characteristic function. Hence, we have to intend

$$\Phi^{in}(\xi) = \exp(-|\xi|^2/2), \tag{26}$$

for the input[1] and

$$\Phi^{ch}(\xi^*,\xi) = \chi_{x_1 x_2}(\alpha_1 = \xi^*, \alpha_2 = \xi)\exp(-|\xi|^2), \tag{27}$$

for the channel. Thus Eq.(25) reads

$$F_{x_1 x_2} = \int \frac{d^2\xi}{\pi} e^{-2|\xi|^2} \chi_{x_1 x_2}(\alpha_1 = \xi^*, \alpha_2 = \xi). \tag{28}$$

---

[1] Since in the expression of the fidelity this appears with the modulus squared, the displacement of a coherent state does not matter.

Using (21), (23), (24) into (28) we finally arrive at

$$F_{0,0} = \frac{1}{p_{00}} \times \frac{1}{(B+1)^2 - B_{12}^2} \times \frac{1}{2 + 2A - 2A_{12} - 2\frac{(C+D)^2}{B+1-B_{12}}}, \qquad (29a)$$

$$F_{0,1} = \frac{1}{p_{01}} \left[\frac{1}{B+1} \times \frac{1}{2 + 2A - 2A_{12} - \frac{(C+D)^2}{B+1}} \right.$$

$$\left. - \frac{1}{(B+1)^2 - B_{12}^2} \times \frac{1}{2 + 2A - 2A_{12} - 2\frac{(C+D)^2}{B+1-B_{12}}} \right], \qquad (29b)$$

$$F_{1,0} = F_{0,1}, \qquad (29c)$$

$$F_{1,1} = \frac{1}{p_{11}} \left[\frac{1}{2 + 2A - 2A_{12}} - \frac{2}{B+1} \times \frac{1}{2 + 2A - 2A_{12} - \frac{(C+D)^2}{B+1}} \right.$$

$$\left. + \frac{1}{(B+1)^2 - B_{12}^2} \times \frac{1}{2 + 2A - 2A_{12} - 2\frac{(C+D)^2}{B+1-B_{12}}} \right], \qquad (29d)$$

The above fidelities should be compared with that coming from the initial squeezed thermal state (8).

$$f = \int \frac{d^2\xi}{\pi} e^{-2|\xi|^2} \exp\left\{ -2|\xi|^2 \left(\frac{\lambda^2 + \lambda}{1 - \lambda^2} + n_T \right) \right\}, \qquad (30a)$$

$$= \frac{1 - \lambda^2}{2\left(1 - \lambda + n_T - \lambda^2 n_T\right)} \qquad (30b)$$

By inspection, it turns out that F_{00}, F_{01} and F_{10} are always below f. On the contrary, F_{11} can be greater than f depending on both μt and λ (besides n_T). However by simply considering $F_{11} - f$ as figure of merit of the entanglement purification process could be misleading. In fact we have to also account for the probability p_{11} of having measurement outcomes 11.

Actually we can introduce an efficiency for the protocol as:

$$\mathcal{E} = \begin{cases} p_{11}\left(F_{11} - f\right) & if \ (F_{11} - f) > 0 \\ 0 & if \ (F_{11} - f) \leq 0 \end{cases} \qquad (31)$$

This quantity is greater than zero in a wide region of parameters μt and λ for $n_T = 0$ (see Fig.2), clearly showing the possibility of entanglement concentration (we speak about entanglement concentration because for $n_T = 0$ the initial state is already pure). Fig. 2 also manifests the existence of a tradeoff between μt and λ to get the maximum of $\mathcal{E}$. The efficiency is affected by the number of thermal photons, however it remains greater than zero even for $n_T \neq 0$ (see Fig. 3) showing the possibility of entanglement purification (in this case the initial state is not pure). It is worth to remark that $\mathcal{E}$ does not decrease exponentially vs n_T, but there exist a cutoff value of n_T at wich it reaches zero [10].

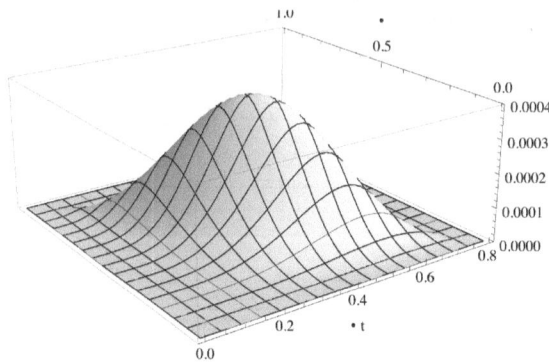

Fig. 2. Efficiency $\mathcal{E}$ vs μt and λ for $n_T = 0$

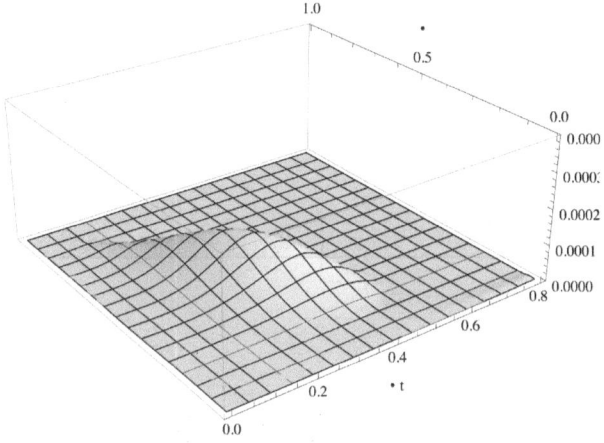

Fig. 3. Efficiency $\mathcal{E}$ vs μt and λ for $n_T = 0.05$

6   Concluding Remarks

We have proposed a realistic scheme for CV entanglement purification. It can
be applied to input optical CV fields, which are then sent to two identical and
independent atomic ensembles. The effective interaction between the optical and
atomic systems is shown to be of the form of either beam splitter or parametric
amplifier. Then supplementing the interaction with an appropriate non-Gaussian
measurement on the atomic ensembles, we have shown the possibility of enhancing the two-mode squeezing in the output. It turns out that by using the effective
Hamiltonian H_{PA} it is impossible to improve the teleportation fidelity [10]. This
is an indication of entanglement monogamy in CV setting [11]. On the contrary,
by using the effective Hamiltonian H_{BS} we have shown that it is possible to
enhance the input entanglement, even if the initial weakly entangled state is

not pure. From the perspective of this effective interaction, our protocol shares analogies with the generation of entangled states through photon subtraction mechanism [12].

Acknowledgments

SR would like to thank European Commission FP6 IST FET QIPC project QAP Contract No. 015848, DEST ISL Grant CG090188. The work of DV and SM is supported by EU through the FET-Open Project HIP (FP7-ICT-221899).

References

1. Braunstein, S.L., van Loock, P.: Rev. Mod. Phys. 77, 513 (2005)
2. Bennett, C.H., et al.: Phys. Rev. Lett. 76, 722 (1996)
3. Eisert, J., Scheel, S., Plenio, M.B.: Phys. Rev. Lett. 89, 137903 (2002)
4. Browne, D.E., Eisert, J., Scheel, S., Plenio, M.B.: Phys. Rev. A. 67, 062320 (2003)
5. Duan, L.M., Lukin, M.D., Cirac, J.I., Zoller, P.: Nature 414, 413 (2001)
6. Holstein, T., Primakoff, H.: Phys. Rev. 58, 1098 (1940)
7. Walls, D.F., Milburn, G.J.: Quantum Optics. Springer, Berlin (2008)
8. Olivares, S., Paris, M.G.A., Bonifacio, R.: Phys. Rev. A 67, 032314 (2003)
9. Pirandola, S., Mancini, S.: Laser Physics 16, 1 (2006)
10. Mancini, S., Morigi, G., Rebic, S., Vitali, D.: in preparation
11. Adesso, G., Illuminati, F.: New J. Phys. 8, 15 (2006)
12. Kitagawa, A., Takeoka, M., Sasaki, M., Chefles, A.: Phys. Rev. A 73, 042310 (2006)

Few Atom Detection and Manipulation Using Optical Nanofibres

Kieran Deasy[1,2], Amy Watkins[2,3], Michael Morrissey[1,2,*], Regine Schmidt[3], and Síle Nic Chormaic[2,3]

[1] Dept. Applied Physics and Instrumentation, Cork Institute of Technology, Bishopstown, Cork, Ireland
[2] Physics Department, University College Cork, Cork, Ireland
[3] Photonics Centre, Tyndall National Institute, University College Cork, Prospect Row, Cork, Ireland
{kieran.deasy,amy.watkins,regine.schmidt, sile.nicchormaic}@tyndall.ie

Abstract. We study the coupling of spontaneously emitted photons from laser-cooled ^{85}Rb atoms to the guided modes of an optical nanofibre to demonstrate the potential such fibres offer as tools for detecting and manipulating cold atoms, even when the number of atoms is very small. We also demonstrate the integration of an optical nanofibre into an absorption spectroscopy setup, showcasing the ability of the evanescent field around nanofibres to interact with atoms in close proximity to the fibre. In principle, trapping of single atoms in engineered optical potentials on the surface of the fibre should facilitate entanglement between distant atoms mediated via the guided modes of the nanofibre.

Keywords: Atom fluorescence, optical nanofibres, evanescent field.

1 Introduction

The significance of single-mode and multimode optical fibres in classical data communications cannot be disputed. Single-mode silica optical fibre forms the backbone of many recent advances in classical data transfer over long distances. It seems inevitable that the functionality of optical fibres in quantum systems and quantum communication schemes should also come to a fore. In recent years, a number of methods for trapping and guiding cold, neutral atoms outside subwavelength optical fibres, i.e. optical nanofibres, has been proposed [1,2]. The strong interest in such techniques stems from the opportunities that arise in quantum information technologies and in studying atom-surface interactions. In order to fully predict the processes involved when a cold atom is close to the surface of an optical nanofibre, a thorough understanding of the spontaneous emission rate of atoms located near the fibre surface is crucial. Earlier theoretical [3] and experimental [4] work shows that the spectral distribution of atomic fluorescence is affected by the presence of the fibre, due to

---

* Current address: IESL, FORTH, P.O. Box 1527, Gr-711 10 Heraklion, Greece.

A. Sergienko, S. Pascazio, and P. Villoresi (Eds.): QuantumCom 2009, LNICST 36, pp. 200–209, 2010.
© Institute for Computer Sciences, Social-Informatics and Telecommunications Engineering 2010

surface interactions resulting in a shift to the atomic transition frequency. In this work, we study the coupling of the spontaneous emission from laser-cooled rubidium atoms to the guided modes of an optical nanofibre to demonstrate the potential such fibres offer as tools for detecting and manipulating cold atoms, even when the number of atoms is very small [5].

As an alternative approach, light can be made propagate through an optical nanofibre, resulting in an intense evanescent field exponentially decaying from the surface of the fibre into the surrounding environment. This phenomenon has been theoretically studied extensively [6,7] and the evanescent light field has applications across a range of areas, in particular as an efficient method for coupling light into whispering gallery mode optical microcavities [8,9]. The evanescent field has also been used as an effective method for investigating the interaction of light and matter, for example in electromagnetically induced transparency (EIT) [10]. Nanofibres present a great opportunity to observe these interactions, providing a dual functionality of both supplying and collecting the probing light. In this work, we demonstrate the use of an optical nanofibre through which light is propagating in order to carry out absorption spectroscopy in a rubidium vapour cell. The interaction between the atoms and the light in the evanescent field yields an absorption signal that can be monitored at one end of the fibre. Currently, the Doppler-broadened peaks of the rubidium spectrum only are visible since a single beam pass has been used through the fibre.

2 Experimental Details and Results

In the work reported, there are two distinct experimental setups being used: one involves the integration of optical nanofibres as passive probes into a cloud of cold ^{85}Rb atoms and the other is the incorporation of an optical nanofibre, through which a very weak pump beam is passing, as an active probe into a Rb vapour cell. The fabrication and characterization of optical nanofibres is briefly described in section 2.1. The cold atom setup is described in section 2.2 and results demonstrating the coupling of spontaneously emitted photons from the laser-cooled atoms into the guided mode of a nanofibre are presented. This provides a unique tool for characterizing the properties of the magneto-optical trap (MOT). In section 2.3 we give details on the incorporation of the optical nanofibre into a Rb vapour cell setup where the eventual aim is to perform saturated absorption spectroscopy using the nanofibre.

2.1 Optical Nanofibres

Optical nanofibres are central to the two different experiments described here and the fabrication of the nanofibres is critical to the success of this work. A schematic diagram of the structure of an optical nanofibre is shown in Figure 1. The fabrication of optical nanofibres is based on heating and pulling a section of standard single mode fibre while ensuring that transmission through the fibre is high. From Figure 1 it can be seen that the fibres go through a transition region from being normal single mode fibre, where light is guided by the core/cladding refractive index difference, to a situation where the core has become negligible in the waist region of the nanofibre and the light is guided by the cladding/vacuum refractive index difference. When light is

(a)

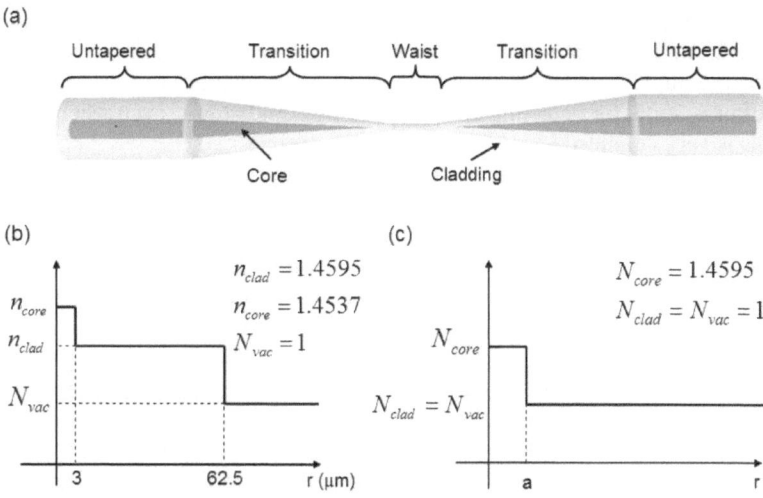

Fig. 1. (a) Diagram of the nanofibre. (b) Refractive index profile of ordinary single mode fibre. (c) Refractive index profile of the nanofibre in the waist region.

guided through the fibre a portion of it will be guided by the evanescent field at the waist region, external to the fibre, and this property is used when studying light-matter interactions with nanofibres. Alternatively, light can also be efficiently coupled into the fibre at the waist region and this property has been exploited by surrounding the fibre with cold atoms and monitoring the spontaneous emission from the atoms into the fibre's guided modes.

There are a number of techniques used for fabricating optical nanofibres [11] and we employ a standard heat and pull technique similar to that described elsewhere [12]. Within a clean environment, stripped, commercially available single-mode fibre is heated using an oxy-butane flame and then the two ends of the fibre are simultaneously drawn apart in opposite directions using a pair of stepper motors while monitoring the fibre transmission. Due to this extension the volume of the fibre within the hot zone is reduced, consequently reducing the fibre diameter [13]. The fibre diameter after the pulling process is a function of both the pull length and the length of the hot zone according to the adiabatic condition [14]. The oxygen-butane mixture is carefully selected to ensure a clean burning flame, eliminating any contaminates which may otherwise be deposited on the fibre, thereby reducing its transmission. A schematic of the pulling rig is shown in Figure 2. Using this design, nanofibres with a diameter of less than 1 μm and a transmission at 780 nm of over 90% are routinely produced.

2.2 Cold Atom Setup

For these experiments a cloud of cold ^{85}Rb atoms is produced using a standard MOT setup [15]. The cooling laser is locked 12 MHz red detuned from the $5S_{1/2}$, F=3→ $5S_{3/2}$, F'=4 transition using saturated absorption spectroscopy (SAS). The repump

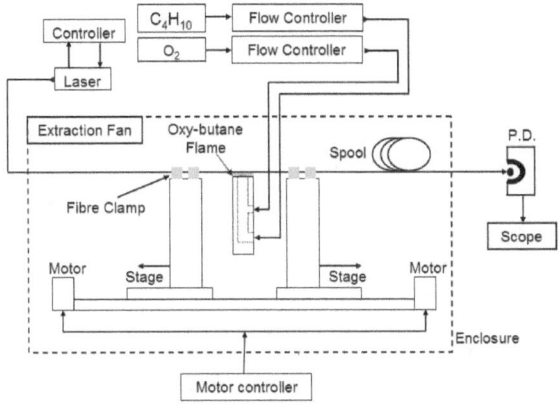

Fig. 2. Schematic diagram of the heat-and-pull rig used to fabricate the optical nanofibres

laser is also locked using SAS to the $5S_{1/2}$, $F=2 \rightarrow 5S_{3/2}$, $F'=3$ transition. Three orthogonal counter-propagating cooling beams intersect with the repumper beam and the zero of an inhomogeneous magnetic field with a field gradient of 10 G/cm at the centre of an octagonal vacuum chamber. Rubidium is supplied into the chamber using resistively heated dispensers, allowing the background Rb vapour to be controlled via the dispenser current.

The nanofibre is mounted on a U-mount and installed vertically into the chamber. The UHV is maintained by feeding the fibre into the chamber through a Teflon ferrule held in place using a Swagelok. The nanofibre is positioned in a manner to ensure the waist is overlapped by the cloud of cold atoms. The fibre we use for this work has a diameter of ~600 nm and a transmission of 85% at 780 nm after fabrication. Small changes to the MOT position can be made by the manipulation of magnetic fields using external coils. This ensures a good overlap of the MOT and the waist region of the fibre, as shown in Figure 3(a). Note that two fibres were installed in the UHV system but the results are all taken using a single fibre (TONF 1).

To monitor the position of the atom cloud a number of CCD cameras are placed around the chamber to provide viewing from all directions. This is to ensure that the overlap is excellent, thereby guaranteeing efficient coupling of spontaneously emitted photons from the atoms into the guided modes of the nanofibre as illustrated in Figure 3(b). The photon count rate is detected at one fibre end using a single photon counting module (SPCM, PerkinElmer, SPC-AQRH-14-FC) connected to a counter (Hamamatsu Counting Unit, C8855). It is also important to compare results obtained using the nanofibre probe with those taken using a more conventional method for analyzing MOT characteristics, such as loading time and lifetime. For this purpose, a fluorescence detection scheme was incorporated, whereby fluorescence from the MOT was focused onto a photodiode using a series of lenses. The signal yields information concerning the number of trapped atoms in the MOT. A schematic of the fluorescence detection scheme is shown in Figure 4.

(a)   (b)

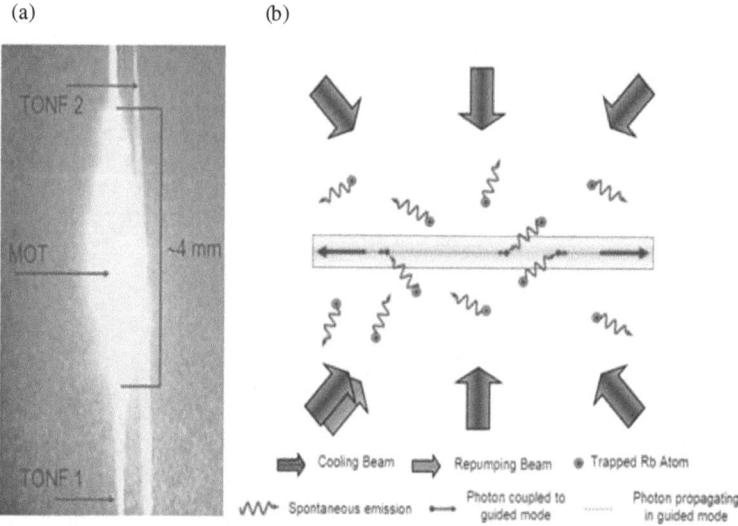

Fig. 3. (a) CCD image of the cloud of laser-cooled ^{85}Rb atoms overlapping the nanofibre. (b) Schematic of the atom cloud overlapping the nanofibre. Spontaneously emitted photons are detected at one end of the fibre.

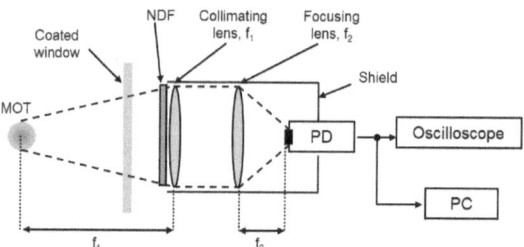

Fig. 4. Schematic diagram of the fluorescence detection system. PD: photodiode, NDF: neutral density filter, PC: personal computer.

2.2.1 Photon Coupling

Efficient coupling of fluorescence photons from the cold atoms into the guided modes of the nanofibre are critical for this work. The first measurement was an analysis of the efficiency of the photon coupling into the nanofibre. Initially, the cooling laser, repumper and anti-Helmholtz coils were switched off to determine the background count rate on the detector. The repump laser was then switched on, followed by the cooling laser and finally the anti-Helmholtz coils were switched on and the cold atom cloud was formed. The components were then switched off in the reverse order. The signal recorded on the single photon counter is shown in Figure 5. From these results, it is clear that the increase in signal obtained when the anti-Helmholtz coils are switched on is significant and this dramatic change in count rate of approximately 4×10^5 s^{-1} is attributed to fluorescence from the trapped atoms.

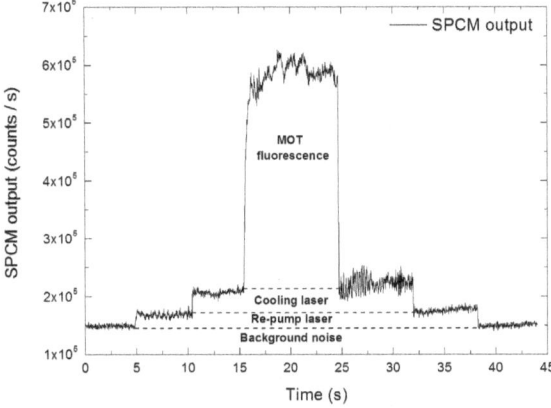

Fig. 5. Coupling of photons into the optical nanofibre from a cloud of ^{85}Rb atoms

To verify this result the photon count rate, η_P, can be determined from:

$$\eta_P = N_{eff}\eta_f\gamma_{sc}\eta_D T, \tag{1}$$

where N_{eff} is the effective number of atoms, η_f is the coupling efficiency of spontaneously emitted photons into the guided modes of the fibre, η_D is the quantum efficiency of the detector at 780 nm, and T is the transmission through the fibre. The effective number of atoms is the number of atoms contributing to the spontaneous emission signal and this is calculated by considering a 2 mm long hollow cylindrical volume extending 300 nm from the fibre surface and using the density of the MOT to determine the number of atoms in this region. Typically six atoms would be contained in this volume. Using the laser intensity and detuning of the cooling laser the atomic scattering rate can be determined as $\gamma_{SC} = 6.5 \times 10^5$ s^{-1}. With these values we calculate the value for $\eta_P = 3.7 \times 10^5$ s^{-1}, in good agreement with the experimental value observed.

2.2.2 Loading Time
The loading time is the $1/e$ time taken for the MOT to load with atoms once the magnetic field is switched on. The MOT is loaded from the background rubidium vapour. Here, we compare the results for the loading time taken from photons coupled into the nanofibre and detected at one end of the fibre on the SPCM for different Rb dispenser currents and these are shown in Figure 6. For comparison, the real-time loading evolution of the MOT as determined by focusing the emitted fluorescence from the atom cloud onto a photodiode (PD) is also shown. The Rb dispenser was turned on and allowed to stabilize for 15 minutes before the magnetic field was switched on. This was to remove any fluctuations due to variances in the Rb vapour pressure for a specific dispenser current. Loading times of ~ 0.5 s were obtained and both detection methods are in reasonably good agreement.

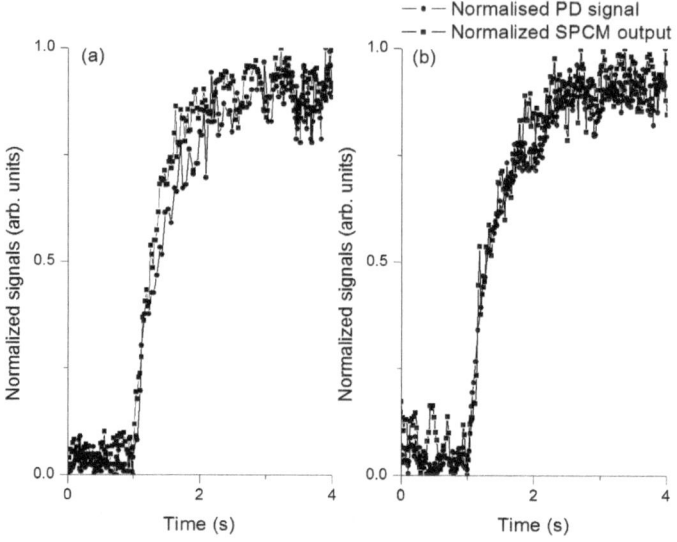

Fig. 6. Loading time of the MOT measured using a PD *(full circles)* and an optical nanofibre *(full squares)*. (a) Getter current of 3.3 A, (b) Getter current of 3.7 A.

2.2.3 Lifetime

The lifetime of the MOT is the $1/e$ time it takes for the atoms to escape from the MOT once the Rb dispenser is switched off. At this point, the loss rate from the trap is greater than the capture rate and the lifetime is determined by measuring the decay the number of trapped atoms. Again, we compare results using fluorescence imaging onto a PD and fluorescence coupling into the nanofibre for two different getter currents and these are shown in Figure 7.

From Figure 7 (a) and (b) it can be seen that the value observed for the lifetime using the PD detection scheme is relatively consistent at 8 - 9 seconds. However, for the nanofibre probe the lifetime varies significantly depending on the getter current. It can be seen that there is a longer lifetime observed using the nanofibre probe compared to the PD signal and this is extended even further when higher getter currents are applied. Two possible explanations for this can be proposed: firstly atoms at the extremities of the atom cloud are hotter than those at the centre of the cloud and, therefore, are lost faster. This will have a greater effect on the fluorescence detected by the PD, which monitors the entire atom cloud as opposed to the nanofibre, which only detects signal from atoms at or near the centre. The second possible reason is due to sensitivity; the nanofibre is a very sensitive device capable of detecting fluorescence for a small number of atoms remaining in its vicinity. This shows a clear advantage for using the nanofibre as a probe in a low density MOT or for detecting few atoms as compared to standard photodiode imaging schemes.

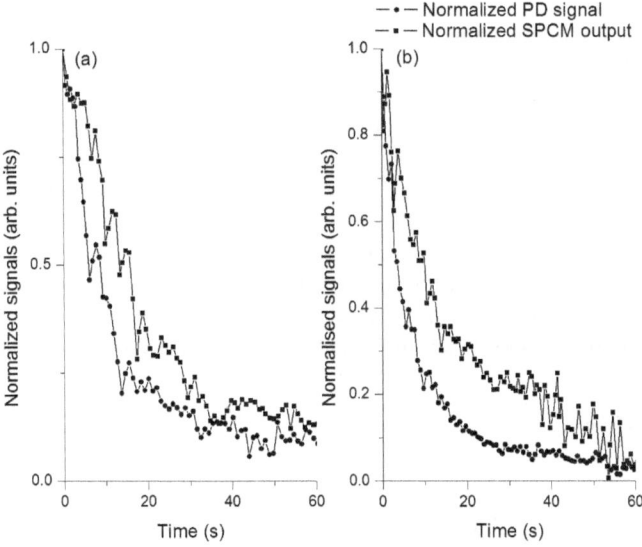

Fig. 7. Plots showing the lifetime of the MOT as taken using a PD *(full circles)* and using the nanofibre *(full squares)*. (a) Getter current of 3.3 Amps, (b) Getter current of 3.7 Amps.

2.3 Rubidium Vapour Cell Setup

For this experiment a rubidium vapour cell is formed in a 4-way vacuum cross pumped down to 10^{-2} mbar using a roughing pump. A Rb dispenser in the chamber

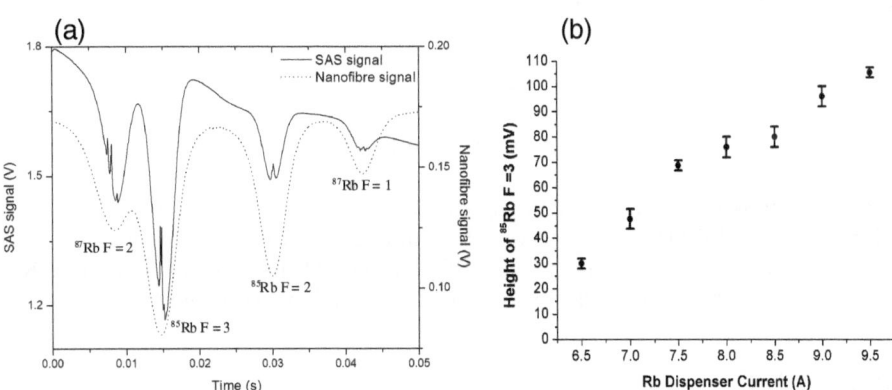

Fig. 8. (a) Absorption signal *(dashed line)* through the optical nanofibre for the D2 lines in ^{85}Rb and ^{87}Rb. For comparison the saturated absorption signal *(solid line)* using a standard setup is also shown. Fibre transmission ~ 2%. (b) Height of the ^{85}Rb F = 3 D2 line as a function of Rb dispenser current. Fibre transmission ~ 5%.

provides both the ^{85}Rb and ^{87}Rb isotopes. An optical nanofibre is mounted on a U-mount and placed in the chamber using the same techniques as in the cold atom setup. A Toptica DL100 laser at around 780 nm is scanned across the Rb transitions and approximately 2 μW of power is coupled into the nanofibre, which has a very low transmission of ~2%. The transmitted signal is monitored using a SensL Silicon Photomultiplier (SPM) Micro3000 Series **2** placed at one end of the fibre and is shown in Figure 8. As a comparison, the saturated absorption spectrum is also shown for the same laser. One can clearly see that the Doppler-broadened absorption dips are obtainable using this technique. In addition, the height of the ^{85}Rb F = 3 dip has been plotted as a function of current through the Rb dispenser and, as expected, it increases with increasing current. The next stage in this work will be the incorporation of a probe beam in order to obtain a saturated absorption signal through the nanofibre.

3 Conclusion

In conclusion, we have demonstrated the use of optical nanofibres as passive probes for cold atom samples and as active probes in a rubidium vapour cell. The nanofibre can be used to determine characteristics associated with the cloud of cold atoms, such as the lifetime and the loading time of the MOT as a function of background vapour pressure. The loading time measurements compare favourably with those made using conventional imaging techniques. However, the lifetime measurements differ significantly for both techniques. This is likely due to the MOT switching from the temperature limited regime to the constant density regime. In order to study this effect in more detail, future work will concentrate on symmetrizing the profile of the atom cloud and increasing MOT loading times by reducing the amount of rubidium dispensed into the vacuum chamber.

We have also observed absorption of the light field through an optical nanofibre when placed in a Rb vapour cell. An initial, Doppler-broadened spectrum has been observed with less than 2 μW of power coupled into the fibre and a very low fibre transmission of 2%. This shows the sensitivity afforded by using optical nanofibres as active probes. The next stage in this work will consist in incorporating a probe beam into the vapour cell in order to observe the saturated absorption spectrum.

Acknowledgments. This work is supported by Science Foundation Ireland under Grant No. 07/RFP/PHYF518 and the HEA through the INSPIRE programme. K.D. and A.W. acknowledge support from IRCSET through the Embark Initiative. R.S. acknowledges support from the ESF CASIMIR research networking programme.

References

1. Le Kien, F., Balykin, V.I., Hakuta, K.: Atom trap and waveguide using a two-color evanescent light field around a subwavelength-diameter optical fiber. Phys. Rev. A 70, 063403 (2004)
2. Sagué, G., Baade, A., Rauschenbeutel, A.: Blue-detuned evanescent field surface traps for neutral atoms based on mode interference in ultra-thin optical fibres. New J. Phys. 10, 113008 (2008)

3. Russell, L., Gleeson, D.A., Minogin, V.G., Nic Chormaic, S.: Spectral distribution of atomic fluorescence coupled into an optical nanofibre. J. Phys. B 42, 185006 (2009)
4. Nayak, K.P., Melentiev, P.N., Moringa, M., Le Kien, F., Balykin, V.I., Hakuta, K.: Optical nanofiber as an efficient tool for manipulating and probing atomic fluorescence. Optics Express 15, 5431 (2007)
5. Morrissey, M.J., Deasy, K., Wu, Y., Chakrabarti, S., Nic Chormaic, S.: Tapered optical fibers as tools for probing magneto-optical trap characteristics. Rev. Sci. Instrum. 80, 053102 (2009)
6. Bures, J., Ghosh, R.: Power density in the vicinity of a tapered fiber. J. Opt. Soc. Am. A 16, 1992 (1999)
7. Tong, L., Lou, J., Mazur, E.: Single-mode guiding properties of subwavelength-diameter silica and silicon wire waveguides. Optics Express 12, 1025–1035 (2004)
8. Cai, M., Painter, O., Vahala, K.J.: Observation of critical coupling in a fiber taper to a silica-microsphere whispering-gallery mode system. Phys. Rev. Lett. 85, 74–77 (2000)
9. Ward, J.M., Féron, P., Nic Chormaic, S.: A taper-fused microspherical laser source. IEEE Photonics Technology Letters 20, 392–394 (2008)
10. Spillane, S.M., Pati, G.S., Salit, K., Hall, M., Kumar, P., Beausoleil, R.G., Shahriar, M.S.: Observation of nonlinear optical interactions of ultralow levels of light in a tapered optical nanofiber embedded in a hot rubidium vapor. Phys. Rev. Lett. 100, 233602 (2008)
11. Tong, L., Gattass, R.R., Ashcom, J.B., He, S., Lou, J., Shen, M., Maxwell, I., Mazur, E.: Subwavelength-diameter silica wires for low-loss optical wave guiding. Nature 426, 816–819 (2003)
12. Ward, J.M., O'Shea, D.G., Shortt, B.J., Morrissey, M.J., Deasy, K., Nic Chormaic, S.: Heat-and-pull rig for fiber taper fabrication. Rev. Sci. Instrum. 77, 083105 (2006)
13. Birks, T., Li, Y.: The shape of fiber tapers. J. Light. Tech. 10, 432 (1992)
14. Love, J.D., Henry, W.M., Stewart, W.J., Black, R.J., Lacroix, S., Gonthier, F.: Tapered single-mode fibres and devices. 1. Adiabaticity criteria. IEEE Proc. 138, 342 (1991)
15. Metcalf, H.J., van der Straten, P.: Laser Cooling and Trapping. Springer, Heidelberg (1999)

An Error Model for the Cirac-Zoller CNOT Gate

Sara Felloni[1,2] and Giuliano Strini[3]

[1] Department of Electronics and Telecommunications, Norwegian University of
Science and Technology (NTNU), NO-7491 Trondheim, Norway
sara.felloni@iet.ntnu.no
[2] UNIK - University Graduate Center, NO-2027 Kjeller, Norway
[3] Dipartimento di Fisica, Università degli Studi di Milano, Via Celoria 16,
20133 Milano, Italy

Abstract. In the framework of ion-trap quantum computing, we develop a characterization of experimentally realistic imperfections which may affect the Cirac-Zoller implementation of the CNOT gate.

The CNOT operation is performed by applying a protocol of five laser pulses of appropriate frequency and polarization. The laser-pulse protocol exploits auxiliary levels, and its imperfect implementation leads to unitary as well as non-unitary errors affecting the CNOT operation.

We provide a characterization of such imperfections, which are physically realistic and have never been considered before to the best of our knowledge. Our characterization shows that imperfect laser pulses unavoidably cause a leak of information from the states which alone should be transformed by the ideal gate, into the ancillary states exploited by the experimental implementation.

Keywords: Ion-trap quantum computing, Cirac-Zoller CNOT gate, decoerence, error models.

1 Decoherence and Quantum Computing by Ion Traps

Quantum computation and quantum communication require special physical environments: Decoherence, noise and experimental imperfections threaten the correctness of quantum computations, as well as the intended functioning of quantum communication protocols. Several experimental implementations have been proposed and explored in order to build systems capable of having limited sensitivity to unwanted perturbations, while allowing both desired interactions among internal components and access from external systems or users.

Ion-trap quantum computation is one of the main and rapidly evolving experimental possibilities. Several new ideas and experimental techniques [1] [2] [3] suggest that ion traps offer a promising architecture for quantum information processing. In this framework, we develop a characterization of experimentally realistic imperfections which may affect the Cirac-Zoller implementation of the CNOT gate [4]. The CNOT operation is performed by applying a protocol of five laser pulses of appropriate frequency and polarization. The laser-pulse protocol exploits auxiliary levels and this could origin unitary as well as non-unitary

A. Sergienko, S. Pascazio, and P. Villoresi (Eds.): QuantumCom 2009, LNICST 36, pp. 210–219, 2010.

errors affecting the CNOT operation. We provide a characterization of such imperfections, which are physically realistic and have never been considered before to the best of our knowledge.

The paper is organized as follows. In Section 2, we briefly review the sequence of five laser pulses which implement the CNOT gate by ion-trap techniques, as proposed by Cirac and Zoller. In Section 3, we explore the physically realistic perturbations which may affect the laser-pulse protocol for the Cirac-Zoller implementation of the CMINUS gate and, consequently, of the CNOT gate. First, in Section 3.1, we describe by diagrams of quantum states the twelve-level Hilbert space in which we reproduce both the ideal and imperfect actions of the CNOT gate. Subsequently, in Section 3.2 we model imperfections in all the three laser pulses which implement the CMINUS gate, by applying perturbations on the parameters determining the physical characteristics of the lasers. Then, in Section 3.3 we formulate our error model by means of the well-known density-matrix formalism and Kraus representation, showing how leakage errors are unavoidable in realistic conditions. Finally, in Section 4 we express our conclusive remarks.

Throughout this paper, we assume the reader is familiar with the basic ion-trap techniques. Useful and detailed physical and computational descriptions of ion-trap quantum computing can be found, for instance, in [5] (pages 544-561) or [6].

2 The Cirac-Zoller CNOT Gate

Ion-trap quantum gates can be obtained by applying to ions implementing qubits the appropriate combination of laser pulses, tuned to the appropriate duration and phase. In controlled two-qubit operations, the motion state of the string of ions (the *phonon* qubit) is exploited as a "bus" to transfer quantum information between two qubits implemented by ions.

Following the proposal of Cirac and Zoller [4], we briefly recall how to obtain a CNOT quantum gate (CNOT$_{CZ}$) acting on a set of N ions, with the ion l as the control qubit and the ion m as the target qubit.

A general state in the computation is described as $|i_l, i_m; i_n\rangle$, where the last position is always filled by the state of the phonon qubit; for the sake of simplicity, we omit all the qubits of the register which are not affected by the protocol. Qubits can be in the ground or excited levels $|g\rangle$ and $|e\rangle$; at some point of the protocol, an auxiliary level $|a\rangle$ is also necessary. The initial state for the computation is $|i_l, i_m; 0\rangle$.

First, a red de-tuned laser acts on the ion l in order to map the quantum information of the control ion onto the vibrational mode. This impulse changes the states of the two-qubit computational basis as follows:

$$\begin{cases} |g_l, g_m; 0\rangle \rightarrow |g_l, g_m; 0\rangle \\ |g_l, e_m; 0\rangle \rightarrow |g_l, e_m; 0\rangle \\ |e_l, g_m; 0\rangle \rightarrow i|g_l, g_m; 1\rangle \\ |e_l, e_m; 0\rangle \rightarrow i|g_l, e_m; 1\rangle. \end{cases} \tag{1}$$

Then, a red de-tuned laser is applied to the ion m. The impulse is now expressed in the basis $\{|g_m; n = 1\rangle, |a_m; n = 0\rangle\}$, thus exploiting the auxiliary level $|a_m\rangle$, and it acts as follows:

$$\begin{cases} |g_l, g_m; 0\rangle \rightarrow |g_l, g_m; 0\rangle \\ |g_l, e_m; 0\rangle \rightarrow |g_l, e_m; 0\rangle \\ i|e_l, g_m; 1\rangle \rightarrow -i|e_l, g_m; 1\rangle \\ i|g_l, e_m; 1\rangle \rightarrow i|g_l, e_m; 1\rangle. \end{cases} \tag{2}$$

Finally, a red de-tuned laser acts once again on the ion l, in order to map the quantum information of vibrational mode back onto the control ion:

$$\begin{cases} |g_l, g_m; 0\rangle \rightarrow |g_l, g_m; 0\rangle \\ |g_l, e_m; 0\rangle \rightarrow |g_l, e_m; 0\rangle \\ -i|g_l, g_m; 1\rangle \rightarrow |e_l, g_m; 0\rangle \\ i|g_l, e_m; 1\rangle \rightarrow -|e_l, e_m; 0\rangle. \end{cases} \tag{3}$$

The global effect of the three laser pulses is to induce a controlled phase-shift gate with a phase-shift of an angle $\delta = \pi$, that is, a CMINUS gate. From the CMINUS gate, the CNOT gate can be obtained by applying a single-qubit Hadamard gate before and after the three laser pulses.

In conclusion, the CNOT$_{\text{CZ}}$ operation between the target qubit m and the control qubit l can be performed by applying a protocol of five laser pulses of appropriate frequency and polarization.

3 Imperfections in the Cirac-Zoller CNOT Gate

In the ion-trap implementation of the CNOT gate proposed by Cirac and Zoller, a key-role is played by the laser protocol implementing the CMINUS operation, which exploits *ancillary levels*. The role of such additional levels is here explored to model unitary as well as non-unitary errors affecting the CMINUS operation and, consequently, the whole implementation of the CNOT gate.

The error model schematically consists in the following steps. First, unitary errors are applied to each impulse gate constituting the CMINUS protocol. Then, a partial trace operation is performed on the collective vibrational motion, whose levels are neglected at the end of both ideal and non-ideal computations. In the ideal protocol, a final step of partial tracing on the ancillary levels would lead to perfectly reproducing the action of the CMINUS operation. However, in a general experimental situation, our error model shows that it is no longer possible to trace over the ancillary level without loss of information: The unitary errors introduced in the three CMINUS laser pulses unavoidably cause an irreversible leak of information in the ancillary states.

3.1 The Hilbert Space

In order to describe the most general perturbation of the CMINUS laser-pulse protocol for the Cirac-Zoller implementation of the CNOT gate, we rely onto a useful representation and ordering of the twelve-level space which reproduces the action of the gate.

This ordering is illustrated by the diagram of states in Figure 1. In the usual way of representing quantum computations by quantum circuits, each horizontal line represents a qubit. In diagrams of states [7], we draw instead a horizontal line for each state of the computational basis, here adding horizontal lines for the additional levels necessary to reproduce the action of the CNOT gate. Similarly to quantum circuits, any sequence of operations in diagrams of states must be read from left (input) to right (output).

We first represent the ideal action of the three laser pulses, that is, we represent how the information contained in the states of the considered twelve-level space is elaborated when the CNOT gate is implemented by a system unaffected by errors or noise. We put the control qubit m, which has two possible levels 0 and 1, in the most significant position; we put the target qubit l, which has two possible standard levels 0 and 1 plus an additional ancillary level 2, in the middle; finally, we put the phonon qubit n, which has two possible levels 0 and 1, in the least significant position (see upper scheme in Figure 1). We order the resulting twelve levels accordingly (see lower diagram of Figure 1). The ordering is such that in the top half of the diagram we have the six levels corresponding to the non-excited state of the phonon qubit, namely, $n = 0$; in the bottom half of the diagram we have the remaining six levels, corresponding to the excited state of the phonon qubit, namely, $n = 1$. Finally, the two bottom levels of each half of the diagram correspond to the ancillary level of the target qubit, namely, $m = 2$.

Since we are considering initial states for which i) the collective vibrational motion is not excited ($n = 0$) and ii) the auxiliary level stores no information, the initial information is stored only in the first four states $|000\rangle$, $|010\rangle$, $|100\rangle$ and $|110\rangle$. The adopted ordering allows us to collect at the top of the diagram the four levels containing the input information, when the collective vibrational motion is not excited and the ancillary states have not yet been exploited. After applying in sequence the three operations corresponding to the three laser pulses previously described, we show at the rightmost end of the diagram how the global transformation affects the twelve levels of the overall system. Again, at the top of the diagram we read on the first four states how the computation reproduces the action of an ideal CMINUS gate.

3.2   Imperfect Impulse Gates

Referring to the Hilbert space previously described, we now model imperfections in the three impulse gates involved in the implementation of the CNOT$_{\text{CZ}}$ gate.

The action of a general unitary matrix acting on a two-qubit system,

$$U = \begin{bmatrix} \cos\frac{\theta}{2}\left(\cos\psi + i\sin\psi\right) & i\sin\frac{\theta}{2}\left(\cos\phi + i\sin\phi\right) \\ i\sin\frac{\theta}{2}\left(\cos\phi - i\sin\phi\right) & \cos\frac{\theta}{2}\left(\cos\psi - i\sin\psi\right) \end{bmatrix}, \tag{4}$$

is determined by the three parameters θ, ψ, ϕ. When considering unitary transformations induced by laser pulses, these parameters are associated with precise

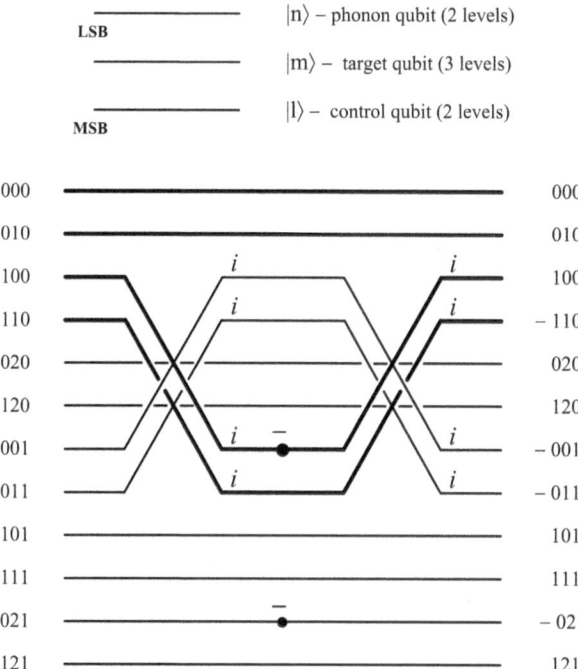

LSB $|n\rangle$ – phonon qubit (2 levels)

$|m\rangle$ – target qubit (3 levels)

$|l\rangle$ – control qubit (2 levels)
MSB

Fig. 1. Diagram of states [7] representing a noisy ion-trap implementation of the CMI-NUS gate, according to the Cirac-Zoller protocol. The control qubit m in set the most significant position, with possible levels 0 and 1; the target qubit l in set the middle, with possible levels 0, 1 and 2 (ancillary); the phonon qubit n in set the least significant position, with possible levels 0 and 1. The ordering of the twelve resulting levels is such that the top six levels of the diagram correspond to the non-excited state of the phonon qubit ($n = 0$), while the bottom six levels of the diagram correspond to the excited state of the phonon qubit ($n = 1$). In each half of the diagram, the two bottom levels correspond to the ancillary state of the target qubit ($m = 2$). The sequence of operations must be read from left (input) to right (output); information flows on the thick lines, while thinner lines correspond to absence of information. Note that the action of the CMINUS gate can here be visualized by simply following the thick lines corresponding to the first four levels, which alone store all the input information at the start of the gate operation.

physical features of the impulses: The parameter $\theta = \Omega t$ denotes the impulse area, where Ω is the intensity and t is the duration of the impulse; the parameter ψ is related to the laser de-tuning; the parameter ϕ denotes the laser phase. By appropriately varying these parameters, we obtain the unitary matrices which correspond to the ideal impulses implementing the CMINUS gate.

The first and the third impulse matrices are obtained by imposing:

$$\{\theta = \pi, \psi = \pi, \phi = 0\}, \qquad U^{(1,3)} = \begin{bmatrix} 0 & i \\ i & 0 \end{bmatrix}, \tag{5}$$

while the second impulse matrix is obtained by imposing:

$$\{\theta = \pi, \psi = \pi, \phi = 0\}, \qquad U^{(2)} = \begin{bmatrix} -1 & 0 \\ 0 & -1 \end{bmatrix}. \tag{6}$$

We apply a perturbation on all the three parameters of all the three impulses, which we will denote from now on with index ι:

$$\{\theta^\iota + \Delta\theta^\iota, \quad \psi^\iota + \Delta\psi^\iota, \quad \phi^\iota + \Delta\phi^\iota\} \qquad \iota = 1, 2, 3. \tag{7}$$

Thus, each noisy impulse is now described by:

$$U^{(\iota)} = \begin{bmatrix} u^\iota_{11} & u^\iota_{12} \\ u^\iota_{21} & u^\iota_{22} \end{bmatrix} \tag{8}$$

where, for $\iota = 1, 2, 3$, we have:

$$u^\iota_{11} = \cos\frac{\theta^\iota + \Delta\theta^\iota}{2}(\cos(\psi^\iota + \Delta\psi^\iota) + i\sin(\psi^\iota + \Delta\psi^\iota)),$$

$$u^\iota_{12} = i\sin\frac{\theta^\iota + \Delta\theta^\iota}{2}(\cos(\phi^\iota + \Delta\phi^\iota) + i\sin(\phi^\iota + \Delta\phi^\iota)),$$

$$u^\iota_{21} = i\sin\frac{\theta^\iota + \Delta\theta^\iota}{2}(\cos(\phi^\iota + \Delta\phi^\iota) - i\sin(\phi^\iota + \Delta\phi^\iota)),$$

$$u^\iota_{22} = \cos\frac{\theta^\iota + \Delta\theta^\iota}{2}(\cos(\psi^\iota + \Delta\psi^\iota) - i\sin(\psi^\iota + \Delta\psi^\iota)). \tag{9}$$

All the three noisy impulse matrices are subsequently embedded in the twelve-level space of the Cirac-Zoller implementation, obtaining respectively the matrices $V^{(1)}$, $V^{(2)}$ and $V^{(3)}$. The entries $\{u^\iota_{i,j}\}$ are given by the parameters defined by equation (9); since the matrices are very sparse, we omit all their null entries to obtain a clearer visualization of each matrix structure:

$$V^{(1)} = \begin{bmatrix}
1 & . & . & . & . & . & . & . & . & . & . & . \\
. & 1 & . & . & . & . & . & . & . & . & . & . \\
. & . & u^1_{11} & . & . & u^1_{12} & . & . & . & . & . & . \\
. & . & . & u^1_{11} & . & . & u^1_{12} & . & . & . & . & . \\
. & . & . & . & 1 & . & . & . & . & . & . & . \\
. & . & . & . & . & 1 & . & . & . & . & . & . \\
. & . & u^1_{21} & . & . & u^1_{22} & . & . & . & . & . & . \\
. & . & . & u^1_{21} & . & . & u^1_{22} & . & . & . & . & . \\
. & . & . & . & . & . & . & 1 & . & . & . & . \\
. & . & . & . & . & . & . & . & 1 & . & . & . \\
. & . & . & . & . & . & . & . & . & 1 & . & . \\
. & . & . & . & . & . & . & . & . & . & . & 1
\end{bmatrix}$$

$$V^{(2)} = \begin{bmatrix}
1 & . & . & . & . & . & . & . & . & . & . & . \\
. & 1 & . & . & . & . & . & . & . & . & . & . \\
. & . & 1 & . & . & . & . & . & . & . & . & . \\
. & . & . & 1 & . & . & . & . & . & . & . & . \\
. & . & . & . & 1 & . & . & . & . & . & . & . \\
. & . & . & . & . & 1 & . & . & . & . & . & . \\
. & . & . & . & . & . & u^2_{11} & . & . & u^2_{12} & . \\
. & . & . & . & . & . & . & 1 & . & . & . & . \\
. & . & . & . & . & . & . & . & 1 & . & . & . \\
. & . & . & . & . & . & . & . & . & 1 & . & . \\
. & . & . & . & . & . & u^2_{21} & . & . & u^2_{22} & . \\
. & . & . & . & . & . & . & . & . & . & . & 1
\end{bmatrix}$$

$$V^{(3)} = \begin{bmatrix} 1 & . & . & . & . & . & . & . & . & . & . & . \\ . & 1 & . & . & . & . & . & . & . & . & . & . \\ . & . & u_{11}^3 & . & . & . & u_{12}^3 & . & . & . & . & . \\ . & . & . & u_{11}^3 & . & . & . & u_{12}^3 & . & . & . & . \\ . & . & . & . & 1 & . & . & . & . & . & . & . \\ . & . & . & . & . & 1 & . & . & . & . & . & . \\ . & . & u_{21}^3 & . & . & . & u_{22}^3 & . & . & . & . & . \\ . & . & . & u_{21}^3 & . & . & . & u_{22}^3 & . & . & . & . \\ . & . & . & . & . & . & . & . & 1 & . & . & . \\ . & . & . & . & . & . & . & . & . & 1 & . & . \\ . & . & . & . & . & . & . & . & . & . & 1 & . \\ . & . & . & . & . & . & . & . & . & . & . & 1 \end{bmatrix} . \tag{10}$$

Finally, we appropriately embed the Hadamard gates in the twelve-level space of the Cirac-Zoller implementation (still denoting by H the resulting matrices of dimension 12×12, for the sake of simplicity).

By multiplying the five matrices in the appropriate order, we obtain the overall matrix of dimension 12×12 which describes the action of the CNOT$_{CZ}$ gate:

$$\text{CNOT}_{CZ} = H \, V^{(3)} \, V^{(2)} \, V^{(1)} \, H. \tag{11}$$

3.3   Density Matrix Evolution and Kraus Operators

We now calculate the density matrix transformation corresponding to the previously described model for imperfections in the impulse gates originating the CNOT$_{CZ}$ gate.

The initial information is stored only in the first four states $|000\rangle$, $|010\rangle$, $|100\rangle$ and $|110\rangle$: This is equivalent to say that all initial information is stored in a density matrix of dimension 4×4, which we denote by $\rho_{4 \times 4}^{in}$. Thus, the overall initial density matrix $\rho_{12 \times 12}^{in}$ has non-zero entries only in the 4×4 upper-and-leftmost positions:

$$\rho_{6 \times 6}^{in} = \begin{bmatrix} \rho_{4 \times 4}^{in} & \mathbf{0}_{4 \times 2} \\ \mathbf{0}_{2 \times 4} & \mathbf{0}_{2 \times 2} \end{bmatrix} ; \qquad \rho_{12 \times 12}^{in} = \begin{bmatrix} \rho_{6 \times 6}^{in} & \mathbf{0}_{6 \times 6} \\ \mathbf{0}_{6 \times 6} & \mathbf{0}_{6 \times 6} \end{bmatrix} . \tag{12}$$

The evolution of the overall initial density matrix caused by the laser-pulse protocol is given by:

$$\rho_{12 \times 12}^{fin} = \text{CNOT}_{CZ} \, \rho_{12 \times 12}^{in} \, \text{CNOT}_{CZ}^{\dagger}. \tag{13}$$

In order to determine the Kraus operators describing the evolution of the reduced density matrix, the unitary matrix CNOT$_{CZ}$ can be decomposed into sub-matrices A, B, C and D, each one of dimension 6×6. Thus, equation (13) can be written as follows:

$$\rho_{12\times12}^{\text{fin}} = \begin{bmatrix} A & B \\ C & D \end{bmatrix} \begin{bmatrix} \rho_{6\times6}^{\text{in}} & \mathbf{0}_{6\times6} \\ \mathbf{0}_{6\times6} & \mathbf{0}_{6\times6} \end{bmatrix} \begin{bmatrix} A^\dagger & C^\dagger \\ B^\dagger & D^\dagger \end{bmatrix}. \tag{14}$$

Since at the end of the process implementing the CNOTCZ gate we neglect the levels corresponding to the collective vibrational motion, we now perform a partial trace operation on the overall final density matrix $\rho_{12\times12}^{\text{fin}}$, in respect to the phonon qubit n:

$$\rho_{6\times6}^{\text{fin}} = \text{Tr}_{n = \text{LSB}} \{\rho_{12\times12}^{\text{fin}}\} = A \, \rho_{6\times6}^{\text{in}} \, A^\dagger + C \, \rho_{6\times6}^{\text{in}} \, C^\dagger. \tag{15}$$

We further decompose the sub-matrices A and C, both having dimension 6×6, into sub-matrices $\{A_i, C_i\}$, for $i = 1 \dots 4$, each one having the appropriate dimension as expressed in the following equation:

$$\rho_{6\times6}^{\text{fin}} = \begin{bmatrix} A1_{4\times4} & A2_{4\times2} \\ A3_{2\times4} & A4_{2\times2} \end{bmatrix} \begin{bmatrix} \rho_{4\times4}^{\text{in}} & \mathbf{0}_{4\times2} \\ \mathbf{0}_{2\times4} & \mathbf{0}_{2\times2} \end{bmatrix} \begin{bmatrix} A1_{4\times4}^\dagger & A3_{4\times2}^\dagger \\ A2_{2\times4}^\dagger & A4_{2\times2}^\dagger \end{bmatrix} +$$

$$+ \begin{bmatrix} C1_{4\times4} & C2_{4\times2} \\ C3_{2\times4} & C4_{2\times2} \end{bmatrix} \begin{bmatrix} \rho_{4\times4}^{\text{in}} & \mathbf{0}_{4\times2} \\ \mathbf{0}_{2\times4} & \mathbf{0}_{2\times2} \end{bmatrix} \begin{bmatrix} C1_{4\times4}^\dagger & C3_{4\times2}^\dagger \\ C2_{2\times4}^\dagger & C4_{2\times2}^\dagger \end{bmatrix}. \tag{16}$$

Consequently, the final reduced density matrix can be expressed as:

$$\rho_{6\times6}^{\text{fin}} = \begin{bmatrix} \rho1_{4\times4}^{\text{fin}} & \rho2_{4\times2}^{\text{fin}} \\ \rho3_{2\times4}^{\text{fin}} & \rho4_{2\times2}^{\text{fin}} \end{bmatrix}, \tag{17}$$

where:

$$\rho1_{4\times4}^{\text{fin}} = A1_{4\times4} \, \rho_{4\times4}^{\text{in}} \, A1_{4\times4}^\dagger + C1_{4\times4} \, \rho_{4\times4}^{\text{in}} \, C1_{4\times4}^\dagger, \tag{18}$$

$$\rho4_{2\times2}^{\text{fin}} = A3_{2\times4} \, \rho_{4\times4}^{\text{in}} \, A3_{4\times2}^\dagger + C3_{2\times4} \, \rho_{4\times4}^{\text{in}} \, C3_{4\times2}^\dagger. \tag{19}$$

Recalling expression (11), we finally explicit the Kraus operators $A1$, $A3$, $C1$ and $C3$:

$$A1_{4\times4} = \begin{bmatrix} 1 & 0 & 0 & 0 \\ 0 & 1 & 0 & 0 \\ 0 & 0 & a_1^1 & a_2^1 \\ 0 & 0 & a_2^1 & a_1^1 \end{bmatrix}, \qquad A3_{2\times4} = \begin{bmatrix} 0 & 0 & 0 & 0 \\ 0 & 0 & 0 & 0 \end{bmatrix}, \tag{20}$$

where:

$$a_1^1 = \frac{1}{2} \left(2u_{11}^1 u_{11}^3 + u_{21}^1 \left(1 + u_{11}^2 \right) u_{12}^3 \right), \qquad a_2^1 = \frac{1}{2} u_{21}^1 \left(-1 + u_{11}^2 \right) u_{12}^3, \tag{21}$$

with parameters $\{u_{i,j}^\iota\}$, for $\iota = 1, 2, 3$, defined by equation (9), and:

$$C1_{4\times4} = \begin{bmatrix} 0 & 0 & c_1^1 & c_1^1 \\ 0 & 0 & c_2^1 & -c_2^1 \\ 0 & 0 & 0 & 0 \\ 0 & 0 & 0 & 0 \end{bmatrix}, \qquad C3_{2\times4} = \begin{bmatrix} 0 & 0 & c^3 & c^3 \\ 0 & 0 & 0 & 0 \end{bmatrix}, \tag{22}$$

where:

$$c_1^1 = \frac{1}{\sqrt{2}} \left(u_{11}^1 u_{21}^3 + u_{21}^1 u_{11}^2 u_{22}^3 \right),$$

$$c_2^1 = \frac{1}{\sqrt{2}} \left(u_{11}^1 u_{21}^3 + u_{21}^1 u_{22}^3 \right), \qquad c^3 = \frac{1}{\sqrt{2}} u_{21}^3 u_{21}^2, \qquad (23)$$

with parameters $\{u_{i,j}^\iota\}$, for $\iota = 1, 2, 3$, once again defined by equation (9).

In conclusion, when the three laser pulses are ideally implemented without perturbation, the overall transformation coincides with the ideal CNOT gate, that is, the upper-and-leftmost final density sub-matrix $\rho_{\text{main}} = \rho 1_{4 \times 4}^{\text{fin}}$ is the result of a perfect CNOT gate application to the density matrix of the four initial states. On the other hand, when we deal with inaccurate laser pulses, the final information is no longer stored only in the evolved four initial states. Errors unavoidably cause a spreading of information, which partially remains stored in the four evolved initial states, and partially leaks in the ancillary levels involved in the action of the second laser pulse.

More precisely, the final density matrix previously calculated shows how information is spread by the sequence of imperfect operations. The density sub-matrix ρ_{main} still holds the information corresponding to the imperfect action of the gate on the desired control and target qubits, while the lower-and-rightmost final density sub-matrix $\rho_{\text{anc}} = \rho 4_{2 \times 2}^{\text{fin}}$ holds the information leaked in the auxiliary levels and thus lost from the control and target qubits.

The traces of sub-matrices ρ_{main} and ρ_{anc} give the probabilities p_{main} and p_{anc} that information spreads and populates the main and ancillary levels, respectively. Obviously, $p_{\text{main}} + p_{\text{anc}} = 1$, as the overall density matrix has unit trace.

4   Conclusive Remarks

We have developed a characterization of experimentally realistic imperfections which may affect the Cirac-Zoller implementation of the CNOT gate.

In the framework of ion-trap quantum computing, the CNOT operation can be performed by applying a protocol of five laser pulses of appropriate frequency and polarization. The laser-pulse protocol exploits auxiliary levels, and this may results in unitary as well as non-unitary errors affecting the CNOT operation. We have provided a characterization of such imperfections, which are physically realistic and have never been considered before to the best of our knowledge. This characterization shows that it is no longer possible to disregard the ancillary levels without loss of information: Unitary errors introduced in the laser pulses unavoidably cause a leak of information in the ancillary states, which can no longer be completely retrieved at the end of the computation.

Modeling experimentally realistic perturbations affecting two-qubit universal operations allow us to explore the impact of two-qubit imperfections on quantum computation and communication protocols. Our previous characterization of general single-qubit errors and its application to a quantum privacy amplification protocol based on entanglement purification [8] proved that different imperfections can affect a quantum protocol very differently. In a similar way, the modeled two-qubit imperfections may have unexpected effects on the quantum computation and communication protocols simulated in their presence. Extensive simulations will highlight which error parameters are most dangerous and which kinds of perturbations should be expected, especially regarding errors on the ancillary levels.

Finally, by addressing general two-qubit errors affecting universal two-qubit operations, we also aim to extend our previously developed general single-qubit error model. Since multi-qubit unitary operations can always be computed by equivalent quantum circuits composed of single-qubit and two-qubit gates, a compact model comprising single-qubit and two-qubit errors would offer a satisfiable characterization of general imperfections. Unfortunately, describing and understanding all parameters which characterize the most general transformation of a two-qubit density matrix can not be achieved by straightforward analytical study. Thus, exploring imperfect implementations of universal two-qubit operations in the main current experimental frameworks can offer a valuable help to determine the role and the physical meaning of significant parameters describing the most general two-qubit noise transformation.

Acknowledgments

Sara Felloni acknowledges support by ERCIM, as this work was partially carried out during the tenure of an ERCIM "Alain Bensoussan" Fellowship Programme, and she wishes to thank Johannes Skaar for useful insights and discussion.

References

1. Cirac, J.I., Zoller, P.: New frontiers in quantum information with atoms and ions. Phys. Today, 38 (March 2004)
2. Ion trap approaches to Quantum Information Processing and Quantum Computing - A Quantum Information Science and Technology Roadmap - part 1: Quantum Computation, section 6.2 (2004), http://qist.lanl.gov/qcomp_man.shtml
3. Haeffner, H., Roos, C.F., Blatt, R.: Quantum computing with trapped ions. Phys. Rep. 469, 155 (2008), arXiv:0809.4368
4. Cirac, J.I., Zoller, P.: Quantum computations with cold trapped ions. Phys. Rev. Lett. 74, 4091 (1995)
5. Benenti, G., Casati, G., Strini, G.: Principles of quantum computation and information. Basic tools and special topics, vol. 2. World Scientific, Singapore (2007)
6. Steane, A.: The ion trap quantum information processor. Appl. Phys. B 64, 623 (1997)
7. Felloni, S., Leporati, A., Strini, G.: Diagrams of states in quantum information: An illustrative tutorial. Int. J. of Unconv. Comp. (2008) (to appear)
8. Benenti, G., Felloni, S., Strini, G.: Effects of single-qubit quantum noise on entanglement purification. Eur. Phys. J. D 38, 389 (2006)

Fiber Coupled Single Photon Detector with Niobium Superconducting Nanowire

Go Fujii[1,2], Daiji Fukuda[1], Takayuki Numata[1], Akio Yoshizawa[1], Hidemi Tsuchida[1], Shuichiro Inoue[2], and Tatsuya Zama[1]

[1] National Instiute of Advanced Industrial Science and Technolagy,
1-1-1 Umezono, Tsukuba, Japan
go-fujii@aist.go.jp
[2] Institute of Quantum Science, Nihon University
1-8-14 Kanda-Surugadai, Chiyoda-ku, Tokyo, Japan

Abstract. We have fabricated Niobium-based superconducting single photon detector (Nb-SSPD) for realizing high detection efficiency and fast reset time. The Nb-SSPD consisted of a 7 nm-thick and 200 nm-wide Nb meander line and exhibited the critical temperature and critical current density of 4 K and 4.6×10^5 A/cm^2, respectively. The Nb-SSPD was coupled to an optical fiber, and the reset time of 2.5 ns was observed with illumination of laser pulses at 1550 nm wavelength.

Keywords: Superconducting nanowire single photon detector, Niobium, Kinetic inductance

1 Introduction

Single photon detectors are indispensable devices in the field of quantum information and communication. Especially in the quantum key distribution (QKD) based on the optical fiber links, it is important to develop the detector at the telecommunication wavelength (1550-nm). Niobium Nitride-based superconducting single photon detectors (NbN-SSPDs) are one of the promising detectors because of their low jitter [1] and low dark count [2]. NbN-SSPDs have been successfully employed in the QKD, boosting both transmission distances and key generation rates [3]. However, the maximum count rate [4] and detection efficiency (DE) [5] of NbN-SSPDs are not adequate in order to further improve the performance of the QKD. The maximum count rate of NbN-SSPD is limited by the long reset time resulting from a large kinetic inductance (KI). The KI can be reduced because of the smaller penetration depth. Recently, from this point of view, several groups have reported on the development of SSPDs by using NbTiN or Nb thin film [6],[7],[8]. Moreover, a sophisticated technique is required for fabricating high-quality NbN films, and the integration of a complete optical cavity structure to improve an absroptance in the nanowire is difficult. Therefore, we employed Nb as the superconducting material, which can be easily fabricated utilizing a standard Josephson junction process. The KI of the Nb-SSPD can be reduced because of the smaller penetration depth. In this paper, we report the fabrication and characterization of the fiber-coupled Nb-SSPD for 1550-nm wavelength.

A. Sergienko, S. Pascazio, and P. Villoresi (Eds.): QuantumCom 2009, LNICST 36, pp. 220–224, 2010.
© Institute for Computer Sciences, Social-Informatics and Telecommunications Engineering 2010

2 Device Fabrication

Figure 1(a) shows a microscope image of the fabricated Nb-SSPD. Thin Nb nanowires were fabricated with a lift off process on a silicon dioxide/silicon substrate. First, a meander pattern was formed by electron beam lithography, and then the Nb was deposited by dc magnetron sputtering. The meander structure consists of a 200 nm-wide lines covering a 10×10 μm^2 area with a filling factor of 50%. Finally, super-conducting Nb leads with a thickness of 50 nm were fabricated using the lift-off process. Figure 1(b) shows the AFM image of the meander structure with the nanowire thickness of 7 nm.

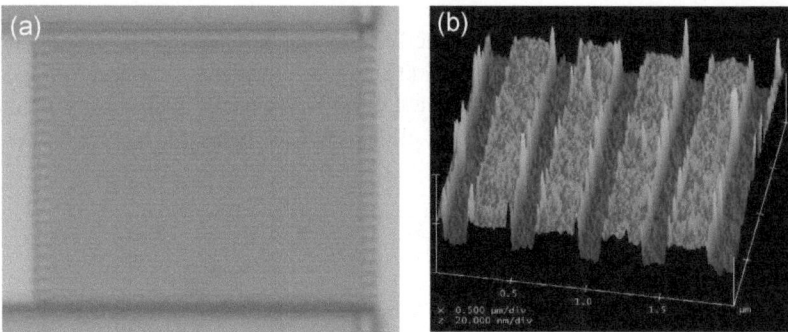

Fig. 1. (a) Micrograph of a meander pattern of 200-nm-wide nanowire covering 10×10 μm^2 area. (b) AFM image of the meander structure.

3 Experiment Setup

The fabricated Nb-SSPD was directly coupled to a single mode optical fiber and was mounted on a cold plate in a liquid-helium refrigerator. A schematic of the readout

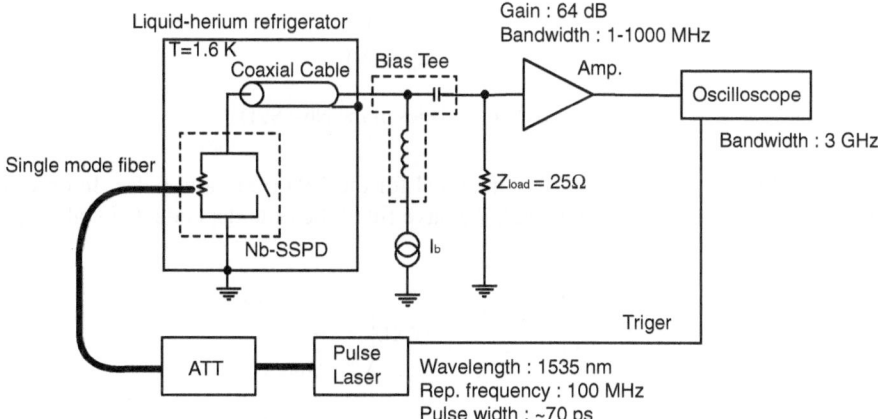

Fig. 2. Schematic of the measurement setup

circuit is shown in Fig. 2. The Nb-SSPD output was connected to a bias-tee placed at room temperature and then, to a 25-Ω parallel shunt resistance. A DC bias current is applied from a low noise voltage source in series with a large bias resistor. The bias-tee output signal was amplified using two cascaded amplifiers with 64 dB total gain, and observed with a 3-GHz bandwidth oscilloscope. We have used a pulsed laser light source at 1535 nm wavelength. The typical pulse width is approximately 70 ps. The repetition frequency of the pulsed laser was 10 MHz. The averaged laser power was strongly attenuated to the level of approximately ten photons in each light pulse.

4 Results

The Nb-SSPD showed the superconductivity at the temperature 4 K. The normal resistance at 10 K of the Nb-SSPD is 75 kΩ. Figure 3 presents a current-voltage (*I-V*) characteristic of the Nb-SSPD at temperature 1.6 K, measured with a constant current mode. From Fig. 3, the critical current I_c of the device is measured to be 5.5 μA.

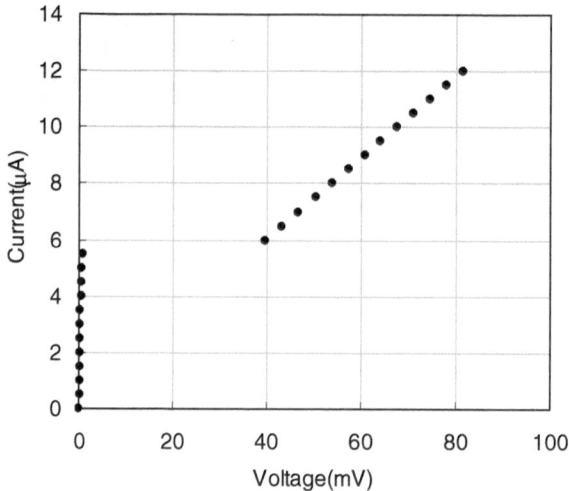

Fig. 3. I-V characteristics of the Nb-SSPD

Figure 4 shows the averaged output signal for the 1550-nm input pulse. In order to obtain the time constant of the signal, we have fitted the time-dependent signal shapes with following formula:

$$f(t) = A\left[\exp\left(-\frac{t-t_0}{\tau_{fall}}\right) - \exp\left(-\frac{t-t_0}{\tau_{rise}}\right)\right],\qquad(1)$$

where A, t_0, τ_{fall}, and τ_{rise} are the pulse height, the laser incident time, the fall time, and the rise time, respectively. The best fitting values of these time constants are $\tau_{rise} = 1.5$

ns and τ_{fall} = 2.5 ns. The response speed of SSPDs are given by following equations, $\tau_{rise} = L_k/(R+R_s)$, $\tau_{fall} = L_k/R_s$, where R is a resistance which derives from hot spot. The value of R = 20 Ω was estimated from the pulse height of the average signal. The KI (L_k) calculated from these equations was approximately 56 nH which is two times lower than that of the NbN-SSPD with the same design.

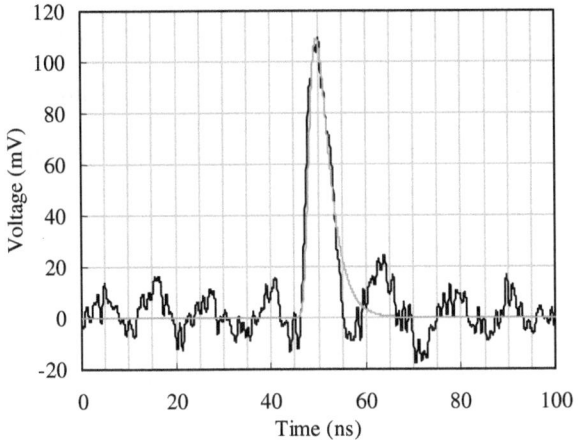

Fig. 4. Averaged signal response of the Nb-SSPD

5   Conclusion

We have reported on the superconducting properties and the signal response of the Nb-SSPDs. The Nb-SSPDs successfully showed the faster response such as τ_{rise} = 1.5 ns and τ_{fall} = 2.5 ns. The superconducting properties of our Nb-SSPD are consistent with the results of other group [8], and will be expected to show much faster reset time with the optimization of the Nb films.

Acknowledgments. This work was supported by KAKENHI (19686010). A part of this work was conducted at the AIST Nano-Processing Facility, supported by "Nanotechnology Network Japan" of the Ministry of Education, Culture, Sports, Science and Technology (MEXT), Japan.

References

1. Gol'tsman, G.N., Okunev, O., Chulkova, G., Lipatov, A., Semenov, A., Smirnov, K., Voronov, B., Dzardanov, A., Williams, C., Sobolewski, R.: Appl. Phys. Lett. 79, 705 (2001)
2. Korneev, A., Matvienko, V., Minaeva, O., Milostnaya, I., Rubtsova, I., Chulkova, G., Smirnov, K., Voronov, V., Gol'tsman, G., Slysz, W., Pearlman, A., Verevkin, A., Sobolewski, R.: IEEE Trans. Appl. Supercond. 15, 571 (2005)

3. Takesue, H., Nam, S.W., Zhang, Q., Hadfield, R.H., Yamamoto, Y.: Nat. Photonics 1, 343 (2007)
4. Kerman, A.J., Dauler, E.A., Yang, J.K.W., Rosfjord, K.M., Anant, V., Berggren, K., Gol'tsman, G.N., Voronov, B.M.: Appl. Phys. Lett. 90, 101110 (2007)
5. Kerman, A.J., Dauler, E.A., Keicher, W.E., Yang, J.K.W., Berggren, K.K., Gol'tsman, G., Voronov, B.: Appl. Phys. Lett. 88, 111116 (2006)
6. Dorenbos, S.N., Reiger, E.M., Perinetti, U., Zwiller, V., Zijlstra, T., Klapwijk, T.M.: Appl. Phys. Lett. 93, 131101 (2008)
7. Miki, S., Takeda, M., Fujiwara, M., Sasaki, M., Otomo, A., Wang, Z.: Appl. Phys. Express 2, 075002 (2009)
8. Annunziata, A.J., Santavicca, D.F., Chudow, J.D., Frunzio, L., Rooks, M.J., Frydman, A., Prober, D.E.: IEEE Trans. Appl. Supercond. 19, 327 (2009)
9. Engel, A., Semenov, A., Hubers, H.W., Il'in, K., Siegel, M.: J. Mod. Opt. 51, 1459 (2004)

Superconducting Nanowire Single-Photon Detectors for Quantum Communication Applications

Chandra M. Natarajan[1], Martin M. Härtig[1], Ryan E. Warburton[1], Gerald S. Buller[1],
Robert H. Hadfield[1], Burm Baek[2], Sae Woo Nam[2], Shigehito Miki[3],
Mikio Fujiwara[3], Masahide Sasaki[3], and Zhen Wang[3]

[1] Heriot-Watt University, Edinburgh, EH14 4AS, United Kingdom
[2] National Institute of Standards and Technology, Boulder, CO 80305, USA
[3] NICT, Tokyo, 184-8795, Japan
cmn10@hw.ac.uk

Abstract. Single-photon detectors are a key enabling technology for optical quantum information processing applications such as quantum key distribution. A new class of single-photon detectors have emerged based on superconducting nanowires. These detectors offer sensitivity at telecommunication wavelengths (1310nm and 1550nm) with low dark counts and excellent timing resolution at an operating temperature of ~4 K. We have integrated four independent fibre-coupled detectors into a practical closed-cycle refrigerator and plan to employ this multichannel detector system in advanced quantum information processing experiments.

Keywords: Superconducting nanowire single photon detectors, SSPD, SNSPD.

1 Introduction

Photons have been widely used in Quantum communication systems due to low noise (or decoherence). High speed single-photon detectors with very low dark counts and high quantum efficiency at the wavelength of interest are ideal requirements for quantum communication systems [1]. At wavelengths below 1000 nm, Silicon Avalanche Photodiodes are the detectors of choice offering high efficiency (~40% at 850 nm) combined with low dark counts (100 Hz). However, for long distance transmission over optical fibre, the ideal operating wavelengths are 1310 nm and 1550 nm. At these wavelengths, Quantum Key Distribution (QKD) experiments have employed InGaAs detectors to achieve distances up to 120km [2]. These detectors typically require cooling to 200 K and offer reduced detection efficiency (~10%). Afterpulsing from InGaAs detectors necessitates long dead times, reducing count rates. Also gated operation is required to reduce the high dark count rate. A new class of single-photon detectors based on superconducting nanowires hold considerable promise for QKD, offering single photon sensitivity at telecom wavelengths, combined with low dark counts, short recovery times (<10 ns) and picosecond timing resolution (<100 ps).

A. Sergienko, S. Pascazio, and P. Villoresi (Eds.): QuantumCom 2009, LNICST 36, pp. 225–232, 2010.
© Institute for Computer Sciences, Social-Informatics and Telecommunications Engineering 2010

2 Detector System Design and Construction

2.1 Detector Technology

The basic operating principle of the superconducting-nanowire single photon detector (SNSPD) is as follows [3]: A 100 nm width wire is defined in a 4 nm thick niobium nitride film. The wire is cooled to ~4 K (below the superconducting transition temperature) and biased close to its critical current. When a visible or infrared photon strikes the wire, the current distribution is perturbed and a short voltage pulse is triggered. To improve the optical coupling efficiency a meander-type geometry is used [4] – the wire is folded back on itself to cover an area up to 20μm x 20μm [5] with a 50% fill factor. The meander pattern is embedded in a 50 Ω coplanar waveguide structure (Fig. 1a). Efforts are also underway to boost the efficiency further using optical cavity designs [6], although these detectors are not yet in widespread use.

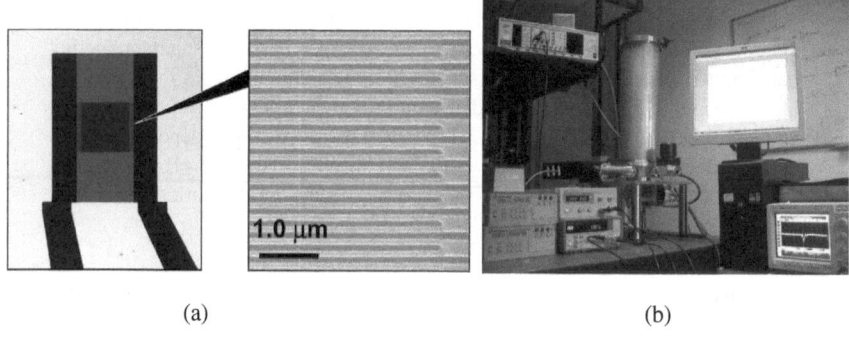

(a)   (b)

Fig. 1. (a) Scanning electron micrograph (SEM) of SNSPD device [5], (b) Superconducting nanowire single-photon detector system based on a closed-cycle Gifford-McMahon refrigerator at Heriot-Watt University, UK

2.2 Closed-Cycle Refrigerator

Over the past decade, commercially available closed-cycle cryocoolers have improved both in terms of attainable base temperature and reliability [7]. The smallest available unit (Gifford-McMahon type), provides 0.1 W of cooling power at 4 K. This runs off a 13A electrical outlet and requires only air cooling. This is sufficient to cool multiple superconducting detector packages [8]. A passive standoff stage with a long thermal time constant is implemented for temperature stability, with a modest increase in base temperature (2.9 K versus the cold head base temperature of 2.5 K) [8]. Several versions of this system have now been constructed by the authors and are now in use in the USA, Japan and Europe. The current version at Heriot-Watt University, UK has four detectors channels (Fig. 1b), and is capable of accommodating up to eight detector channels – required for advanced quantum information processing experiments.

2.3 Optical Alignment

Photons are fibre-coupled to the meander nanowire devices. Proper alignment is essential for the efficiency of the system as in any other optics setup. The challenge lies in aligning the device at room temperature and subsequently cooling the package down to ~3K without affecting the alignment. It is important that two parameters, firstly the separation between the fibre ferrule and the device (d) and secondly the x-y alignment, are precisely controlled and maintained over repeated thermal cycling in order to avoid the need for realignment after every thermal cycle.

In our setup (Fig 2.a) white light interferometer [9] is used to study and control the separation between the fibre ferrule and the device. Light from a broad spectrum light source is launched through a fibre coupler (50:50 @ 700 nm) that splits it into two halves, one of which is dumped and the other half is coupled to the device using a polished ferrule. A fraction (~4%) of the light is reflected back from the polished end of the ferrule, 33% of the remaining light is reflected back from the device surface and the rest of the light gets transmitted through the device. The two reflected signals interfere at the coupler and a spectrometer (350-1000nm) is used to study the interference intensity profile over the broad spectrum. The separation (d) can be retrieved from the frequency of the interfered signal using the equation (Fig. 2b)

$$\Delta\lambda = \frac{\lambda^2}{2d}. \tag{1}$$

While packaging the device and the ferrule, precisely machined shims are introduced to vary d. At room temperature, d is set to ~50µm and as the package is cooled down to ~3K, it contracts and d reduces to about ~25µm (Fig. 2b).

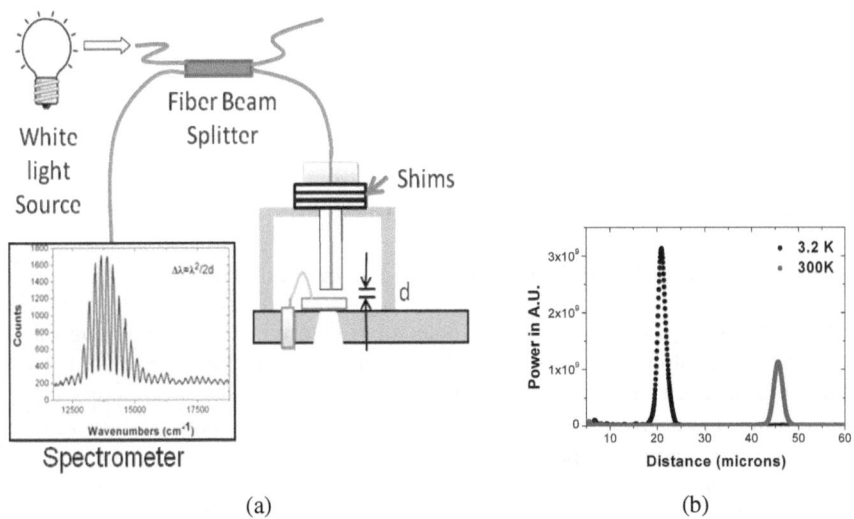

Fig. 2. (a) White light interferometry setup, (b) distance (d) at 300K and 3.2K

In order to achieve accurate x-y alignment, the sample holder and the ferrule holder are secured onto a platform and a sub-micron x-y alignment stage respectively (Fig. 3). A 1550nm beam from a diode laser is sent through the fibre ferrule, the beam emanating from the ferrule is transmitted through the device and is viewed using a long working distance near-infrared objective (50x). Acceptable alignment is achieved via careful positioning of the x-y alignment. Thereafter the ferrule holder and sample holder are fixed together using screws. The alignment is performed at room temperature and the package is attached to the standoff stage in the cryostat, which is maintained at about 3.2K. This robust packaging scheme enables the alignment to be preserved over repetitive thermal cycling between room temperature and cryogenic temperatures.

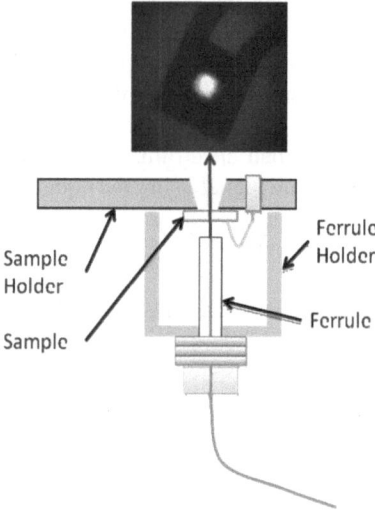

Fig. 3. Fibre ferrule – device alignment setup

3   Practical Detector System Performance

3.1   System Detection Efficiency at 830nm, 1310nm and 1550nm

System detection efficiency (η) of the detector is determined by calculating the ratio of number of photons detected to number of photons sent into the system, without subtracting the losses in the optical path inside the system. The schematic setup used to determine η at a given wavelength for a fibre coupled detector, is shown in figure 4. A laser diode, gain switched by a pulse pattern generator (PPG), generates optical pulses at the drive frequency f, which are attenuated using two programmable digital attenuators. Each attenuator is calibrated upto 60dB, therefore resulting in a range of 0-120dB attenuation when used in series. The device is current biased at various bias points and the corresponding dark counts (D) of the device are measured. An average number of photons per pulse (μ) is controlled by varying the output power of the diode and the attenuation. The incident photon flux is varied and the corresponding response from the device is recorded as R counts per second. η is extracted by fitting the data to the following curve [8]:

$$R = D + f\left(1 - e^{-\eta\mu}\right).$$
(2)

η improves at shorter wavelengths due to higher photon energy (Fig. 5a). As the bias current approaches the critical current of the nanowire, the device become more sensitive, picking up electronic and thermal fluctuations resulting in higher dark count rates. The device can be operated at an appropriate bias point depending on the signal-to-noise and detection efficiency requirements of the experiment.

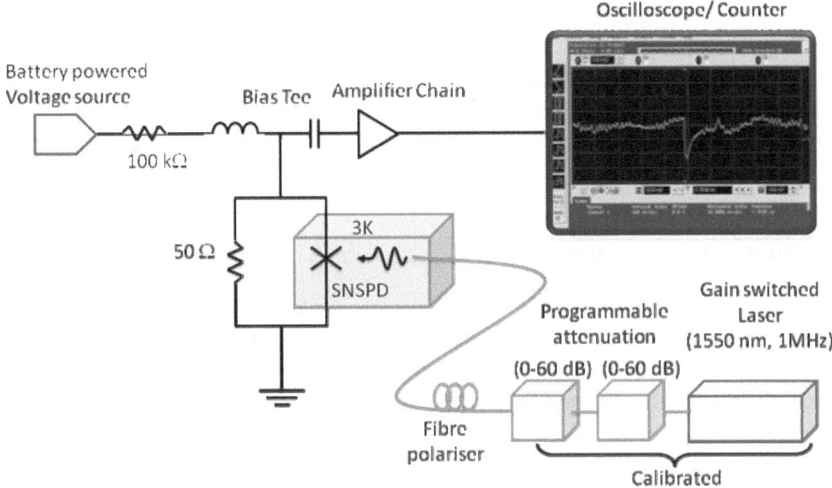

Fig. 4. System detection efficiency characterization setup

3.2 Polarization Dependence

The meander SNSPD device is observed to possess noticeable polarization sensitivity. It has been reported [10] that the count rate from the detector displays a maximum and a minimum value depending on the orientation of the electric field being either parallel or perpendicular to the orientation of the nanowire. In our setup, the polarization dependence is observed by varying the polarization of the incident photons using a fibre polarizer. The fibre polarizers are adjusted either to high or low count rate and the η is subsequently measured. The experimental ratios of the high count rate η_{high} to that of low count rate η_{low} are well in accordance with results from simulations (Fig. 5b). However, the measurements from the 830nm experiment were taken using the same single mode telecom fibre. This could affect the polarization control of the 830nm photons owing to the possible multimode nature of the fibre below 1200 nm.

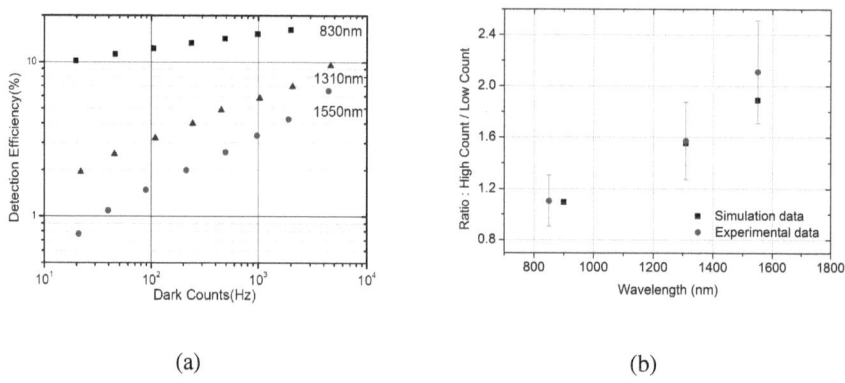

(a)   (b)

Fig. 5. (a) Practical SNSPD system detection efficiency η versus ungated dark count rate D for a fibre-coupled detector at wavelengths 830nm, 1310nm and 1550nm, (b) Comparison of experimentally-determined polarization sensitivity (η_{high}/η_{low}) versus wavelength, plotted with simulations

3.3 Detector Timing Jitter

In QKD experiments, timing jitter of the detectors plays a major role in determining the total key transmission rate [11]. Jitter is usually given by the full width half maximum (FWHM) of the histogram of the detection time of photons. In our setup (Fig. 6), a picosecond diode laser (1550nm) driven by a pulse/pattern generator is used to generate photon pulses. The pulses are attenuated to single photon per pulse regime and coupled to the SNSPD. Sync pulses from the laser driver and output pulses from the detector are used to start and stop the time-correlated single-photon counting (TCSPC) card respectively. The TCSPC card generates a histogram of the start-stop intervals (Fig. 7). The data was recorded with a TCSPC card with a timing resolution of 10ps and the binning width of 4ps. The histogram has a Gaussian profile and a FWHM of ~70ps.

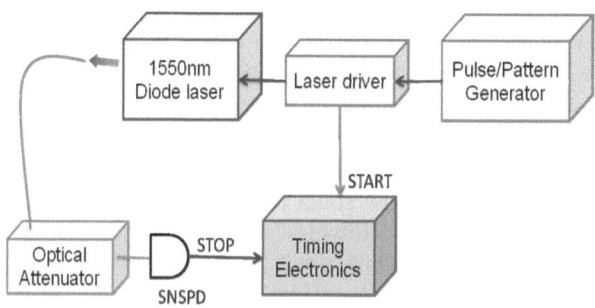

Fig. 6. Instrument response curve of a fibre-coupled detector (1550nm laser diode excitation)

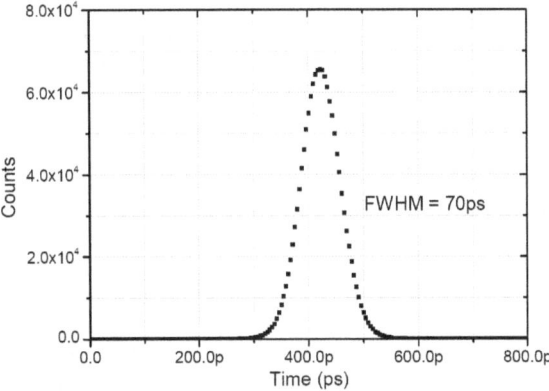

Fig. 7. Instrument response curve of a fibre-coupled SNSPD (1550nm picosecond laser diode excitation)

4 Conclusion and Outlook

Superconducting nanowire single-photon detectors (SNSPDs) are a promising new technology for telecom wavelength quantum key distribution. The detectors can be implemented into closed-cycle refrigerator systems for practical operation. Record bit rates and transmission distances have been achieved in QKD experiments exploiting this detector technology [11-15]. In the near future, there are plans to employ these devices in various experiments such as (i) characterization of single photon sources at telecom wavelengths [16], (ii) long distance, high bit rate QKD experiments and (iii) to operate these devices with quantum waveguide circuits [17]. Considerable improvements in practical detector performance are anticipated [6] and we expect these detectors to play a significant future in the field of quantum information processing.

Acknowledgements

CMN and RHH acknowledge support from the UK Engineering and Physical Sciences Research Council. RHH is also support by a Royal Society of London University Research Fellowship. The authors thank V. Anant (MIT) for providing simulation results in support of the SNSPD polarization sensitivity measurements.

References

1. Gisin, N., Ribordy, G., Tittel, W., Zbinden, H.: Quantum cryptography. Rev. Mod. Phys. 74, 145–196 (2002)
2. Gobby, C., Yuan, Z.L., Shields, A.J.: Quantumkey distribution over 122 km of standard telecom fiber. Appl. Phys. Lett. 84, 3762 (2004)
3. Gol'tsman, G.N., Okunev, O., Chulkova, G., Lipatov, A., Semenov, A., Smirnov, K., Voronov, B., Dzardanov, A., Williams, C., Sobolewski, R.: Picosecond superconducting single-photon optical detector. Appl. Phys. Lett. 79, 705–707 (2001)

4. Verevkin, A., Zhang, J., Sobolewski, R., Lipatov, A., Okunev, O., Chulkova, G., Korneev, A., Smirnov, K., Gol'tsman, G.N., Semenov, A.: Detection efficiency of large-active-area NbN single-photon superconducting detectors in the ultraviolet to near-infrared range. Applied Physics Letters 80(25), 4687–4689 (2002)
5. Miki, S., Fujiwara, M., Sasaki, M., Baek, B., Miller, A.J., Hadfield, R.H., Nam, S.W., Wang, Z.: Large sensitive-area NbN nanowire superconducting single-photon detectors fabricated on single-crystal MgO substrates. Appl. Phys. Lett. 92, 061116–061118 (2008)
6. Rosfjord, K.M., Yang, J.K.W., Dauler, E.A., Kerman, A.J., Anant, V., Voronov, B.M., Gol'tsman, G.N., Berggren, K.K.: Nanowire Single-photon detector with an integrated optical cavity and anti-reflection coating. Opt. Express 14, 527–534 (2006)
7. Radebaugh, R.: Refrigeration for Superconductors. Proceedings of the IEEE 92(10), 1719–1734 (2004)
8. Hadfield, R.H., Stevens, M.J., Gruber, S.S., Miller, A.J., Schwall, R.E., Mirin, R.P., Nam, S.W.: Single photon source characterization with a superconducting single photon detector. Opt. Express 13, 10846–10853 (2005)
9. Schmitt, J.M.: Optical Coherence Tomography, A Review. IEEE J. Select. Topics Quantum Electron. 5, 1205–1215 (1999)
10. Anant, V., Kerman, A., Dauler, E., Yang, J., Rosfjord, K., Berggren, K.: Optical Properties of Superconducting Nanowire Single-Photon Detectors. Opt. Express 16, 10750–10761 (2008)
11. Baek, B., Ma, L., Mink, A., Tang, X., Nam, S.W.: Detector performance in long-distance quantum key distribution using superconducting nanowire single-photon detectors. In: Proc. of SPIE, vol. 7320, p. 73200D (2009)
12. Collins, R.J., Hadfield, R.H., Fernandez, V., Nam, S.W., Buller, G.S.: Low timing jitter detector for gigahertz quantum key distribution. Electronics Letters 43(2) (2007)
13. Takesue, H., Nam, S.W., Zhang, Q., Hadfield, R.H., Honjo, T., Tamaki, K., Yamamoto, Y.: Quantum key distribution over 40 dB channel loss using superconducting single-photon detectors. Nature Photonics 1(6), 343 (2007)
14. Honjo, T., Nam, S.W., Takesue, H., Zhang, Q., Kamada, H., Nishida, Y., Tadanaga, O., Asobe, M., Baek, B., Hadfield, R.H., Miki, S., Fujiwara, M., Sasaki, M., Wang, Z., Inoue, K., Yamamoto, Y.: Long distance entanglement-based quantum key distribution over optical fiber. Optics Express 16(23), 19119 (2008)
15. Rosenberg, D., Peterson, C.G., Harrington, J., Rice, P.R., Dallman, N., Tyagi, K.T., McCabe, K.P., Nam, S.W., Baek, B., Hadfield, R.H., Hughes, R.J., Nordholt, J.E.: Practical long distance quantum key distribution system using decoy levels. New Journal of Physics 11, 045009 (2009)
16. Ward, M.B., Karimov, O.Z., Unitt, D.C., Yuan, Z.L., See, P., Gevaux, D.G., Shields, A.J., Atkinson, P., Ritchie, D.A.: On-demand single-photon source for 1.3 μm telecom fiber. Appl. Phys. Lett. 86, 201111 (2005)
17. Politi, A., Cryan, M.J., Rarity, J.G., Yu, S., O'Brien, J.L.: Silicon-on-silicon waveguide Quantum circuits. Science 320, 646–649 (2008)

Interferometric Technique for Density Matrix Reconstruction by On/Off Detectors

G. Brida[1], M. Genovese[1], M. Gramegna[1], P. Traina[1], L. Ciavarella[1],
S. Olivares[2], and M.G.A. Paris[3]

[1] INRIM; strada delle Cacce 91, 10135 Torino, Italy
[2] CNISM Udr Milano, I-20133 Milano, Italia
Dipartimento di Fisica dell'Università di Milano, I-20133 Milano, Italia
[3] Dipartimento di Fisica dell'Università di Milano, I-20133 Milano, Italia
CNISM, Udr Milano, I-20133 Milano, Italia
ISI Foundation, I-10133 Torino, Italia
p.traina@inrim.it

Abstract. The density matrix provides the most complete description of a quantum optical state and a scheme addressed to its reconstruction for arbitrary sources can be fundamental for several applications, ranging from quantum information to the foundations of quantum mechanics and quantum optics. We demonstrate an innovative state reconstruction technique, which provides the density matrix of a field mode and requires only avalanche photodetectors (coupled with phase modulation with respect to a local oscillator), without any phase or amplitude discrimination power. It represents an alternative of simpler implementation to quantum homodyne tomography.

Keywords: Nanomechanical quantum systems and their interaction with nonclassical light.

1 Introduction

Gaining knowledge on the density matrix of an arbitrary quantum state is fundamental for several applications, ranging from quantum information [1] to the foundations of quantum mechanics [2] and quantum optics [3, 4]. Quantum tomography [5, 6, 7], which requires the measurement of a continuous set of field quadrature, can deliver such kind of information, but since it is based on homodyne detection, it turns out to be rather a complicate technique when applied to short pulses in photo-counting regime. Since also reliable photo-counting detectors are not available at the moment[8, 9, 10, 11, 12, 13] (in the sense that photo-detectors that can operate as photon counters are rather rare and affected either by a low quantum efficiency or require cryogenic conditions), various theoretical studies [14, 15, 16, 17] have been addressed to obtain a reconstruction of the (diagonal) elements of the density matrix by using the information that one can achieve by means of realistic detectors, more precisely, the frequencies of the *off* events at different quantum efficiencies of the detector. The technique

A. Sergienko, S. Pascazio, and P. Villoresi (Eds.): QuantumCom 2009, LNICST 36, pp. 233–241, 2010.

found a favourable experimental test in [18, 19, 20, 21, 22, 23], where a very satisfactory reconstruction of the statistics of mono-partite and bi-partite quantum optical states was obtained by using on/off detectors following the method of Ref.[16, 17].

Furthermore, full state reconstruction is possible by a suitable processing of the data obtained with on/off detectors, such as single-photon avalanche photodiodes coupled with some phase modulation. This modulation of the signal, corresponds to the application of a coherent displacement by mixing the state under investigation with a known coherent reference in a Mach-Zehnder interferometer [24], and measuring or reconstructing the photon distribution for different values of the modulation one has enough information for a full state reconstruction of the original state[25].

In this paper, we present an experimental application of this method addressed to the full reconstruction of the density matrix for two possible optical states (coherent or pseudo-thermal) by using on/off detection coupled to phase modulation. The work is structured as follows: the first section reviews the method to reconstruct the density matrix and gives a simple description of the experiment. In the second section we explain in details our setup. In the third section we present the results concerning a coherent and a pseudo-thermal signal. The fourth section closes the paper with some concluding remarks.

2   Reconstruction Method

In this section we briefly introduce the density matrix reconstruction method proposed in Ref. [25]. Let ρ be the density matrix of the state we want to reconstruct. The reconstruction is based on the measurement of the photon distribution of the displaced state $\rho(\beta) \equiv D(\beta)\,\rho\,D^{\dagger}(\beta)$, $\beta = |\beta|\,e^{i\varphi}$, for fixed $|\beta|$ and different values of φ, i.e., $p_n(\varphi) = \langle n|\rho(\beta)|n\rangle$. The density matrix elements of the reconstructed state ϱ are then given by [25]:

$$\langle m + s|\varrho|m\rangle = \frac{1}{\mathcal{N}_\varphi} \sum_{l=1}^{\mathcal{N}_\varphi} \sum_{n=0}^{\bar{n}} F_{m,n}^{(s)}\, e^{is\varphi_l}\, p_n(\varphi_l), \tag{1}$$

where $\mathcal{N}_\varphi$ is the number of phase φ_l considered, $l = 1, \ldots, \mathcal{N}_\varphi$, $\bar{n}$ is such that $\varrho_{kh} \equiv \langle h|\varrho|k\rangle$ can be neglected if $h, k > \bar{n}$, and $F_{m,n}^{(s)}$ are functions of $|\beta|$ whose analytic expression is reported in Ref. [25].

The challenging task is now the measurement of the photon statistics (*i. e.* the amplitudes of the diagonal elements in the density matrix expressed in Fock basis), since this for a direct measurement requires number-resolving photodetectors. An alternative method has been given so far in Ref. [16, 17], where it was shown that the photon distribution p_n for a quantum optical state can be retrieved via on/off detectors. Assuming negligible dark counts, the *off* probability is related to the p_n's thanks to the relation:

$$P_0(\eta) = \sum_n (1 - \eta)^n p_n, \tag{2}$$

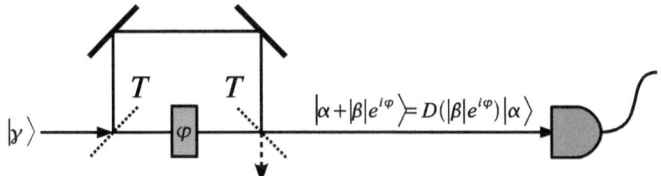

Fig. 1. Simplified sketch of the experiment aimed to reconstruct the density matrix of a coherent state $|\alpha\rangle$. A signal excited to the coherent state $|\gamma\rangle$, γ real, enters a Mach-Zehnder interferometer with two beam splitters with transmissivity T. The output field can be written as $D(|\beta|e^{i\varphi})|\alpha\rangle$, where $|\beta| = T\gamma$, $\alpha = R\gamma$ and φ is the phase difference between the signals added by the interferometer. The other output is discharged.

η being the quantum efficiency of the detector. Measuring a given signal for different quantum efficiencies η_ν, $\nu = 1, \ldots, N$, starting form Eq. 2 we obtain a statistical model for the positive parameters p_n, which can be solved by *Expectation Maximization Algorithm* (EM) [16, 17].

The additional phase information needed by Eq. (1) can be obtained following the interferometric scheme proposed in [25]. As a first step toward a test of that, here we analyze the reconstruction of the density matrix of a coherent state $|\alpha\rangle$ and, without loss of generality, we assume the amplitude α as real. Fig. 1 shows a simplified version of the scheme used in our experiment. The amplitude of the coherent state and the one of the displacement can be also modified by reducing the power along one of the optical paths. This scheme allows to vary φ by suitably tuning the interferometer and, then, the $p_n(\varphi)$ in Eq. (1) are finally retrieved by on/off measurements onto the output field and EM algorithm.

3   Experimental Setup

The setup for the implementation of our experiment for the coherent state is depicted in Fig. 2.

The output of a He-Ne laser ($\lambda = 632.8$ nm) is lowered to single photon regime by neutral filters. The spatial profile of the beam is purified by a spatial filter realized by two converging lenses and a 100 μm diameter wide pinhole. The laser cavity is also preserved by backreflections, which may cause instability by means of an optical isolator consisting in a Faraday rotator between two orthogonal polarizers.

After a beam-splitter, part of the beam is addressed to a control detector in order to monitor the laser amplitude fluctuations, while the remaining part, which is the signal to be reconstructed, is sent to the interferometer, its main structure consisting in a single invar block custom designed and developed at INRIM. Fig. 3 shows the interferometer itself and Fig. 4 reports on the expected stability of the system against thermal noise as simulated in Ansys software environment. The interferometer also features a piezo-movement system, which

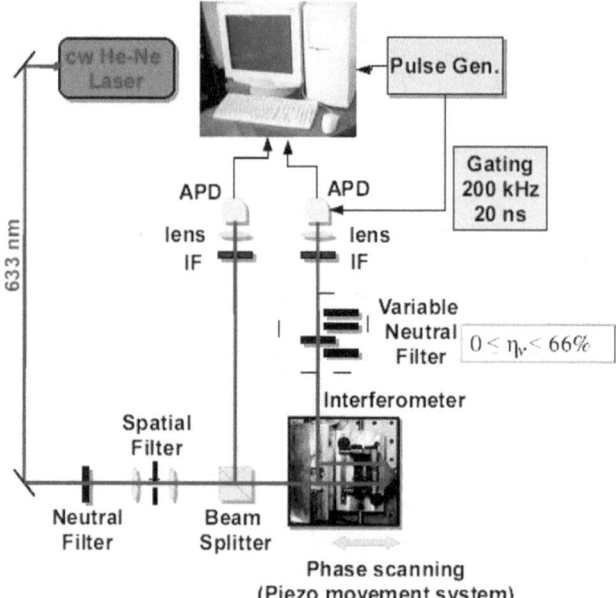

Fig. 2. Setup for the reconstruction of the density matrix for a coherent state. The emission of a He-Ne laser ($\lambda = 632.8$ nm) is lowered to single photon regime by neutral filters. A spatial filter realized by two converging lenses and a $100\mu m$ diameter-wide pinhole purifies the shape of the beam and allows to select a single spatial mode. A beam-splitter reflects part of the beam to a control detector used to monitor the laser amplitude fluctuations, while the remaining part, which is the signal to be reconstructed, is sent to the interferometer. The phase between the "short" and "long" paths in the interferometer can be changed by driving the position of the reflecting prism by means of a piezo-movement system. A set of variable neutral filters allows to collect photons for different values of the quantum efficiency. The detectors used are Perkin-Elmer Single Photon Avalanche Photodiode(SPCM-AQR) gated by a 20 ns wide time window with (repetition rate = 200 kHz). A single run consists of 5 repetitions of 4 seconds acquisitions and events are recorded by a NI-6602 PCI counting module.

allows to scan the phase between the "short" and "long" paths by driving the position of the reflecting prism with nanometric resolution and high stability.

For each position of the prism, the "no-click" probabilities are collected for different sets of neutral filters in front a Perkin-Elmer Single Photon Avalanche Photodiode (SPCM-AQR).

The detector is gated by a 20 ns wide time window with a repetition rate of 200 kHz. In order to obtain a reasonable statistics, a single run consists of 5 repetitions of 4 seconds acquisitions. Events are recorded by a NI-6602 PCI counting module.

In a second acquisition the signal to be reconstructed was a pseudo-thermal state simulated by inserting an Arecchi's rotating glass in the signal arm of the interferometer.

Fig. 3. Side view (left) and top view (right) of the realized interferometer

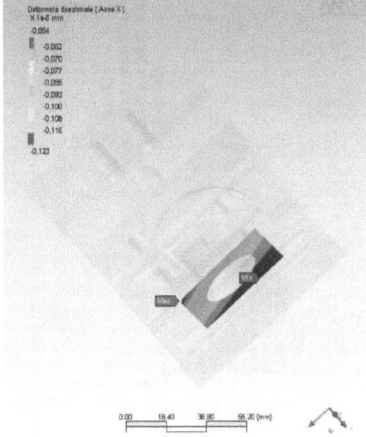

Fig. 4. Ansys simulation of the longitudinal stability of the systems for different external conditions and materials. The simulation shows an expected maximum expansion for the system below the order of the nanometer.

4   Experimental Results

In Fig. 5 the interference fringes at the output of the interferometer are shown. We have chosen $\mathcal{N}_\varphi = 11$ different phases and the average energy E has been obtained from the reconstructed $p_n(\varphi)$. Even if the visibility is not very high, for coherent signals this is not an issue, since this can be controlled by rescaling the experimental data without affecting the nature of the reconstructed density matrix.

In the bottom part of Fig. 6 we report the reconstructed density matrix in the Fock representation (diagonal and subdiagonal) for a coherent state with real amplitude $\alpha \simeq 1.8$ and a thermal state with average number of photons equal to $n_{th} \simeq 1.4$.

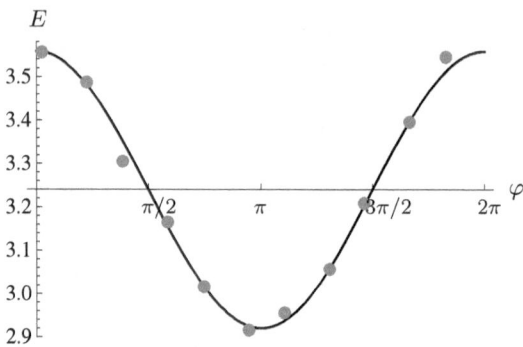

Fig. 5. Interference fringes: experimental data (gray disks) and fit (solid line)

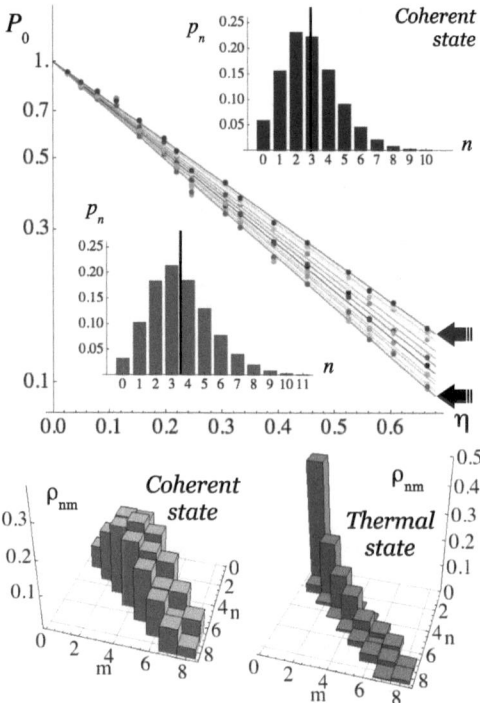

Fig. 6. (Color online) Upper plot: *off* frequencies as a function of the quantum efficiency when the signal is a coherent state and for different phase-shifts. The two insets show the reconstructed photon distributions for the two phase-modulated versions of the signal corresponding to maximum and minimum visibility at the output of the Mach-Zehnder interferometer. The vertical black bars denote the mean value of the photon number for the two distributions, $\langle a^\dagger a \rangle = 3.5$ and $\langle a^\dagger a \rangle = 2.9$. lower left plot: (left) the corresponding reconstructed density matrix in the Fock representation (diagonal and subdiagonal elements). lower right plot: density matrix for the signal excited in a thermal state.

As it is apparent from the plots the off-diagonal elements are correctly re-produced in both cases despite the limited visibility. Here the raw data are frequencies of the *off* event as a function of the detector efficiency, taken at dif-ferent phase modulations, ϕ, whereas the intermediate step corresponds to the reconstruction of the photon distribution for the phase-modulated signals. In our experiments we used $|\alpha|^2 = 0.01$ for the coherent state and $|\alpha|^2 = 1.77$ for the thermal state. The use of a larger $\mathcal{N}_\varphi$ would allow the reliable reconstruction of far off-diagonal elements, which is not possible in the present configuration. In the insets of Fig. 6 we report the reconstructed distributions at the minimum and maximum of the interference fringes.

The evaluation of uncertainties on the reconstructed states involves the contri-butions of experimental fluctuations of on/off frequencies as well as the statisti-cal fluctuations connected with photon-number reconstruction. For our purposes this implies that neither large displacement amplitudes may be employed, nor states with large field and/or energy may be reliably reconstructed, although the mean values of the fields measured here are definitely non-negligible. For the regime of weak field or low energy, the density matrix can be safely evaluated from experimental data (see Fig. 7).

Notice also that any uncertainty in the nominal efficiency η_{max} of the involved photodetectors does not substantially affect the reconstruction [16].

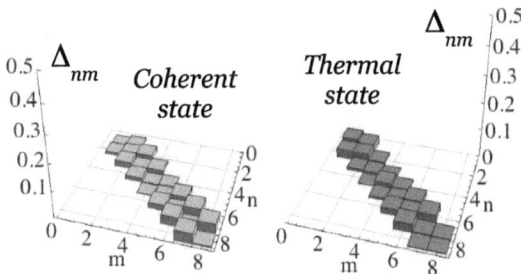

Fig. 7. (Color online) Difference $\Delta_{nm} = |\varrho_{nm}^{exp} - \varrho_{nm}^{th}|$ between reconstructed and theo-retical values of the density matrix elements for the coherent (left) and thermal (right) states used in our experiments

We point out that our scheme allows to estimate the uncertainty in real time, that it is suitable both for pure and mixed states and that statistic information is obtained by sampling a discrete matrix rather than measuring a continuous distribution in phase space as homodyne quantum tomography schemes.

In order to improve the efficiency of the reconstruction scheme, the next efforts will be aimed in the directions of optimizing the visibility of the interference fringes and of extending the reconstruction to elements in density matrix farther from the diagonal by increasing the number of phase steps.

5   Concluding Remarks

We demonstrated the partial reconstruction of the density matrix both of a coherent state and a pseudo-thermal one following Ref. [25]. Our results show the validity of the method and prompt to further efforts in order to provide complete reconstruction of the density matrix for arbitrary quantum optical sources by using on/off detection coupled to phase measurements.

Acknowledgments

This work has been supported by Compagnia di San Paolo Foundation, EU project QuCandela, by Regione Piemonte (E14) and by the CNR-CNISM convention.

References

[1] Bouwmeester, D., Ekert, A.K., Zeilinger, A.: The Physics of Quantum Information: Quantum Cryptography, Quantum Teleportation, Quantum Computation. Springer, New York (2000)

[2] Genovese, M.: Physics Reports 413(6) (2005)

[3] Perina, J., Hradil, Z., Jurco, B.: Quantum Optics and Fundamental Physics. Kluwer, Dordrecht (1994)

[4] Mandel, L., Wolf, E.: Optical Coherence and Quantum Optics. Cambridge Univ. Press, Cambridge (1995)

[5] Munroe, M., Boggavarapu, D., Anderson, M.E., Raymer, M.G.: Phys. Rev. A 52, 924–927 (1995)

[6] Zhang, Y., Kasai, K., Watanabe, M.: Opt. Lett. 27, 1244–1246 (2002)

[7] Raymer, M., Beck, M.: Quantum States Estimation. Lect. Not. Phys. 649, 235–295 (2004)

[8] Zambra, G., Bondani, M.: Rev. Sci. Instrum. 75, 2762–2765 (2004)

[9] Kim, J., Takeuchi, S., Yamamoto, Y.: Appl. Phys. Lett. 74, 902–904 (1999)

[10] Peacock, A., Verhoeve, P., Rando, N., van Dordrecht, A., Taylor, B.G., Erd, C., Perryman, M.A.C., Venn, R., Howlett, J., Goldie, D.J., Lumley, J., Wallis, M.: Nature 381, 135–137 (1996)

[11] Zappa, F., Lacaita, A.L., Cova, S.D., Lovati, P.: Opt. Eng. 35, 938–945 (1996)

[12] Achilles, D., Silberhorn, C., Liwa, C., Banaszek, K., Walmsley, I.A.: Opt. Lett. 28, 2387–2389 (2003)

[13] Di Giuseppe, G., Sergienko, A.V., Saleh, B.E.A., Teich, M.C.: Quantum Information and Computation. In: Proc. SPIE, vol. 5105, pp. 39–50 (2003)

[14] Mogilevtsev, D.: Opt. Comm. 156, 307–310 (1998)

[15] Mogilevtsev, D.: Acta Phys. Slov. 49, 743–748 (1999)

[16] Rossi, A.R., Olivares, S., Paris, M.G.A.: Phys. Rev. A 70, 055801 (2004)

[17] Rossi, A.R., Paris, M.G.A.: E. Phys. Jour. D 32, 223–226 (2005)

[18] Zambra, G., Andreoni, A., Bondani, M., Gramegna, M., Genovese, M., Brida, G., Rossi, A., Paris, M.G.A.: Phys. Rev. Lett. 95, 063602/1-4 (2005)

[19] Gramegna, M., Genovese, M., Brida, G., Bondani, M., Zambra, G., Andreoni, A., Rossi, A.R., Paris, M.G.A.: Laser Physics. 16, 385–392 (2006)

[20] Brida, G., Genovese, M., Gramegna, M., Paris, M.G.A., Predazzi, E., Cagliero, E.: Open Systems & Information Dynamics 13, 333–341 (2006)

[21] Brida, G., Genovese, M., Piacentini, F., Paris, M.G.A.: Optics Letters 31, 3508 (2006)

[22] Brida, G., Genovese, M., Paris, M.G.A., Piacentini, F., Predazzi, E., Vallauri, E.: Optics and Spectroscopy 103, 95 (2007)

[23] Brida, G., Genovese, M., Meda, A., Olivares, S., Paris, M.G.A., Piacentini, F.: Journ. Mod. Opt. 56, 196–200 (2009)

[24] Paris, M.G.A.: Phys. Lett. A 217, 78 (1996)

[25] Opatrný, T., Welsh, D.G.: Phys. Rev. A 55, 1462 (1997)

Simulating BB84 Protocol in Dephasing Qubit Channel

Xiao-yu Chen

Zhejiang Gongshang University, Hangzhou 310013, China
xychen@zjsu.edu.cn

Abstract. BB84 protocol of quantum key distribution had been proved to be absolutely secure. We simulate the rate of secure key distribution in dephasing channel on a classical computer with the method of event-by-event simulation. Theoretically, the private classical capacity of dephasing channel can be obtained in principle, since the channel is degradable. We give the formula of capacity with respect to the dephasing coefficient. The simulation meets the theory well.

Keywords: Dephasing channel, quantum capacity, BB84 protocol, degradable.

1 Introduction

Quantum communications on fiber or free space come into a reality today. The rate of transmitting quantum information faithfully down noisy channel is upper bounded by the quantum capacity of the channel. Quantum capacity is closely related to the private classical capacity in quantum cryptograph. For a given protocol such as BB84, non-ideal factors such as the control precision in communication process, noise in the transmission and dark counting at the receiving end can be modelled as the effects of a channel.

Unfortunately, quantum capacity exhibits a kind of nonadditivity that makes it extremely hard to deal with [1]. The quantum capacity of the most common quantum channel, the depolarizing channel, is not obtained yet. It requires a regulation process of the maximization of the coherent information that can hopelessly be solved with brute force. It has been proved that once the channel is degradable, single letter formula of the quantum capacity is available [2]. Thus for some special classes of channels, quantum capacity can be obtained analytically or numerically. Moreover, quantum capacity is equal to the private classical capacity for degradable channel [3]. The first example with calculable quantum capacity is quantum erasure channel[4]. Other examples are dephasing qubit channel[2], amplitude damping qubit channel[5], and continuous variable lossy channel[6], where the channels are either degradable or anti-degradable[7].

To provide a guideline for experimenters with the communication rate of faithfully transmitting of quantum information, one need the quantum capacity theory. One one side, quantum capacity usually can not be calculated exactly. One

A. Sergienko, S. Pascazio, and P. Villoresi (Eds.): QuantumCom 2009, LNICST 36, pp. 242–248, 2010.

the other hand, practical quantum optical communication systems are costly and systematical experimental data is not available. To fill the gap between the theory and the experiment, there is the method of classical computer simulation.

We in this paper will present event by event simulation of BB84 protocol in dephasing qubit channel. The reasons to simulate such a system are: (i) phase decoherence is one of the main obstacles in quantum information transmission, it is stronger than amplitude decay for many systems, the decay time of dephasing is 10-100 times shorter than the decay time of amplitude damping[8], (ii) protocols such as BB84 neglect the missing photons in the late treatment, so there is no need to treat amplitude damping separately, (iii) quantum capacity formula can be analytically worked out for dephasing qubit channel, providing a clear upper bound to the faithfully quantum information transmitting rate. (iv)BB84 protocol is well studied and widely used one. Its security has been proved [9]. Thus the properties of the simulation can be verified with theoretical and experimental results. The parameters such as length of the date package can be determined for the simulation.

2   Quanutm Capacity of Qubit Phase Damping Channel

Quantum capacity is one of the main issues in quantum information theory. It is concerned with the transmission ability of unknown quantum state on a given quantum channel. The critical quantity involved in the quantum capacity is the coherent information (CI) $I_c(\sigma, \mathcal{E}) = S(\mathcal{E}(\sigma)) - S(\sigma^{QR'})$ [10] [11]. Here $S(\varrho) = -\text{Tr}\varrho \log_2 \varrho$ is the von Neumann entropy, σ is the input state, the application of the channel $\mathcal{E}$ results the output state $\mathcal{E}(\sigma)$; $\sigma^{QR'} = (\mathcal{E} \otimes \mathbf{I})(|\psi\rangle \langle\psi|)$, with R referred to the 'reference' system[10] (the system under process is Q system with annihilation and creation operators a and $a^\dagger$, we denote σ^Q as σ for simplicity), $|\psi\rangle$ is the purification of σ. The quantum channel capacity is[12][13][14]

$$Q = \lim_{n\to\infty} \sup_{\sigma_n} \frac{1}{n} I_c(\sigma_n, \mathcal{E}^{\otimes n}). \tag{1}$$

Since phase damping channel (or dephasing channel) is degradable, the quantum capacity can be expressed with single letter formula,

$$Q = \sup_{\sigma} I_c(\sigma, \mathcal{E}_p), \tag{2}$$

where $\mathcal{E}_p$ refers to the superoperator of phase damping channel.

In the computational basis $|0\rangle, |1\rangle$ of the Hilbert space $\mathcal{H}_Q$, the most general qubit state input can be parameterized as follows

$$\rho = \begin{bmatrix} 1-p, \alpha^* \\ \alpha, \quad p \end{bmatrix}, \tag{3}$$

where $p \in [0,1]$ is the probability associated with the state $|1\rangle$ and $|\alpha| = \sqrt{p(1-p)}$ is a coherent term. The purification of ρ can be the state

$$|\Psi\rangle = \sqrt{1-p}|0\rangle \otimes |R_0\rangle + \sqrt{p}|1\rangle \otimes |R_1\rangle, \tag{4}$$

where the 'reference' qubit system $\mathcal{H}_R$ is introduced, and $|R_0\rangle, |R_1\rangle$ are the unit vectors of $\mathcal{H}_R$. Since $\rho = Tr_R |\Psi\rangle \langle\Psi| = \sqrt{1-p}|0\rangle\langle 0| + \sqrt{p(1-p)}(\langle R_1 | R_0\rangle |0\rangle \langle 1| + \langle R_0 | R_1\rangle |1\rangle \langle 0|) + \sqrt{p}|1\rangle\langle 1|$, we should put

$$\langle R_0 | R_1\rangle = \frac{\alpha}{\sqrt{p(1-p)}}. \tag{5}$$

The channel output state of a phase damping channel is

$$\mathcal{E}_p(\rho) = \begin{bmatrix} 1-p, \gamma\alpha^* \\ \gamma\alpha, \quad p \end{bmatrix}, \tag{6}$$

where the coefficient $\gamma \in [0,1]$ characterizes the phase damping channel. The eigenvalues of the output state $\mathcal{E}_p(\rho)$ is

$$\lambda_{1,2} = \frac{1}{2}[1 \pm \sqrt{1 - 4(p(1-p) - |\alpha|^2 \gamma^2)}], \tag{7}$$

The joint output state $\rho^{QR'} = \mathcal{E}_p \otimes I_R(|\Psi\rangle\langle\Psi|)$ can be expressed in the computational basis $\{|00\rangle, |01\rangle, |10\rangle, |11\rangle\}$ of $\mathcal{H}_Q \otimes \mathcal{H}_R$, with

$$|R_0\rangle = |0\rangle_R, \tag{8}$$

$$|R_1\rangle = \frac{\alpha}{\sqrt{p(1-p)}}|0\rangle_R + \sqrt{1 - \frac{|\alpha|^2}{p(1-p)}}|1\rangle_R. \tag{9}$$

Thus

$$\rho^{QR'} = \begin{bmatrix} 1-p, & 0, \gamma\alpha^*, & \gamma\sqrt{p(1-p)-|\alpha|^2} \\ 0, & 0,0, & 0 \\ \gamma\alpha, & 0, \frac{|\alpha|^2}{(1-p)}, & \alpha\frac{\sqrt{p(1-p)-|\alpha|^2}}{1-p} \\ \gamma\sqrt{p(1-p)-|\alpha|^2}, & 0, \alpha^*\frac{\sqrt{p(1-p)-|\alpha|^2}}{1-p}, & p - \frac{|\alpha|^2}{(1-p)} \end{bmatrix}. \tag{10}$$

Although $\rho^{QR'}$ is a matrix function of the coherent term $|\alpha|$, the eigenvalues of $\rho^{QR'}$ do not rely on $|\alpha|$. Two of the eigenvalues are

$$\Lambda_{1,2} = \frac{1}{2}[1 \pm \sqrt{1 - 4p(1-p)(1-\gamma^2)}], $$

the other two eigenvalues $\Lambda_{1,2}$ are 0. The von Neumann entropy of the output state is

$$S(\mathcal{E}_p(\rho)) = H_2(\lambda_1), \tag{11}$$

where $H_2(x) = -x \log_2 x - (1-x) \log_2(1-x)$ is the binary entropy function. The maximum of $H_2(\lambda_1)$ should be achieved when $\lambda_1 = \frac{1}{2}$, that is, $|\alpha| = 0$, $p = \frac{1}{2}$. Thus, the maximization conditions of entropy of the output state $S(\mathcal{E}_p(\rho))$ are $|\alpha| = 0, p = \frac{1}{2}$. Meanwhile, the entropy of the joint output state is

$$S(\rho^{QR'}) = H_2(\Lambda_1), \tag{12}$$

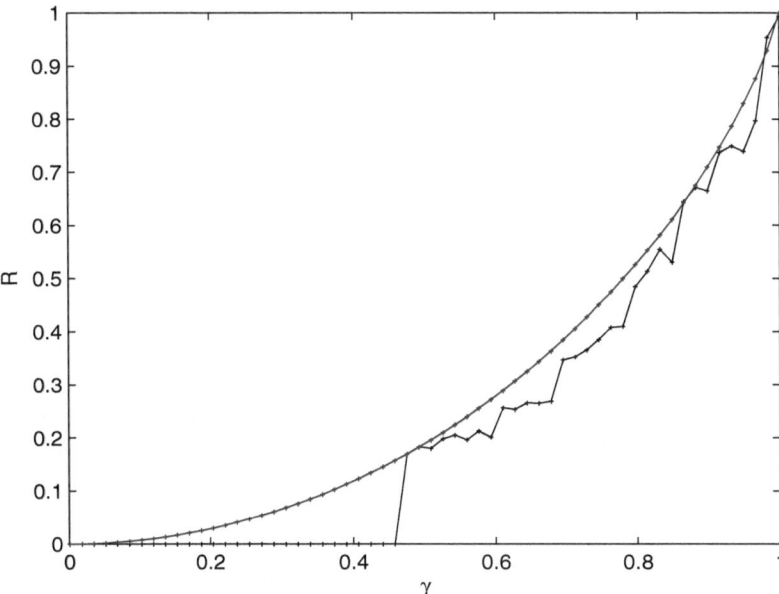

Fig. 1. The comparison of simulation result with quantum capacity, γ is the dephasing coefficient, R is the secure key transmission rate

it is an decreasing function of Λ_1 for $\Lambda_1 \geq \frac{1}{2}$. The minimum should be achieved with maximal Λ_1, which is only possible when $p = \frac{1}{2}$,regardless of the off-diagonal element α of the input state. Hence the coherent information $I_c(\rho, \mathcal{E}_p) = S(\mathcal{E}_p(\rho)) - S(\rho^{QR'})$ achieves its maximum at $|\alpha| = 0$, $p = \frac{1}{2}$. The maximum of I_c is just the quantum capacity since the channel is degradable. We have the quantum capacity

$$Q = 1 - H_2(\frac{1}{2}(1 + \gamma)). \tag{13}$$

For a degradable channel, it has been proved that private classical capacity is equal to quantum capacity[3]. The depasing channel is degradable, thus the private classical capacity of the dephasing channel is

$$C_p = Q = 1 - H_2(\frac{1}{2}(1 + \gamma)). \tag{14}$$

the achievable private information transmission rate R is upper bounded by private classical capacity C_p. The theoretical private classical capacity formula (14) is shown in Figure 1 as the smooth curve.

3   Simulation with Classical Computer

The standard BB84 protocol use polarized quantum state ensemble $\{|-\rangle, |\,|\rangle, |/\rangle ,|\backslash\rangle\}$ as the source. The polarizations are horizontal ($0°$), vertical ($90°$), diagonal up (45°), diagonal down (135°), respectively. Denote $|\psi_0\rangle = |-\rangle, |\psi_1\rangle = |\rangle, |\psi_2\rangle = |/\rangle, |\psi_3\rangle = |\backslash\rangle$, we have $|\psi_{2,3}\rangle = \frac{1}{\sqrt{2}}(|\psi_0\rangle \pm |\psi_1\rangle)$. In the basis of $\{|\psi_0\rangle, |\psi_1\rangle\}$, the density matrices of the four states are

$$\rho_0 = \begin{bmatrix} 1 & 0 \\ 0 & 0 \end{bmatrix}, \rho_1 = \begin{bmatrix} 0 & 0 \\ 0 & 1 \end{bmatrix}, \rho_2 = \frac{1}{2}\begin{bmatrix} 1 & 1 \\ 1 & 1 \end{bmatrix}, \rho_3 = \frac{1}{2}\begin{bmatrix} 1 & -1 \\ -1 & 1 \end{bmatrix}. \tag{15}$$

The output density matrices are

$$\rho_0' = \begin{bmatrix} 1 & 0 \\ 0 & 0 \end{bmatrix}, \rho_1' = \begin{bmatrix} 0 & 0 \\ 0 & 1 \end{bmatrix}, \rho_2' = \frac{1}{2}\begin{bmatrix} 1 & \gamma \\ \gamma & 1 \end{bmatrix}, \rho_3' = \frac{1}{2}\begin{bmatrix} 1 & -\gamma \\ -\gamma & 1 \end{bmatrix}. \tag{16}$$

Thus if horizontal and vertical basis is chosen to convey classical bit in BB84 protocol, the dephasing channel takes no effect to the bit, and if diagonal (up and down) basis is chosen, the dephasing channel does affect the bit transmitted. At the sending end, Alice randomly chooses the horizontal and vertical basis or diagonal basis. Alice converts the binary string into qubit and sends the polarization states to Bob through the quantum channel. At the receiving end, Bob measures the received polarizations and converts them into binary string. They sift measured polarization by declaring the basis chosen with public classical channel and get the raw key, if the error bit rate is larger than a threshold, they abort. Even if the communication is effective, they can not use the raw key directly, the bit error rate is still very large. They have to perform error correction and privacy amplification.

In the simulation, random numbers are needed at the sending process and at the measurement in order to choose the horizontal and vertical basis or diagonal basis.

3.1   Data Reconciliation

The process which performs error correction in the public channel is known as data reconciliation. The requirements of data reconciliation are as follows: (a) to reduce the bit error rate to an appropriate value for use. (b) to reduce the information Eve obtained in this process as far as possible; (c) to maintain the useful data as many as possible; (d) to be fast and to save resources as far as possible.

In this paper, we will use bipartition protocol to correct errors. The method is simple and practical, although it can not correct errors with even number bit flips. We improved the original protocol, increased the error correction capability. The detailed steps are as follows: (a) Alice and Bob rearrange their sequence according to the same random sequence. The purpose is to make the errors uniformly distributed. (b) The data is divided into small packets, with the length of each packet being 11 bits (according to our empirical data); (c) Alice and Bob detect the parity of each packet data respectively and compare the results through some public channel. If they have different results, it is to say that the packet has odd error bits. We divide this packet into two further groups, detect the parity once more. If Alice and Bob have different results, they abort this group; On the contrary, they abort the last bit of this group. Furthermore, they calculate the bit error rate q_1 of the first round; (d) After the first round error correction, bit errors may still exist, we rearrange the bit sequence with some other random sequence which is the same to Alice and Bob and divide the data into packets, the length of each packet in this step is $k \approx 1/(3q_1)$. Repeat the step (c). If $q_1 > 10^{-5}$, repeat this step until the bit error rate is much lower, such that $.q_1 < 10^{-5}$.

3.2   Privacy Amplification

Privacy amplification is a method of extracting a secret key from a string which is partially-known to an eavesdropper.This method is at the cost of reducing the information which legitimate users obtained to improve the security of the data in public channel.

We will discuss the privacy amplification in our simulation algorithm.We suppose that Alice and Bob have l bits after error correction, they estimate Eve knows t bits, choose s as security parameters. We use Hash function F, $f \in F, F : \{0,1\}^l \to \{0,1\}^r , r = l - t - s$.After data reconciliation, the mutual information $I(A, B)$ will reduce from l to r, the mutual information $I(E, A)$ will reduce from t to less than $2^{-s}/\ln 2 \approx 1.443 \times 2^{-s}$, where r is the length of the final key. In this paper, the value of t is chosen in a reasonable range, e.g., from 0 to 5. Furthermore, we set $s = 30$.

The Hash function we used is a $l \times r$ matrix only containing 0 and 1. We apply the mod-2 operation to the l bits sequence and the Hash matrix. Finally, we can get the final key with unconditional security.

4   Conclusions

We simulate BB84 protocol of quantum key distribution in dephasing channel with classical computer. With properly chosen of the length of the first round package in the data reconciliation and the precision of leaked information to the environment, we calculation the yield of the protocol. The transmission rate with respect to the dephasing coefficient is given. Theoretically, the private classical capacity of the channel is also deduced and shown in the figure as a comparison. We can see that the simulation are quite in agreement with theory except in the severe depasing end.

References

1. DiVincenzo, D.P., Shor, P.W., Smolin, J.A.: Phys.Rev. A 57, 830 (1998)
2. Devetak, I., Shor, P.W.: Comm. Math. Phys. 256, 287 (2005)
3. Simth, G.: Phys. Rev. A 78, 022306 (2008)
4. Bennett, C.H., Divincenzo, D.P., Smolin, J.A.: Phys. Rev. Lett. 78, 3217 (1997)
5. Giovannetti, V., Fazio, R.: Phys. Rev. A 71, 032314 (2005)
6. Wolf, M.M., Perez-Garcia, D., Giedke, G.: Phys. Rev. Lett. 98, 130501 (2007)

7. Caruso, F., Giovannetti, V.: Phys. Rev. A 74, 062307 (2006)
8. Ioffe, L., Marc Mzard, M.: Phys. Rev. A 75, 032345 (2007)
9. Shor, P.W., Preskill, J.: Phys. Rev. Lett. 85, 441 (2000); PRL85 (2000)
10. Schumacher, B.: Phys. Rev. A 54, 2614 (1996); Schumacher, B., Nielsen, M.A.: Phys. Rev. A 54, 2629 (1996)
11. Lloyd, S.: Phys. Rev. A 55, 1613 (1997)
12. Devetak, I.: IEEE Trans. inf. Theory 51, 44 (2005); Devetak, I., Winter, A.: Proc. R. Soc. Lond. A 461, 207 (2005)
13. Barnum, H., Knill, M., Nielsen, M.A.: IEEE Trans. Inf. Theory 46, 1317 (2000)
14. Horodecki, M., Horodecki, P., Horodecki, R.: Phys. Rev. Lett. 85, 433 (2000)

Using Multi-particle Entanglement in Secure Communication Scenarios

Mosayeb Naseri[*]

Islamic Azad University, Kermanshah Branch, Kermanshah, Iran
Sepehr1976@yahoo.com

Abstract. The Quantum Entanglement lies at the heart of the new field of quantum communication and computation. Recently, quantum information theory has shown the tremendous importance of quantum correlations for the formulation of new methods of information transfer and for algorithms exploiting the capabilities of quantum computers. This paper describes the application and the importance of the multi-particle entangled states in secure communication scenarios. We show how to make communication secure against eavesdropping using entanglement-based quantum communication, and how to apply this communication protocols in real life situations.

1 Introduction

Over the last 20 years quantum information theory and specially quantum communication has established itself as promising applications of quantum physics. In this way, one of the important scopes is to evaluate the practical applicability of quantum information science in real life. In real life every day people have to make important decisions that should remain secret. Protecting the privacy of those decisions, if their results are to be communicated, can be a challenging problem. Some examples of such decisions are voting and auction.

Recent research on quantum computation and quantum information allowed using it for describing real life communication scenarios. Financial market phenomena, has been considered in quantum information by researchers. Quantum information has extended the scope of game theory for the quantum world [1-4]. Also quantum game theory has been used for describing financial market phenomena [5, 6].

An auction is one of the basic businesses in commerce. The protection of bidder privacy and the prevention of bidder default are the key problems needed to be urgently solved. There are few quantum auction protocols, where the applications of quantum game theory have been extended to quantum version of auctions [5, 6]. An auction usually has three transactional types: traditional English auction, Dutch auction and sealed-bid auction. Traditional English auction is also known as public bid auction, wherein each bidder casts his/her own bid, and the bid must be higher than the bottom price. The bottom price is adjusted upwards after a round. The auction goes on until there is only one bidder left who is willing to offer the price. Dutch

---

[*] Corresponding author.

A. Sergienko, S. Pascazio, and P. Villoresi (Eds.): QuantumCom 2009, LNICST 36, pp. 249–257, 2010.
© Institute for Computer Sciences, Social-Informatics and Telecommunications Engineering 2010

auction is similar to traditional English auction, but it begins with the top price, and then the price goes down round after round until the first bidder decides to offer the price. Unlike the previous two kinds of auctions, all the customers who are willing to name their bids are gathered in a sealed-bid auction, where each bidder submits their own bids to the auctioneer. After the opening phase, the auctioneer makes all the bids public and determines the winner. Recently we have presented a secure quantum sealed-bid auction protocol, in which, a quantum auction is considered as a communication process and it has been designed using a quantum secure direct communication based on GHZ states protocol [7].

On the other hand, a quantum telephone protocol including the dialing process and the talking one has proposed by Xiaojun Wen, Yun Liu, Nanrun Zhou in 2007 [8]. In this protocol in the dialing process, with their respective secret keys, the legitimate communicators Alice and Bob can pass the authentication by Charlie acting as a telephone company. In the talking process, Charlie provides the authenticated Alice and Bob with a quantum channel sequence, on which Alice and Bob can communicate with each other directly and privately by virtue of some encoding operations. Unfortunately, it has been shown that the quantum telephone protocol in its original form is not as secure as it claimed. In our recent work we have shown that the dishonest server can obtain full information of the communication with zero risk of being detected [9].

The aim of this work is to show how to use quantum information in designing a real life communication protocols.

2   Quantum Telecommunication

Quantum key distribution (QKD) is an ingenious application of quantum mechanics, in which two remote legitimate users (Alice and Bob) establish a shared secret key through the transmission of quantum signals and use this key to encrypt (decrypt) the secret messages. Since Bennett and Brassard presented the pioneering work in 1984 [10], a variety of QKD protocols have been proposed. Quantum key distribution has attracted much attention of the researchers. Quantum secure direct communication (QSDC) [11-22] is a branch of quantum cryptography, which allows that the sender transmits directly the secret (not a random key) to the receiver in a deterministic and secure manner. Quantum encryption algorithm has also been investigated [23, 24]. The goal of quantum encryption algorithm and classical encryption algorithm is consistent, i.e. to protect secret information or keep communications private. Quantum secret sharing (QSS) [25, 26] is another important application of quantum mechanics, which allows a secret to be shared among many participants in such a way that only the authorized groups can reconstruct it. Boström and Felbinger put forward a ping-pong QSDC scheme by using Einstein-Podolsky-Rosen (EPR) pairs [13]. Based on the idea of a ping-pong QSDC scheme, Nguyen proposed a quantum dialogue scheme (the quantum dialogue is actually two-way communication) by using EPR pairs [27]. However, an eavesdropper who adopts the intercept-and-resend attack strategy can steal the secret messages without being detected.

2.1 Quantum Dialogue

To get information from Alice, Bob prepares two qubits $\left|\phi_{kl}\right\rangle_{ht}$ in one among the four mutually orthogonal Bell states,

$$\left|\phi_{00}\right\rangle_{ht} = \frac{1}{\sqrt{2}}(\left|00\right\rangle + \left|11\right\rangle),$$

$$\left|\phi_{01}\right\rangle_{ht} = \frac{1}{\sqrt{2}}(\left|00\right\rangle - \left|11\right\rangle),$$

$$\left|\phi_{10}\right\rangle_{ht} = \frac{1}{\sqrt{2}}(\left|01\right\rangle + \left|10\right\rangle),$$

$$\left|\phi_{11}\right\rangle_{ht} = \frac{1}{\sqrt{2}}(\left|01\right\rangle - \left|10\right\rangle),$$

where, h and t stand for "home" and "travel" qubits, respectively. Then he sends qubit "t" to Alice while stores qubit "h" with himself. Alice decides to use qubit "t" as the message mode (MM) or the control mode (CM) randomly. In the MM, Alice encodes her information by performing a unitary operation I or σ_z on qubit "t" corresponding to her message bit 0 or 1, then he pongs it back to Bob, who can obtain Alice's information by a Bell measurement. In the CM, Alice performs a measurement in the basis, $B_z = \left|0\right\rangle, \left|1\right\rangle$ and sends the result to Bob via a public classical channel. Bob then also switches to the CM and performs a measurement in the same basis B_z. Comparing his own result with that of Alice, Bob can detect the presence of Eve [27] .

2.2 Quantum Telephone

A quantum telephone protocol including the dialing process and the talking one has proposed by Xiaojun Wen , Yun Liu , Nanrun Zhou in 2007 [8]. In this protocol in the dialing process, with their respective secret keys, the legitimate communicators Alice and Bob can pass the authentication by Charlie acting as a telephone company. In the talking process, Charlie provides the authenticated Alice and Bob with a quantum channel sequence, on which Alice and Bob can communicate with each other directly and privately by virtue of some encoding operations. Unfortunately, it has been shown that the quantum telephone protocol in its original form is not as secure as it claimed .i.e., recently Y. Sun et al. in [28] have shown that an attacker could eavesdrop on the communicator's conversation without introducing any error by an attack with fake particles and local operations. At the same time, very recently, we have realized that a dishonest server, an eavesdropper, can gain full information of the communication with zero risk of being detected by using fake entangled particles [29]. Then the authors of the both papers [28, 29] have presented a modification procedure to avoid the vulnerability of the protocol against the possible presented attacks.

It is apparently that the modifications presented in [28, 29], improve the original protocol against the eavesdropping of the secure information, but I think the main theoretical source of insecurity of the protocol still remains! Since I see here the main source of theoretical insecurity, let me spend some more words on the theoretical condition for the security. As a matter of fact, each and every secure quantum communication protocol, in fact, the efficiency of transportation was bounded by Holevo quantity [30], which shows that n qubits cannot be used to transmit more than n bits of classical information in a 2-level system. Obviously, in secure quantum telephone protocol, Alice and Bob can transmit 4 bits secret message (two for Alice and two for Bob) via per EPR pair in the above communication. Whereas, Gao et al. [28] pointed out that among the 4-bit information only 2 bits are transmitted securely. Due to Bob's declaration, everyone (of course the Eve) can infer that there would be four possibilities for operations performed by Bob and Alice. Assuming four possibilities having equal probability, the channel contains only two bits of secret information for Eve.

$$- \sum_i p_i \log p_i = -4(\frac{1}{4}) \log_2 \frac{1}{4} = 2.$$

In other words, 2-bits secret has been leaked to Eve. Since, this capacity has been exceeded in secure quantum telephone protocol, it is undoubtedly insecure. However, it seems that this theoretical limitation has been tacitly mentioned by the authors of reference [28] (Ref. [28] page 2280 second column second paragraph).

3 Quantum Finance (Quantum Auction)

Recent research on quantum computation and quantum information allowed using it for describing financial market phenomena. Quantum information has extended the scope of game theory for the quantum world [1-3]. Also quantum game theory has been used for describing financial market phenomena [5,6]. An auction is one of the basic businesses in commerce. The protection of bidder privacy and the prevention of bidder default are the key problems needed to be urgently solved. There are few quantum auction protocols, where the applications of quantum game theory have been extended to quantum version of auctions [31]. An auction usually has three transactional types: traditional English auction, Dutch auction and sealed-bid auction, traditional English auction is also known as public bid auction, wherein each bidder casts his/her own bid, and the bid must be higher than the bottom price. The bottom price is adjusted upwards after a round. The auction goes on until there is only one bidder left who is willing to offer the price. Dutch auction is similar to traditional English auction, but it begins with the top price, and then the price goes down round after round until the first bidder decides to offer the price. Unlike the previous two kinds of auctions, all the customers who are willing to name their bids are gathered in a sealed-bid auction, where each bidder submits his/her own bids to the auctioneer. After the opening phase, the auctioneer makes all the bids public and determines the winner.

The crucial issue of any auction protocol is its security. Each secure auction protocol includes an auctioneer, a third party or an auction host and many bidders (sometimes an auctioneer plays the role of third party in auction protocols). Essentially the most important requirements of secure auction can be summarized as follows [32]:

(1) Anonymity: all bidders can keep anonymity in an auction, even if the bid is opened, i.e., no one can gain access to other bidder's information, except the auction host. In addition, only the auction host stores the bidder's information, there- fore, it can maintain anonymity in the auction, even after the bid is opened.

(2)   Public verifiability: all the bidding prices and the winning prices can be verified by anyone, i.e., everybody should be able to see all the bids and verify that the auctioneer chosen the biggest one to prevent the dishonest auction host or auctioneer cheating bidders and performing a conspiracy with a malicious bidder.

(3)   Accountability of bidder: the auction cannot be interrupted by any malicious bidders with a dishonest bid without being detected. That is to say, the auction host can verify each bid when the bidder casts a bid.

(4)   Fairness: all sealed-bids are opened at the same time, and the third party or the auctioneer cannot collude with a malicious bidder to cheat the other bidders.

(5)   Non-repudiation: the property of non-repudiation is that both the bidder cannot deny having cast his/her bid and the auction host cannot deny that he has received the bid from the bidder.

(6)   Traceability: the winning bidder can be identified when the auction is finished.

There are few quantum auction protocols, where the applications of quantum game theory have been extended to quantum version of auctions. Recently, we proposed a quantum sealed-bid auction protocol using quantum secure direct communication based on GHZ states [7]. Here, the protocol of quantum sealed-bid auction protocol is reviewed.

3.1   Quantum Sealed-Bid Auction

The auction model consists of one buyer agent Alice, who is the auctioneer that needs a particular L items (service or product) and a fixed number of n seller agents Bob, Charlie . . . and Zach, who are the bidders. The scheme can be explained as follows:

- At the beginning of the auction, the buyer, Alice announces her request items (which consist of the items desired characteristics and the auction protocol) by classical channel. In this step all parties agree on that Auctioneer, Alice can perform the four unitary operations I; σ_x; i σ_y; σ_z to encode two bits classical information 00, 01, 10, 11. Also they agree on that all bidders can only perform the two unitary operations I; σ_x to encode classical bits information 0, 1, where I; σ_x; i σ_y and σ_z are Pauli matrixes [7].

- The auctioneer Alice generates M groups N-partite GHZ states. For example, suppose that the i-th GHZ state is in the form of

$$|\phi\rangle^i_{abc\,...\,z} = \frac{1}{\sqrt{2}}\left(\left|0\,s^i_{\,1}s^i_{\,2}\,...\,s^i_{\,N-1}\right\rangle + \left|1\,\overline{s}^i_{\,1}\,\overline{s}^i_{\,2}\,...\,\overline{s}^i_{\,N-1}\right\rangle\right)_{abc\,...\,z}.$$

Where $s^i_{\,j} = 0$ or 1, $\overline{s}^i_{\,j} = s^i_{\,j} + 1$ modulo 2, and the subscripts a; b; c; . . . ; z represent the particles belonging to Alice, Bob, Charlie . . . and Zach, respectively. Then Alice sends b particles to Bob, c particles to Charlie . . . z particles to Zach.

- Bob, Charlie . . . and Zach confirm Alice that they have received all the particles b, particles c . . . and particles z, respectively. Afterwards, Bob (anyone of the N − 1 bidders Bob, Charlie, . . ., Zach, we say, Bob) selects randomly a sufficiently large subset of particles from the M groups b particles, which we call the T groups b particles, and measures each of these particles using one of the two measuring bases $|0\rangle, |1\rangle$ or $|+\rangle, |-\rangle$ randomly and tells all other bidders the position, the measuring basis and the measurement result for each of the T groups b particles via classical channel and ask them to measure T groups c particles, . . ., z particles using the same measuring bases, respectively. Then all of the bidders tell Alice their measurement results for each of the particles. Also Alice measure T groups a particles. According to the measurement results of Bob, Charlie. . . Zach and herself, Alice can determine whether there is any eavesdropping. If there is an error, Alice concludes that the channel is not secure, and halts the auction. Otherwise, Alice, Bob, Charlie . . . and Zach continue the next step.

- If no error happens, the N- 1 bidders Bob, Charlie . . . and Zach encode their secret bids by applying operators I; $i\sigma_y$ on their leftover particles and return their particles to Alice.

- After receiving all the particles from the N- 1 bidders, for each GHZ state, Alice encodes her final message with one of the four unitary operations I; σ_x; i σ_y and σ_z on her particle. Then she performs N-particle GHZ-basis measurements on the remaining groups a; b; c; . . . ; z and publicly announces the initial states and the measurement results of the GHZ states leftover. According to the initial states and the measurement results of the GHZ states leftover, Alice can read out the secret bids of Bob, Charlie . . . and Zach. Also every bidder can deduce the secret bids encoded by the other bidders. Concluding the auction winner can be determined simply.

Very recently, the Quantum Sealed-Bid Auction has been considered by different research groups and it has been shown that be shown that a dishonest bidder can obtain all the other one's secret bids by double CNOT attack or using fake entangled particles attack and according to these secret bids values he can encode the better bids on his particles and win the auction [33-33].

However, it is apparently that with the presented improvement, the original protocol does complete the task of a sealed-bid auction fairly. But I think one of the main sources of insecurity of the original protocol still remains!, in fact, in any secure auction protocol the possibility of collusion of the bidders with the auctioneer should be mentioned and the protocol should contains a security method which is guarantees that if one of the bidders or a group of the bidders decide to collude with the auctioneers, they would not succeed. It is clear, the original article, does not present such a method to bind the collusion of the bidders with the auctioneer. It is noticeable that every secure communication protocol, whether quantum or classical, needs an authenticated channel. User authentication (also called user identification) makes it possible for a communicator to prove his/her identity, often as the first step to log into a system. If there is no authenticated channel, then a man-in-the-middle attack is always possible, resulting in a complete loss of security. In fact, the perfect security of a quantum communication protocol stands and falls with the integrity of the public channel. Usually the classical channel is tacitly assumed to be as the one and only trusted channel of the protocol. Since in the original protocol it is not mentioned how the public communication is realized, the reader must assume that it is not necessarily the auction host or the auctioneer who act as the public channel throughout the entire protocol. Concluding, it is possible to use the auction host or auctioneer as the one and only trusted channel of the protocol, doing so would make the protocol more straightforward, because it is in fact the auction host who establishes the quantum channels between the bidders in the auction, and thus actively performs an authentication of the legitimate parties (not of himself, though). Also, engaging the auction host as the authenticated channel would avoid the vulnerability of the protocol against the collusion of the dishonest auction host or auctioneer with the bidders.

4 Quantum Voting

Recent research on quantum computation and quantum information makes it possible to using it for voting protocols [35, 36]. The advantage of quantum voting protocols is that security that is based on the laws of quantum mechanics rather than assumptions about computational complexity.

Reliable voting protocols should satisfy a number of conditions [37], three of which are: (i) security, (ii) verifiability, and (iii) privacy. The security condition guarantees that all users can influence the result only by casting a *single* valid vote. That is, each voter can vote just once (non-reusability), only legitimate users can vote (eligibility) and no one can learn any intermediate result (fairness). The strongest version of the verifiability requirement is that each voter can verify the correctness of the result; however none of the voters is able to prove how he or she voted. This prevents vote buying. A voting scheme satisfying all properties except the privacy condition is easy to implement. Privacy is related to the secrecy of the ballots, or equivalently to the anonymity of the voters. Ideally, no one should be able to tell how any of the voters voted.

In 2006 Mark Hillery et al., proposed quantum protocols that guarantee the anonymity of participants in voting procedures and can be used in several complex communication tasks. The advantage of quantum voting protocols is that security that is based on the laws of quantum mechanics rather than assumptions about computational complexity [35]. In this protocol quantum voting is based on voters applying local operations to an entangled state, in which a method of preventing voters from cheating is presented.

On the other hands, quantum protocols for ensuring the anonymous voting in a number of different scenarios have been presented by J. A. Vaccaro et al. [36].

5 Conclusion

Quantum computing and communications are novel ways of engineering quantum systems and are proving to dramatically change the way we think about computation, complexity, information, and communication. We have presented some applications of multi-particle entanglement quantum communication, in real life scenarios. Undoubtedly, quantum technologies will play a very important role in the future, and already to date, several companies are commercializing quantum communication systems.

Acknowledgement

It is a pleasure to thank Soheila Gholipour and Yasna Naseri for their interests in the present work. This work is supported by Islamic Azad University, Kermanshah Branch, Kermanshah, IRAN.

References

1. Meyer, D.A.: Quantum Strategies. Phys. Rev. Lett. 82, 1052 (1999)
2. Eisert, J., Wilkens, M., Lewenstein, M.: Quantum Games and Quantum Strategies. Phys. Rev. Lett. 83, 3077 (1999)
3. Piotrowski, E.W., Sladkowski, J.: Quantum Game Theory. In: Mathematical Physics Frontiers. Nova Science Publishers, Inc., Berlin (2004)
4. Piotrowski, E.W., Sladkowski, J.: An invitation to quantum game theory. Int. J. Theor. Phys. 42, 1089 (2003)
5. Piotrowskia, E.W., Sladkowski, J.: Quantum English Auctions. Physica A 318, 505 (2003)
6. Piotrowskia, E.W., Sladkowski, J.: Quantum Auctions: Facts and Myths. Physica A 387, 3949 (2008)
7. Naseri, M.: Secure quantum sealed-bid auction. Opt. Commun. 282, 1939 (2009)
8. Wen, X., et al.: Secure Quantum Telephone. Optics Communications 275, 278–282 (2007)
9. Naseri, M.: Eavesdropping on secure quantum telephone protocol with dishonest server. Opt. Commun. (2009), doi:10.1016/j.optcom.2009.05.012
10. Bennett, C.H., Brassard, G.: Proceedings of the IEEE International Conference on Computers, Systems and Signal Processings, Bangalore, India, p. 175. IEEE, New York (1984)
11. Beige, A., Engler, B.G., Kurtsiefer, C., Weinfurter, H.: Secure communication with a publicly known key. Acta Phys. Pol. A 101, 357 (2002)
12. Long, G.L., Liu, X.S.: Theoretically efficient high-capacity quantum-key-distribution scheme. Phys. Rev. A 65, 32302 (2002)
13. Bostrom, K., Felbinger, T.: Deterministic secure direct communication using entanglement. Phys. Rev. Lett. 89, 187902 (2002)
14. Deng, F.G., Long, G.L., Liu, X.S.: Two-step quantum direct communication protocol using the Einstein-Podolsky-Rosen pair block. Phys. Rev. A 68, 042317 (2003)
15. Deng, F.G., Long, G.L., Li, X.H., Wen, K., Wang, W.Y.: Improving the security of secure direct communication based on the secret transmitting order of particles. Front. Phys. China 2(3), 251 (2007)
16. Zhang, Z.J., Man, Z.X., Li, Y.: Improving Wójcik's eavesdropping attack on the ping–pong protocol. Phys. Lett. A 333, 46 (2004)
17. Deng, F.G., Long, G.L.: Secure direct communication with a quantum one-time pad. Phys. Rev. A 69, 052319 (2004)
18. Zhang, Z.J., Li, Y., Man, Z.X.: Improved Wójcik's eavesdropping attack on ping-pong protocol without eavesdropping-induced channel loss. Phys. Lett. A 341, 385 (2005)
19. Wang, C., Deng, F.G., Long, G.L.: Multi-step quantum secure direct communication using multi-particle Green–Horne–Zeilinger state. Opt. Commun. 253, 15 (2005)
20. Zhang, Z.J., Man, Z.X., Li, Y.: The Improved Bostrom–Felbinger Protocol Against Attacks Without Eavesdropping. Int. J. Quantum Inform. 2, 521 (2005)
21. Man, Z.X., Zhang, Z.J., Li, Y.: Quantum Dialogue Revisited. Chin. Phys. Lett. 22, 18 (2005)

22. Deng, F.G., Long, G.L.: Secure direct communication with a quantum one-time pad. Phys. Rev. A 69, 052319 (2004)
23. Wojcik, A.: Eavesdropping on the "Ping-Pong" Quantum Communication Protocol. Phys. Rev. Lett. 90, 157901 (2003)
24. Zeng, G.H., Keitel, C.H.: Arbitrated quantum-signature scheme. Phys. Rev. A 65, 042312 (2002)
25. Boykin, P.O., Roychowdhury, V.: Optimal encryption of quantum bits. Phys. Rev. A 67, 042317 (2003)
26. Hillery, M., Buzek, V., Berthiaume, A.: Quantum secret sharing. Phys. Rev. A 59, 1829 (1999)
27. Karlsson, A., Koashi, M., Imoto, N.: Quantum entanglement for secret sharing and secret splitting. Phys. Rev. A 59, 162 (1999)
28. Nguyen, B.A.: Quantum Dialogue. Phys. Lett. A 328, 6 (2004)
29. Sun, Y., et al.: Improving the security of secure quantum telephone against an attack with fake particles and local operations. Opt. Commun. 282, 2278 (2009)
30. Naseri, M.: Eavesdropping on secure quantum telephone protocol with dishonest server. Optics Commun. 282, 3375–3378 (2009)
31. Nielsen, M.A., Chuang, I.L.: Quantum Computation and Quantum Information. Cambridge University Press, Cambridge (2000)
32. Liu, X.S., et al.: General scheme for superdense coding between multiparties. Phys. Rev. A 65, 022304 (2002)
33. Subramanian, S.: Design and verification of a secure electronic auction protocol. In: Proc. IEEE 17th Symposium on Reliable Distributed Systems, Washington DC, USA, pp. 204–210 (1998)
34. Yang, Y.G., et al.: Improved Secure Quantum Sealed-Bid Auction. Optics Communications 282, 4167–4170 (2009)
35. Qin, S.-J., et al.: Cryptanalysis and improvement of a secure quantum sealed-bid auction. Optics Communications 282, 4014–4016 (2009)
36. Hillery, M., et al.: Towards quantum-based privacy and voting. Physics Letters A 349, 75–81 (2006)
37. Vaccaro, J.A., et al.: Quantum protocols for anonymous voting and surveying. Physical Review A 75, 012333 (2007)
38. Schneier, B.: Applied Cryptography, p. 125. Wiley, New York (1996)

Subband Tunneling and Coulomb Effects in Coupled Quantum Wells

H. Cruz

Departamento de Fisica Basica, Universidad de La Laguna,
38204 La Laguna, Tenerife, Spain
hcruz@ull.es

Abstract. We have solved in space and time the effective-mass non-linear Schrödinger equation for an electron-hole gas in a semiconductor superlattice. Considering a Coulomb interaction between both electron-hole gases and two subbands, we have obtained a time-varying dipole moment in the heterostructure. In this way, we have shown the possibility of having another kind of terahertz electromagnetic radiation emerging from a double quantum dot system.

Keywords: Electron states in low-dimensional structures, Tunneling, Electronic transport phenomena in thin films.

1 Introduction

In recent years, Coulomb effects in semiconductor quantum dots has attracted significant interest in semiconductor physics. [1] Semiconductor quantum dots are nanoestructures which allow confinement of carriers in all directions within dimensions smaller than their De Broglie wavelength. The tunneling dynamics of an electron into a semiconductor quantum dot can be blocked by the classical Coulomb repulsion of the electrons already in the dot. We notice that one remaining key question is to extend the Coulomb effect analysis to a semiconductor superlattice. In principle, photoexited electron-hole pairs can be created in a well-defined superlattice quantum well. We know that if a large number of carriers are localized in a superlattice quantum well, the electrical properties can be strongly modified due to charge-buildup effects. Then, the diffusion process could be modified due to Coulomb effects at high carrier sheet density values. In this work, we study the time-dependent evolution of both wave packets by considering electron-electron Coulomb interactions. The method of calculation is based on discretization of space and time for carrier wave functions. We show that the diffusion process can be strongly modified due to Coulomb interactions.

2 Model

In order to study the charge density dynamics in the structure growth direction, we need to solve the time-dependent Schrödinger equation associated with

A. Sergienko, S. Pascazio, and P. Villoresi (Eds.): QuantumCom 2009, LNICST 36, pp. 258–260, 2010.

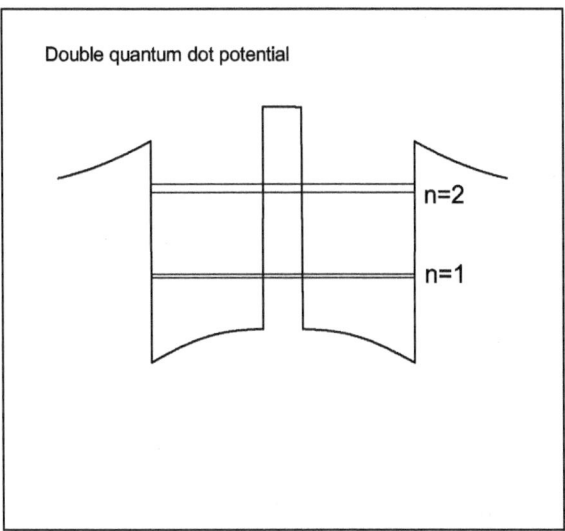

Fig. 1. Double quantum dot potential. It is shown both electron subbands. The level splitting in the first subband is much smaller than in the second subband due to the double quantum dot potential.

a spinless electron in a double quantum well potential. We separate the total electron wave function into the motion along z and the in-plane motion of the electron. The $\psi_{n=1}$ and hole $\psi_{n=2}$ wave functions in the z axis will be given by the nonlinear Schrödinger equations[2]$^-$[4]

$$i\hbar\frac{\partial}{\partial t}\psi_{n=1}(z,t) = \left[-\frac{\hbar^2}{2m_e^*}\frac{\partial^2}{\partial z^2} + V_e(z) + V_H\left(\mid\psi_{n=1}\mid^2,\mid\psi_{n=2}\mid^2\right)\right]\psi_{n=1}(z,t),$$
(1)

$$i\hbar\frac{\partial}{\partial t}\psi_{n=2}(z,t) = \left[-\frac{\hbar^2}{2m_e^*}\frac{\partial^2}{\partial z^2} + V_e(z) + V_H\left(\mid\psi_{n=1}\mid^2,\mid\psi_{n=2}\mid^2\right)\right]\psi_{n=2}(z,t),$$
(2)

where the subscripts $n = 1, 2$ refer to both subbands, respectively, and $V_e(z)$ is the potential due to the quantum wells. The m_e^* is the electron effective mass. V_H is the Hartree potential given by the electron interaction in the heterostructure region. Such a many-body potential is given by the Poisson's equation[3]

$$\frac{\partial^2}{\partial z^2}V_H(z,t) = -\frac{e^2}{\varepsilon}\left[n_{n=1}\mid\psi_{n=1}(z,t)\mid^2 + n_{n=2}\mid\psi_{n=1}(z,t)\mid^2\right],$$
(3)

where ε is the GaAs dielectric constant and $n_{n=1,n=2}$ is the carrier sheet density in each subband.

3   Results and Discussion

The numerical integration in time allows us to obtain the probability of finding both electron and hole charge densities at any time t. We have numerically

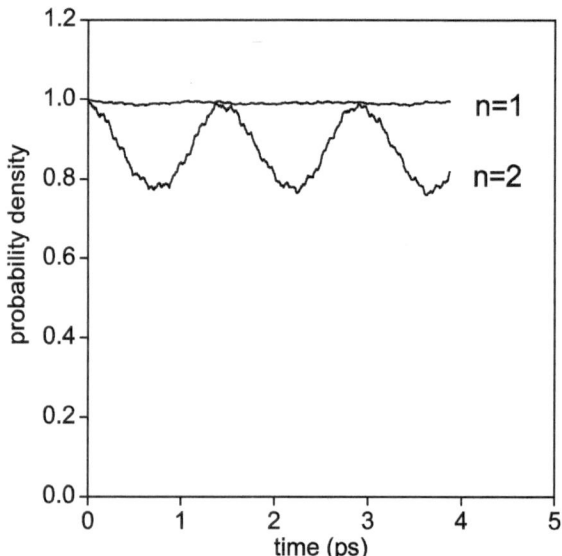

Fig. 2. Probability density versus time. We show tunneling oscillations in the second subband. The tunneling oscillations in the first subband are suppresed due to electron-electron interactions. The level splitting in the first subband is much smaller than the $n = 2$ level splitting. The quantum energy levels in the first subband are easily decoupled with the Coulomb interaction, and then, the tunneling oscillations suppresed.

integrated in space and time the effective-mass Schrödinger equation for an electron gas in a quantum dot. Due to both nonlinear effective mass equations, it is found that the charge dynamically trapped in both wells produces a reaction field that modifies the system resonant condition. In principle, an experimental observation of a such process is possible. In this way, we have shown the possibility of having a different kind of electromagnetic emission emerging from a semiconductor quantum well after an optical excitation of the sample.

Notes and Comments. This work was supported in part by FEDER and *Ministerio de Educación y Ciencia* in Spain, project FIS2006-10268-C03-03.

References

1. Alhassid, Y.: Rev. Mod. Phys. 74, 895 (2000); Reimann, S.M., Manninen, M.: Rev. Mod. Phys. 74, 1283 (2002)
2. Luis, D., Cruz, H., Capuj, N.E.: Phys. Rev. B 59, 9787 (1999)
3. Cruz, H.: J. Appl. Phys. 93, 1620 (2003)
4. Cruz, H., Luis, D.: J. Appl. Phys. 94, 3264 (2003)

Generation of Non-Gaussian Quantum State in Telecommunication Band

Naoto Namekata[1], Yuta Takahashi[2], Go Fujii[1], Daiji Fukuda[3], Sunao Kurimura[4], and Shuichiro Inoue[1]

[1] Institute of Quantum Science, Nihon University, 1-8-14 Kanda-Surugadai, Chiyoda-ku, Tokyo 101-8308, Japan
nnao@phys.cst.nihon-u.ac.jp
[2] Department of Applied Physics, Waseda University, 3-4-1 Okubo, Shinjuku-ku, Tokyo 169-8555, Japan
[3] National Institute of Advanced Industrial Science and Technology, 1-1-1 Umezono Tsukuba, Ibaraki 305-8568 Japan
[4] National Institute for Materials Science, 1-1 Namiki, Tsukuba, Ibaraki 305-0044, Japan

Abstract. We have implemented the non-Gaussian operation on the pulsed squeezed vacuum at telecommunication wavelengths. The non-Gaussian operation based on photon subtraction was carried out using a Titanium-based superconducting transition-edge-sensor that can resolve the photon number. We observed a dip in reconstructed Wigner functions of generated quantum states, which is the clear evidence that a non-Gaussian operation was realized.

Keywords: Non-Gaussian operation, pulsed squeezed light, telecomm-band, transition-edge-sensor, photon-number resolving detector.

1 Introduction

Non-Gaussian operation is one of the essential resources for continuous-variable based quantum information processing. It is well-known that we cannot realize entanglement distillation [1-3] and universal quantum computation without non-Gaussian operation [4]. In a practical application of quantum communication such as teleportation and dense cording, we must share highly entangled quantum states between distant parties using optical fibers. Because of unavoidable losses in a fiber, it is essential to distill the entangled states.

The conventional information theory gives transmission capacity bounds that depend on the number of channels and their noises [5]. These bounds are so-called "Shannon limit". However, quantum information theory allows us to overcome the Shannon limit and indicate novel bounds, "Holevo limit" [6, 7]. The implementation of ultimate-high-capacity optical communication which can reach the Holevo limit requires a quantum computation for coherent light pulses propagating in an optical fiber [8].

A. Sergienko, S. Pascazio, and P. Villoresi (Eds.): QuantumCom 2009, LNICST 36, pp. 261–266, 2010.

Non-Gaussian operation can be realized by utilizing the third or higher order optical nonlinearity. However, the higher order optical nonlinearity in the existing materials is so small for a single-photon that it is difficult to use the nonlinearity in materials to implement quantum gates. Alternatively Gottesman et al., have proposed the cubic-phase gate that consists of Gaussian operations (e.g., the squeezer, the displacement operation and the homodyne detection) and the photon-number resolving detector (PNRD) [9]. The measurement-induced nonlinearity with the PNRD, which is effective as a non-Gaussian operation in the cubic-phase gate, because the rest of cubic-phase gate is substantially composed of Gaussian operations. Recently several experiments on the non-Gaussian operation have been demonstrated in the visible light region [10-12]. However, all of them have implemented non-Gaussian operations by using avalanche photodiodes instead of the PNRD. Using avalanche photodiodes, we cannot measure the number of photons and can only know the arrival of photons. They are inefficient in detecting two or more photons, because the corresponding number of detectors is necessary and each detector must catch only a single photon.

In this paper, we present an experimental demonstration of non-Gaussian operations at a telecommunication wavelength. Employing the PNRD based on a superconducting transition-edge-sensor (TES)[14, 15], the one- or two-photon-subtracted squeezed state at a telecommunication wavelength was generated. As mentioned above, the non-Gaussian operation with the PNRD at a telecommunication wavelength is one of the challenging tasks, and the experimental demonstration will lead to the practical application of the quantum information technologies to the conventional optical communication systems.

2   Experimental Setup

Schematic diagram of the experimental setup is shown in Fig. 1. The pulsed squeezed vacuum was generated by the spontaneous parametric downconversion (SPDC) process in a Type-0 periodically poled lithium niobate adhered-ridge-waveguide (PPLN-ARW) [15]. The Nearly Fourier-limited ∼5 ps laser pulse from

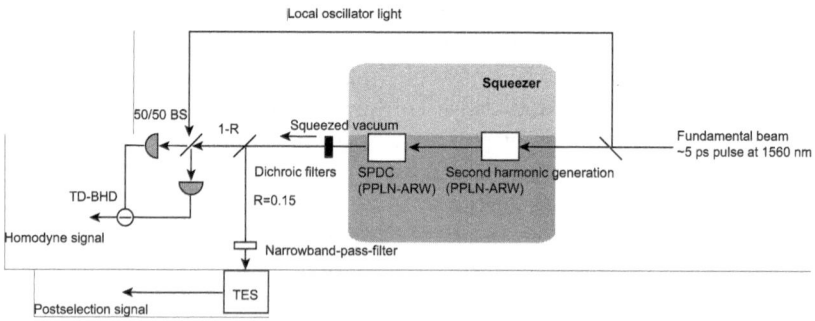

Fig. 1. Experimental setup

an Erbium-doped fiber laser was used as a primary source of the fundamental beam at 1560 nm. It is also used as the local oscillator light for the time-domain balanced homodyne detector (TD-BHD: the overall homodyne efficiency was ~ 0.3). The fraction ($R = 0.15$) of the squeezed vacuum beam was taken out from the homodyne detection mode. The subtracted photons (reflected beam) led to the Ti-TES after passing through the (single-mode) fiber based narrow band pass filter with a pass-band of 0.5 nm. The Ti-TES has a quantum efficiency of 65% and an energy resolution of 0.3 eV with which the Ti-TES can resolve the photon number at 1560 nm (0.8 eV) well. The dark count rate of the Ti-TES was ~ 1000 cps when the discrimination level at a discriminator was set to the level for one-photon detection. To improve the signal-to-noise ratio, the 100 ns gate synchronized with the laser pulse was applied to the DSC. In this way, the dark count rate was decreased to 40 cps, corresponding to a dark count probability of 10^{-4} per 100 ns gate. As regards discrimination level for two-photons detection, a dark count probability was $< 10^{-5}$ per 100 ns gate. For every postselection signals from the Ti-TES, a digital oscilloscope registered homodyne signals. Such postselection procedure provides the non-Gaussian operation. In this experiment, the conditional homodyne detections were carried out in case of not only one-photon but also two-photon subtraction, since the Ti-TES can resolve the photon number. The data segment was stored one after another until the 4000 data segment filled up the oscilloscope's memory. The data acquisition time was $1 \sim 10$ seconds since the trigger photon detection rate was $0.4 \sim 5$ kHz, which depended on the threshold for the Ti-TES. The relative phase between the LO light and the measured beam was changed from 0 to π during the single data acquisition by the piezo transducer (PZT) located in 50/50 beamsplitter (BS). Since the interferometer in the setup was substantially stable during the data acquisition time, it is not stabilized actively. For each data segment, the quadrature amplitude was evaluated from a peak value of the homodyne signal and the time when the homodyne signal was registered, then the Wigner function was reconstructed from $\sim 10,000$ of the quadrature amplitudes using the iterative maximum-likelihood estimation algorithm [16].

3   Experimental Results

Figure 2 shows experimental Wigner functions (left panel), their contour plots (middle panel), and the theoretical Wigner functions (right panel), when the initial quantum state was a -1.25 dB squeezed vacuum (anti-squeezed level was 4.1 dB). The experimental Wigner functions are not corrected for measurement imperfections, that is, the homodyne efficiency and dark counts of the detectors. Figure. 2(a) show a Wigner function of a measured quantum state when the threshold level was set to one-photon detection. We observed the dip in the center of the Wigner functions, which obviously indicates that the non-Gaussian operation was realized. Figure. 2(b) shows a Wigner function when the threshold level was set to two-photon detection. We observed the Wigner function that was clearly different from the case of one-photon detection as shown in figure.

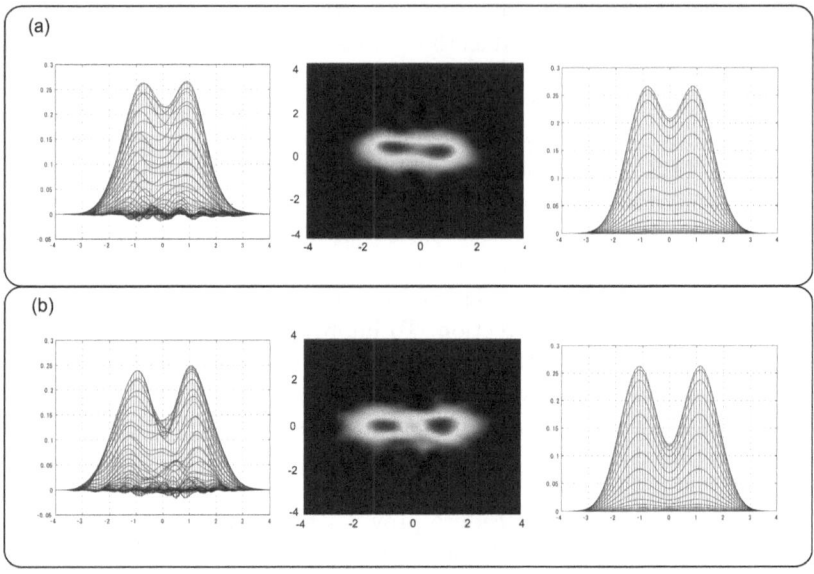

Fig. 2. Experimental results. (a) Wigner functions of one-photon-subtracted squeezed state. (b) Wigner functions of two-photon-subtracted squeezed state.

In our experiments, although we could demonstrate the presence of the dips into the measured Wigner functions, we could not observe the negative value in the measured Wigner functions. In general, the prepared state before the homodyne measurement can be evaluated by correcting the homodyne efficiency. We were not able to observe the negative value in Wigner functions even after the correction of the measurement imperfections. It turns out that the modal purity (the mode matching efficiency of the trigger photon against the mode of LO light for the homodyne measurement) was low to get the ideal Wigner function. We cannot observe the negative value with a modal purity of less than ∼0.7, even if the homodyne efficiency is perfect. If the modal purity is imperfect, a fraction of the trigger photons is out of the mode for the homodyne measurement. Such photons become noise photons that trigger the attenuated squeezed vacuum. Therefore, the fidelity of the generated photon-subtracted squeezed state is degraded. The right panels of Fig.2 (a) and (b) show the theoretical Wigner functions of one- and two-photon-subtracted squeezed states, respectively, taking a modal purity of 0.69, and other experimental conditions into account. The shapes of the experimental Wigner functions are in good agreement with these theoretical ones. According to the numerical result, the modal purity in this experiment is not perfect and the negative values can not be obtained in the corrected Wigner functions.

To improve the fidelity of the non-Gaussian operation, it is essential to improve the modal purity. Using a narrower band-pass-filter in front of the trigger detectors, the modal purity would be improved. In the condition, the trigger detection rate becomes lower. Therefore, the signal-to-noise ratio of the trigger detector must be improved to observe the negative value in the Wigner functions. In use of TES for the trigger detector at a telecommunication wavelength, the signal-to-noise ratio will be improved by reducing the noise counts mainly originating from the black body radiation in the fiber that guides the trigger photons to TES. To get a pure non-Gaussian state, the purity of initial squeezed state should be improved as well. Specifically, the homodyne efficiency and optical losses of setups must be improved.

4   Conclusion

In conclusion, we have demonstrated the non-Gaussian operation against a pulsed squeezed vacuum at a telecommunication wavelength using the photon-number-resolving detector based on the titanium transition-edge-sensor. Improving the imperfect modal purity and homodyne efficiency, the high-purity non-Gaussian operation will be realized, and eventually, it will contribute to progress of a squeezing enhancement and the entanglement purification for a quantum repeater of continuous-variable-based quantum key distribution system over optical fiber networks.

Acknowledgement

This work was partly supported by the National Institute of Information and Communications Technology (NICT).

References

1. Fiurasek, J.: Phys. Rev. Lett. 89, 137904 (2002)
2. Eisert, J., Scheel, S., Plenio, M.B.: Phys. Rev. Lett. 89, 137903 (2002)
3. Giedke, G., Cirac, J.I.: Phys. Rev. A 66, 032316 (2002)
4. Lloyd, S., Braunstein, S.L.: Phys. Rev. Lett. 82, 1784 (1999)
5. Shannon, C.E.: Bell System Tech. J. 27, 379 (Part I) and 623 (Part II) (1948)
6. Holevo, A.S.: IEEE Trans. Inform. Thoery 44, 269 (1998)
7. Schumacher, B., Westmoreland, M.D.: Phys. Rev. A 56, 131 (1997)
8. Giovannetti, V., Guha, S., Lloyd, S., Maccone, L., Shapiro, J.H., Yuen, H.P.: Phys. Rev Lett. 92, 027902 (2004)
9. Gottesman, D., Kitaev, A., Preskill, J.: Phys. Rev. A 64, 012310 (2001)
10. Wenger, J., Brouri, R.T., Grangier, P.: Phys. Rev. Lett. 92, 153601 (2004); Ourjoumtsev, A., Brouri, R.T., Laurat, J., Grangier, P.: Science 312, 83–86 (2006)
11. Neergaard-Nielsen, J.S., Melholt Nielsen, B., Hettich, C., Mlmer, K., Polzik, E.S.: Phys. Rev. Lett. 97, 083604 (2006)
12. Wakui, K., Takahashi, H., Furusawa, A., Sasaki, M.: Opt. Express 15, 3568–3574 (2007)
13. Miller, A.J., Nam, S.W., Martinis, J.M., Sergienko, A.V.: Appl. Phys. Lett. 83, 793 (2003)

14. Fukuda, D., Fujii, G., Yoshizawa, A., Tsuchida, H., Damayanthi, R.M.T., Takahashi, H., Inoue, S., Ohkubo, M.: J. Low Temp. Phys. 151, 100–105 (2008)
15. Kurimura, S., Kato, Y., Maruyama, M., Usui, Y., Nakajima, H.: Appl. Phys. Lett. 89, 191123 (2006)
16. Lvovsky, A.I.: J. Opt. B: Quantum Semiclass. Opt. 6, S556 (2004)
17. Fukuda, D., Fujii, G., Numata, T., Yoshizawa, A., Tsuchida, H., Fujino, H., Ishii, H., Itatani, T., Inoue, S., Zama, T.: Metrologia 46, S288 (2009)

Efficiency of the Eavesdropping in B92 QKD Protocol with a QCM

Michael Siomau[1] and Stephan Fritzsche[2,3]

[1] Max-Planck-Institut für Kernphysik, Postfach 103980,
D-69117 Heidelberg, Germany
siomau@physi.uni-heidelberg.de
[2] Department of Physical Sciences, P.O. Box 3000,
Fin-90014 University of Oulu, Finland
[3] Frankfurt Institute for Advanced Studies, D-60438 Frankfurt am Main, Germany

Abstract. Success of any eavesdropping attack on a quantum crypto-graphic protocol can be reduced by the legitime users if they partially compare their data. It is important to know for the legitime users what is (necessary and enough) amount of data which should be compared to ensure that (possible) illegitime user has an arbitrary small information about the rest of data. To obtain such amount the legitime users need to know efficiencies of all possible attacks for particular cryptographic protocol. In this work we introduce the eavesdropping attack on Ben-nett's B92 protocol for quantum key distribution (QKD) with a quantum cloning machine (QCM). We demonstrate efficiency of suggested attack and compare it with efficiencies of alternative attacks proposed before.

Keywords: B92 QKD protocol, QCM, mutual information, discrepancy.

1 Introduction

Security of quantum cryptography [1] implies that legitime users (say Alice and Bob) *always* can provide such error-correction and privacy amplification codes [2] that final information of an eavesdropper (Eve) about the message transmitted from Alice to Bob is as low as it was set by legitime users before applying the codes. During the codes realization Alice and Bob compare the values of some of their bits from the message. The rigorous problem for Alice and Bob is to construct the codes so that, from one side, the number of compared bits is minimal and, from another side, final Eve's information about the message is an arbitrary small. Thus it is important to estimate the maximal amount of information that Eve may take in an eavesdropping attack.

Efficiency of an eavesdropping attack depends from particular realization of cryptographic protocol and method of Eve's intervention. Since quantum key distribution (QKD) protocols [3]-[8] takes central place in cryptography, we pay our attention on particular QKD scheme – Bennett's B92 protocol [5]. In the B92 protocol only two nonorthogonal quantum states are utilized in order to encode

A. Sergienko, S. Pascazio, and P. Villoresi (Eds.): QuantumCom 2009, LNICST 36, pp. 267–274, 2010.

and transmit information about the cryptographic key. As usual, we suppose that information is sent from Alice to Bob by means of a quantum communication channel. At the beginning of the key distribution protocol, Alice encodes each logical bit, 0 or 1, into two nonorthogonal states, which can be parameterized as

$$|u\rangle = \cos\frac{\theta}{2}|0\rangle + \sin\frac{\theta}{2}|1\rangle ,$$

$$|v\rangle = \sin\frac{\theta}{2}|0\rangle + \cos\frac{\theta}{2}|1\rangle , \tag{1}$$

where $|0\rangle$ and $|1\rangle$ introduce computational basis. The overlap of the states $O(\theta) \equiv |\langle u | v \rangle|^2 = \sin^2\theta$ gives the distance between states $|u\rangle$ and $|v\rangle$ in geometric sense.

Prepared qubits (1) are sent to Bob who measures them randomly either in the standard $\oplus$ or the diagonal $\otimes$ basis. If, for example, Bob has chosen the standard basis $\oplus$ and measured the state $|1\rangle$, he concludes that the state $|+\rangle$ was sent by Alice (because $|0\rangle$ state is orthogonal to measured $|1\rangle$ state). Similarly, if Bob had chosen the diagonal basis $\otimes$ and has measured $|-\rangle$, he concludes that the state $|0\rangle$ was sent. Only this described cases are conclusive for Bob. In fact, in order to obtain the maximum information about received qubits, Bob should minimize the probability to obtain some inconclusive result. As discussed in detail by Ekert *et al.* [9], this can be achieved by a positive operator-valued measurement (POVM), and the best operators for that are

$$G_1 = \frac{1}{1 + \langle u | v \rangle} \left(1 - |u\rangle\langle u|\right) , \tag{2}$$

$$G_2 = \frac{1}{1 + \langle u | v \rangle} \left(1 - |v\rangle\langle v|\right) , \tag{3}$$

$$G_3 = 1 - G_1 - G_2 , \tag{4}$$

where $|u\rangle$ and $|v\rangle$ denote nonorthogonal states (1) that are used in order to encode the information.

After all the qubits have been sent (and measured), Bob tells to Alice numbers of conclusive measurements via a public channel, which can be monitored but not modified by possible eavesdropper. Only those bits (obtained in Bob's conclusive measurements) can be used to construct the key, while all the rest need to be discarded because no definite conclusion can be drawn from the outcome of Bob's measurement.

Although an eavesdropping of the (quantum) communication between Alice and Bob is seriously hampered by the well-known impossibility to produce exact copies of quantum information (non-cloning theorem), the generation of approximate copies with a quantum cloning machine (QCM) [10]-[12] may enlarge the success rate for such an attack. A QCM is a device represented as unitary transformations which provides approximate copies of the input qubit states with certain fidelity

$$F = \langle\psi|\rho|\psi\rangle , \tag{5}$$

where ρ presents a density matrix of each copy and $|\psi\rangle$ is the input state. The QCM is called *symmetric* if the output states are identical to each other, and are said to be *nonsymmetric* otherwise. We restrict ourself by considering only symmetric QCM's which provides two copies from a single input state $1 \to 2$, because such QCM's provide copies with maximal fidelity in comparison with $1 \to N$ QCM's, where $N > 2$, as it was shown by Werner [13].

We always consider *individual* eavesdropping attack in which Eve interacts individually and in the same way with the qubit states traveling from Alice to Bob: Eve copies a single qubit intercepted from the communication channel, and sends one copy to Bob keeping another. Thus, in this work we consider individual eavesdropping attack on B92 protocol with symmetric $(1 \to 2)$ QCM's.

The paper is organized as follows. In the next section we introduce necessary mathematical approach – discrepancy and mutual information – to describe the success of Eve within a cryptographic protocol. In section 3 we analyze efficiencies of the eavesdropping with different symmetric $(1 \to 2)$ QCM's and show that meridional QCM [14] is optimal for eavesdropping in B92 protocol. In section 4 we compare efficiency of the eavesdropping with meridional QCM with efficiencies of alternative attacks discussed by Ekert *et al.* [9] and conclude that suggested eavesdropping attack is optimal for particular choice of states (1) which are used to encode information. We finish this work with conclusions.

2   Discrepancy and Mutual Information

To describe efficiency of an eavesdropping attack it is sufficient to know two values: disturbance of a qubit received by Bob and the mutual information between participants of the cryptographic protocol.

A disturbance in the transmission of the qubits can have very different reasons. Apart from an eavesdropper, the quantum control during the preparation or transmission of the qubits might be incomplete for a given realization of the quantum channel. For all practical realizations of QKD protocols, therefore, a certain error rate (discrepancy) need to be accepted, and an eavesdropper might be successful in extracting information even if the protocol is inherently secure in the ideal case. To quantify discrepancy in the transmission of a single qubit, a convenient measure is the probability that Bob detects an error. If Bob would know the state $|s\rangle$ of one or several qubits in advance, that were sent to him by Alice, he can easily test for a possible eavesdropping attack. In this case, he will receive in general the qubits no longer in a pure but a mixed state that have to be described in terms of their density matrix ρ. Then, the discrepancy that is detectable by Bob is given by

$$D = 1 - \langle s \,|\, \rho | s \rangle \,, \tag{6}$$

Since Bob knows the maximal discrepancy D_{max} for the given channel, he could recognize an eavesdropping attack for $D > D_{max}$ and discard the key accordingly. Later we assume, for simplicity, that the quantum control of the given transmission is complete. With this assumption all discrepancy detected by Bob is caused by Eve.

A central question for Eve is of how much information she can extract from the transmission of the key and what is the price in terms of discrepancy. From the initial agreement between Alice and Bob about the basis states which are to be chosen randomly, Eve might know that Alice prepares the qubits in one of several states with proper probabilities. This initial agreement has to be made typically independent of a particular protocol. Before Eve has measured a given qubit, her (degree of) ignorance is given by Shannon's entropy. After the measurement, she increased her knowledge about the system by decreasing this entropy, a measure that is called the mutual information I_{AE} that Eve has acquired about Alice's message due to the measurement. Obviously, Eve will try to obtain as much information as possible keeping the discrepancy $D < D_{max}$.

Since Eve copy (attack) each qubit independently within B92 protocol as they are sent from Alice to Bob, as output of her cloning transformation, she then obtains two copies of one of the two possible states $\rho_{|0\rangle}$ and $\rho_{|+\rangle}$ (which just correspond to the two input states $|0\rangle$ and $|+\rangle$) with a fidelity as defined by the given QCM. While Eve transmits one of her copies further to Bob, she could measure the second copy following the same procedure as Bob.

To calculate the mutual information between Alice and Eve that is to be extracted from the eavesdropping, we can follow the procedure as described in Ref. [15]. Using the POVM elements (2)-(4), the probability for Eve to obtain the outcome μ is

$$P_{\mu i} = \text{Tr}(G_\mu \rho_i), \tag{7}$$

and where the operators ρ_i refer to the two possible states $\{\rho_{|0\rangle}, \rho_{|+\rangle}\}$ of her copy. After the measurement, when she has obtained a particular outcome μ, the *posterior* probability $Q_{i\mu}$ that ρ_i was prepared by Alice is

$$Q_{i\mu} = \frac{P_{\mu i} p_i}{q_\mu}, \tag{8}$$

where $q_\mu = \sum_j P_{\mu j} p_j$, and $p_j = 1/2$ is the probability for sending the states $|0\rangle$ and $|+\rangle$ within the B92 protocol. With these probabilities, the Shannon entropy (which was $H = -\log_2(1/2) = 1$ initially), becomes

$$H_\mu = -\sum_i Q_{i\mu} \log Q_{i\mu}, \tag{9}$$

once the result μ was obtained, and hence the mutual information is

$$I_{AE} = H - \sum_\mu q_\mu H_\mu. \tag{10}$$

The mutual information between Alice and Bob I_{AB} can be obtained in the same manner as it described above. In case of the eavesdropping attack with a symmetric QCM the mutual information between Alice and Bob is equal to the mutual information between Alice and Eve $I_{AE} = I_{AB}$, since Eve and Bob receive the copies with equal fidelities.

With the help of discrepancy (6) and mutual information (10) values we may now describe efficiency of the eavesdropping in B92 QKD protocol with a QCM.

3   Optimal QCM for the Eavesdropping

QCM's have found a remarkable application in quantum cryptography, since they introduce optimal attacks for particular QKD protocols [12]. For example, *universal* QCM [10], that provides copies with constant fidelity $F = 5/6 \approx 0.83$ for an arbitrary input state, is optimal for the eavesdropping in six-state protocol [6]. Although only states from a one-dimensional subspace of original two-dimensional Hilbert state space are used to encode information in BB84 protocol [3], *equatorial* QCM [11], that realize copying of the states form the subspace with fidelity $F = 1/2 + \sqrt{1/8} \approx 0.85$, is optimal for the eavesdropping within this protocol. Obviously, optimal QCM for the eavesdropping in B92 protocol should create copies with higher fidelity than universal and equatorial QCM's for restricted set of states which covers states (1). Recently presented *meridional* QCM [14] is optimal QCM for the eavesdropping in B92 protocol, since it provides copies with fidelity

$$F(\theta) = \frac{9}{10} - \frac{1}{5}\sin^2\theta + \frac{1}{5}\sin\theta, \tag{11}$$

for the input states $|s\rangle = \cos\frac{\theta}{2}|0\rangle + \sin\frac{\theta}{2}|1\rangle$, where $0 \leq \theta \leq \pi$. All states $|s\rangle$ belong to Eastern meridian (that includes the three states $\{|0\rangle, |1\rangle, |+\rangle\}$) of the main circle on the Bloch sphere. It is easy to see that for the input states $|s\rangle$ fidelity (11) varies a little $0.90 \leq F \leq 0.95$ and may favor this region to produce quantum copies with fidelity higher than in case of universal and equatorial quantum copying. Figure 1 displays the behavior of this fidelity for states $|s\rangle$ from Eastern meridian.

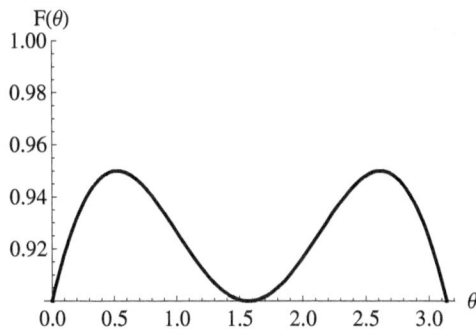

Fig. 1. Fidelity (11) of meridional QCM as function of the angle θ for the states from Eastern meridian on the Bloch sphere

4   Efficiency of the Eavesdropping with Meridional QCM

In the previous section we have shown that meridional QCM is optimal QCM
for the eavesdropping in B92 protocol. With the help of definitions of the mutual
information (10) and discrepancy (6) values, taking into account behavior of the
fidelity function (11), we may analyze the success of Eve's attack with meridional
QCM within B92 protocol.

The mutual information between Alice and Eve as function of parameter
θ of states (1) is shown at Figure 2 with solid line. The mutual information
monotonically decrease from the value $I(0) = 0.5310$ (what correspond to the
case when states (1) are orthogonal) to $I(\pi/2) = 0$ (when the states coincide
with each other). In order to show the upper bound of information which can
be achieved by Eve within B92 protocol, we also display (with dashed line) the
mutual information in the case of Eve's intervention with (unreal) 'ideal' QCM
that provides two exact copes of the given input.

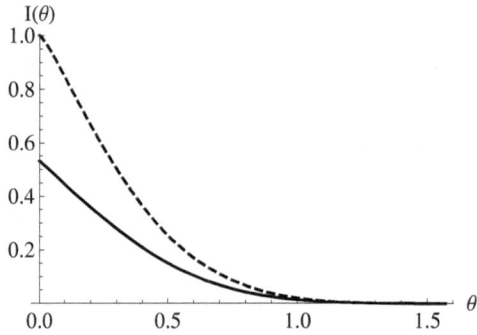

Fig. 2. This plot gives the mutual information between Alice and Eve $I(\theta)$ as function of the angle θ. Solid line introduces the mutual information in the case of Eve's
intervention with meridional QCM. Dashed line introduce unreal situation when Eve
attacks the qubits with 'ideal' QCM.

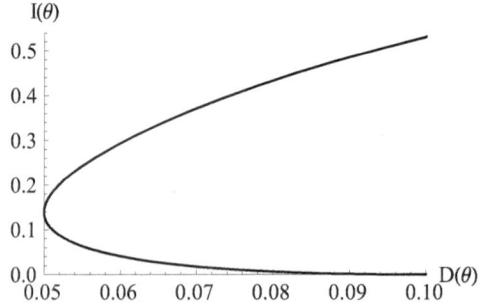

Fig. 3. This plot gives parametric dependence of the mutual information between Alice
and Eve $I(\theta)$ form discrepancy $D(\theta)$. See text for further discussion.

Discrepancy $D(\theta)$ of the qubit received by Bob as function of the parameter θ of states (1) is given by equation $D(\theta) = 1 - F(\theta)$ (where $F(\theta)$ denotes the fidelity of the copy received by Bob), as it follows from definitions (5) and (6).

Having found functions $I(\theta)$ and $D(\theta)$ we present at Fig. 3 parametric dependence of the mutual information between Alice and Eve from discrepancy. The mutual information equals zero when the two states are coincide (and the overlap $O(\pi/2) = 1$). In this case Alice always prepares *one state* and sends it to Bob via communication channel. Formally, Eve introduce discrepancy $D = 0.10$ in the eavesdropping (as it follows from Eqn. (11)), but single (known) state can not be used to construct a key. With decreasing of the overlap the mutual information increases and discrepancy becomes lower due to the fidelity function (11) behavior. When the overlap between states (1) equals $O(\pi/6) = 1/4$ the mutual information is $I_{AE}(\pi/6) = 0.1384$ and discrepancy has value $D(\pi/6) = 0.05$. With further decreasing of the overlap the mutual information still increase, but discrepancy also increase.

To bring our analysis to the end we now compare efficiency of the eavesdropping with meridional QCM and efficiencies of alternative attacks proposed before [9]. In particular, Ekert *et al.* showed that *intercept-resend* attack, when Eve measures intercepted from communication channel qubit and sends to Bob qubit prepared in state $|u\rangle$ or $|v\rangle$ according to the outcome of her measurements, is optimal attack if states $|u\rangle$ and $|v\rangle$ has small overlap $O \ll 1$. In case of large overlap $O \approx 1$ the eavesdropping with *entangled probe* gives advantage for Eve to obtain maximal information about intercepted qubits causing minimal discrepancy. However, for a wide range of the overlap $0.067 \leq O \leq 0.5$, the eavesdropping with meridional QCM is more effective than both mentioned attacks. For example, if the overlap between the states $O(\pi/6) = 1/4$, discrepancy caused by Eve is $D(\pi/6) = 0.05$ and the mutual information between Alice and Eve is equal to the mutual information between Alice and Bob.

5   Conclusions

We have analyzed the eavesdropping attack on the simplest QKD scheme, where the key encoding based on two nonorthogonal quantum states, with a QCM. From this analysis, it is found that Eve, the eavesdropper, can obtain more information from meridional than from universal or equatorial QCM's causing lower disturbance to intercepted qubits. We have also demonstrated that the attack supported with meridional QCM is the most effective (i.e. optimal) attack on B92 protocol comparing to intercept-resend attack and attack with entangled probe for particular choice of the states.

Since eavesdropping attack on B92 with meridional QCM is optimal (for particular choice of the states), it introduces the maximal amount of information that Eve may take in an eavesdropping attack. Thus, obtained result should be taken into account in construction of error-correction and privacy amplification codes for B92 QKD protocol.

References

1. Lo, H.-K., Chau, H.-F.: Unconditional security of quantum key distribution over arbitrarily long distances. Science 283, 2050–2056 (1999); Shor, P.W., Preskill, J.: Simple proof of security of the BB84 quantum key distribution protocol. Phys. Rev. Lett. 85, 441–444 (2000)
2. Nielsen, M.A., Chuang, I.L.: Quantum computation and quantum information. Cambridge University Press, Cambridge (2000); references therein; Kilin, S.Ya., Choroshko, D.B., Nizovtsev, A.P.: Quantum Cryptography. Belorus science, Minsk (2007); references therein
3. Bennett, C.H., Brassard, G.: Quantum cryptography: Public key distribution and coin tossing. In: Proceedings of IEEE International Conference on Computers, Systems and Signal Proceeding, Bangalore, India, pp. 175–179 (1985)
4. Ekert, A.K.: Quantum cryptography based on Bell's theorem. Phys. Rev. Lett. 67, 661–663 (1991)
5. Bennett, C.H.: Quantum cryptography uning any two nonorthogonal states. Phys. Rev. Lett. 68, 3121–3124 (1992)
6. Bruß, D.: Optimal eavesdropping in quantum cryptography with six states. Phys. Rev. Lett. 81, 3018–3021 (1998)
7. Cerf, N.J., Levy, M., Van Assche, G.: Quantum distribution of gaussian keys using squeezed states. Phys. Rev. A 63, 052311 (2001); Grosshans, F., Grangier, P.: Continuous variable quantum cryptography using coherent states. Phys. Rev. Lett. 88, 057902 (2002)
8. Scarani, V., Acin, A., Ribordy, G., Gisin, N.: Quantum cryptography protocols robust against photon number splitting attacks for weak laser pulse implementation. Phys. Rev. Lett. 92, 057901 (2004)
9. Ekert, A.K., Huttner, B., Palma, G.M., Peres, A.: Eavesdropping on quantum-cryptographical systems. Phys. Rev. A 50, 1047–1056 (1994)
10. Bužek, V., Hillery, M.: Quantum copying: Beyond the non-cloning theorem. Phys. Rev. A 54, 1844–1852 (1996)
11. Bruß, D., Cinchetti, M., D'Ariano, G.M., Macchiavello, C.: Phase covariant quantum cloning. Phys. Rev. A 62, 012302 (2000)
12. Scarani, V., Iblisbir, S., Gisin, N.: Quantum cloning. Rev. Mod. Phys. 77, 1225–1256 (2005)
13. Werner, R.F.: Optimal cloning of pure states. Phys. Rev. A 58, 1827–1832 (1998)
14. Siomau, M., Fritzsche, S.: High-fidelity copies from a symmetric $1 \rightarrow 2$ quantum cloning machine (2009), ArXiv:0906.1453v1
15. Peres, A.: Quantum Theory: Concepts and Methods. Kluwer Acad. Publ., Dordrecht (2002)

Improvement of Lattice-Based Cryptography Using CRT

Thomas Plantard, Mike Rose, and Willy Susilo

School of Computer Science and Software Engineering
University of Wollongong, Wollongong NSW, Australia
{thomaspl,mrose,wsusilo}@uow.edu.au

Abstract. In this paper, we first critically analyze two existing lattice-based cryptosystems, namely GGH and Micciancio, and identify their drawbacks. Then, we introduce a method for improving the implementation of GGH using the Chinese Remainder Theorem (CRT). Furthermore, we also propose another cryptosystem optimized for CRT, drawing on the strengths of both cryptosystems. We provide a fair comparison between our scheme and the existing ones.

1 Introduction and Motivation

With the continuous advancements in the field of quantum computing, the security of many existing asymmetric key cryptosystems has been demonstrated to be broken in the theoretical sense, with computational insecurity of these schemes being dependent only on technological advancements. As a result, new one-way trapdoor functions must be developed that will remain secure after quantum computers become available. One promising avenue of research in this direction is lattice-based cryptography.

Recent advancements in the field of lattice-based cryptography have brought a sustained interest in producing a lattice-based cryptosystem that runs in a similar space and time complexity as existing conventional asymmetric key cryptosystems. Two specific related cryptosystems showing much promise are the cryptosystems introduced by Goldreich, Goldwasser and Halevi [5] (GGH), and its modification and improvement by Micciancio [9].

While the security of any cryptosystem in an information theoretic sense relies on the space and time complexity of the 'trapdoor function' being used, we need a stronger definition to have the cryptosystem implemented and deemed practical for use. It is insufficient to define operations in polynomial time and non-polynomial time when discussing practical implementations of the cryptosystem.

Even though the cryptosystem may be provably secure, if the implementation speeds involved are too slow for an acceptable security parameter, then it is clear that alternatives must be sought. After careful analysis of both lattice-based cryptosystems analyzed in this paper, we found that both have significant space or time complexity drawbacks, possibly preventing widespread adoption.

A. Sergienko, S. Pascazio, and P. Villoresi (Eds.): QuantumCom 2009, LNICST 36, pp. 275–282, 2010.
© Institute for Computer Sciences, Social-Informatics and Telecommunications Engineering 2010

In this paper we propose a method for improving the speed of Babai's Round-Off CVP approximation algorithm [1] in lattices using the Chinese Remainder Theorem (CRT). We then formulate a new lattice-based cryptosystem with implementation as a prime consideration. This new cryptosystem is based on the work done by Goldreich et al. and Micciancio but has much faster encryption and decryption speeds when implemented on common hardware platforms.

2   Lattice Theory and Lattice-Based Cryptography

Definition 1 (Lattice). *A lattice $\mathcal{L}$ is a discrete sub-group of $\mathbb{R}^n$, or equivalently the set of all the integral combinations of $d \leq n$ linearly independent vectors over $\mathbb{R}$.*

$$\mathcal{L} = \mathbb{Z} b_1 + \cdots + \mathbb{Z} b_d, \quad b_i \in \mathbb{R}^n.$$

$B = (b_1, ..., b_d)$ is called a basis of $\mathcal{L}$ and d, the dimension of $\mathcal{L}$.
We will refer $\mathcal{L}_B$ as a lattice of basis B.

For a given lattice $\mathcal{L}$, there exists an infinite number of bases. However, the Hermite Normal Form basis (Definition 2) is unique [2].

Definition 2 (HNF). *Let $\mathcal{L}$ be an integer lattice of dimension d and $H \in \mathbb{Z}^{d,n}$ a basis of $\mathcal{L}$. H is a Hermite Normal Form basis of $\mathcal{L}$ if and only if*

$$\forall 1 \leq i, j \leq d \quad H_{i,j} \begin{cases} = 0 & \text{if } i > j \\ \geq 0 & \text{if } i \leq j \\ < H_{j,j} & \text{if } i < j \end{cases}$$

The HNF basis can be computed from a given basis in a polynomial time [6].
    Many lattice theory problems are based on distance minimization, as determined via the euclidean norm.

Definition 3 (Euclidean norm). *Let w be a vector of $\mathbb{R}^n$. The euclidean norm is the function $\|.\|$ defined by $\|w\| = \sqrt{\sum_{i=1}^{n} w_i^2}$.*

There are many different problems based on the minimization of distance in lattice theory. In this paper we focus on the one used by the cryptosystem we analyzed, namely the *Closest Vector Problem.*

Definition 4 (γ-CVP). *Let w be a vector in a lattice $\mathcal{L}$. The Closest Vector Problem is to find a vector $u \in \mathcal{L}, \forall v \in \mathcal{L}, v \neq u, \|w - u\| \leq \gamma\|v - u\|.*

CVP (for $\gamma = 1$) has been demonstrated to be NP-hard by Emde Boas [3]. However, by limiting ourselves to a special instance of CVP, we will be able to use a good basis to solve CVP using two existing algorithms proposed by Babai [1]. In 1986, Babai proposed two polynomial methods to solve CVP: the *nearest plane* and the *round-off* methods. These algorithms solve CVP within $\gamma = 2^{d/2}$ and $\gamma = 1 + 2d \left(\frac{9}{2}\right)^{d/2}$, respectively. Babai's algorithms use an LLL-reduced basis [7].

2.1   Lattice-Based Cryptography

In 1996, Goldreich, Goldwasser and Halevi (GGH) [5] proposed an efficient way
to use lattice theory to build a cryptosystem inspired by McEliece cryptosystem
[8] and based on the Closest Vector Problem (CVP). Their practical proposition
of a cryptosystem was attacked and broken severely by Nguyen in 1999 [10].
However, the general idea is still viable. Until then, the other propositions were
made using the same principle [9]. In the following, we briefly review the GGH
cryptosystem. A GGH cryptosystem comprises the following algorithms.

- **KeyGenerate:** Compute a "good basis" A and a "bad basis" B of a lattice
  $\mathcal{L}$, $\mathcal{L}(A) = \mathcal{L}(B)$. Provide B as public and keep A secret.
- **Encrypt:** To encrypt a plaintext vector $p \in \mathbb{Z}^n$: use the bad basis to create
  a vector v in the lattice $\mathcal{L}$, $v = pB$. Publish the encrypted message which is
  the addition of this vector with a random error vector $e \in \mathbb{Z}^n$: $c = v + e$.
- **Decrypt:** Use the good basis to find the closest vector in the lattice of the
  encrypted message c. The closest vector of the encrypted message c is the
  message vector v^1. Using this, obtain the plaintext vector p.

The important points for the security and efficiency of those cryptosystems are
defined as follows.

i) It is easy to compute a "bad basis" from a "good basis", but it is difficult
   to compute a "good basis" from a "bad basis".
ii) It is easy to create a random vector of a lattice even with a "bad basis".
iii) It is easy to find the closest vector with a "good basis" but difficult to do
   so with a "bad basis".

In 2001, Micciancio [9] proposed some major improvements of the speed and the
security of GGH. In this scheme, the public key uses a Hermite Normal Form
(HNF) for the bad basis. In this scheme, to encrypt a message, we perform a
modulo lattice reduction on the plaintext vector (Algorithm 1). In effect, this is
done by using the small plaintext vector as the error vector in the GGH scheme,
i.e., $c = rB + p$ where r is some minimal vector. The advantage of this scheme
is that we are able to use any basis inverse to recover the plaintext and hence
do not need to store a transformation matrix, unlike GGH's scheme.

   To perform decryption, Micciancio's scheme uses Babai's nearest-plane CVP
approximation, as this provides a better approximation than the Round-Off
algorithm [1].

2.2   Drawbacks of Existing Schemes

GGH. A thorough test implementation of the GGH cryptosystem revealed two
major shortcomings. Firstly, GGH suffers from slow decryption speeds. Using
a theorized computationally secure dimension of 1000, decryption speeds on
a conventional, modern PC were around 2 seconds, compared to conventional

---

[1] Under the supposition that the norm of e is sufficiently small.

Algorithm 1. Vector Reduction modulo a HNF Basis

Input : $B \in \mathbb{Z}^{n,n}$ a HNF basis and $p \in \mathbb{Z}^n$ a vector.
Output: $c \in \mathbb{Z}^n$ such that $(c - p) \in \mathcal{L}(B)$ and $\forall i, 0 \le c_i < B_{i,i}$.
begin
    $c \leftarrow p$
    for $i \leftarrow n - 1$ **to** 0 **do**
        $q \leftarrow \lfloor c_i / B_{i,i} \rfloor$
        $c \leftarrow c - q \times B_i$
    end
end

---

cryptosystems operating orders of magnitude faster. This significantly limits the practical applicability of this cryptosystem. Secondly, the storage requirements of public keys for GGH are enormous. In a presumed computationally secure dimension of 1000, public key sizes were seen to be over 290Mb. This is impractical in a public key infrastructure with a large number of clients, such as the Internet or large internal networks.

Micciancio. We found that implementing the Micciancio cryptosystem also revealed some shortcomings. Firstly, the use of Nearest-Plane CVP recovery in the Micciancio cryptosystem's decryption phase gave a much lower decryption speed than GGH's round-off CVP recovery method (Figure 1). Secondly, we found that Micciancio requires a large amount of memory to perform key generation, which became larger than the 4Gb maximum on our 32-bit test PC at higher dimensions, preventing key generation at these dimensions on common, consumer hardware.

2.3   Chinese Remainder Theorem

The Chinese Remainder Theorem allows the representation of a large, variable-precision integer by its residue modulo some small moduli. This technique is often used to replace arithmetic on large integers with operations over these small moduli.

Theorem 1 (Chinese Remainder Theorem). *Let $p_i \in \mathbb{N}$ n coprimes integers, $P = \prod_{i=1}^{n} p_i$ and $P_i = P/p_i$. Then, for any n-tuple a_i there exists an unique integer $0 \le A < P$ such that $a_i = A \bmod p_i$,*

$$A = \sum_{i=1}^{n} a_i (P_i^{-1} \bmod p_i) P_i \bmod P.$$

Motivation. We apply the Chinese Remainder Theorem to lattice-based cryptosystems by operating the cryptosystem in an integer ring with an order greater than the largest element, for several reasons. Firstly, performing our calculations in these 'small' finite fields allows us to take advantage of the various platform optimizations for integer arithmetic, since working with variable precision integers larger than the implementation machine's word-size imposes a significant overhead. Secondly, we reduce the need to rely on using a highly optimized implementation of variable-precision integer and matrix arithmetic, specific for each particular platform. While such libraries are readily available for common PC computing platforms, we felt that providing a platform-independent variable precision optimization would be valuable, especially for embedded platforms such as smart cards and cellular telephones. Thirdly, since each finite field calculation is independent of the others it is indeed possible to perform each finite field operation in parallel. While this has not been implemented in this library due to the lack of thread support in Shoup's NTL library [11], we see no intrinsic property of the cryptosystems discussed that would prevent this from operating as intended.

Construction. We can construct a sufficiently sized ring in the following way. First, we assess the bound of the size of the coefficients to be calculated. If the coefficients involved include negative coefficients, we must double this bound in order to cover these values. Next, we construct many small finite fields of prime order $p < 2^b$, where b is our target platform's word size, until the product of these primes is above our coefficient bound. We perform our calculations independently in these fields and once these have completed, we perform a simple CRT reconstruction to calculate the final value.

A significant hurdle to adopting this approach for all lattice problems, however, is that it is often difficult to find an appropriate tight bound of the size of the coefficients involved. Due to this, we design our new scheme around this problem, creating a private basis with which we are able to calculate a satisfactorily tight bound on the coefficients to allow for faster key generation.

Memory usage. Since each finite field is independent, it is possible to decrypt a plaintext via CRT serially with respect to the key, i.e., only loading each matrix over some finite field into memory as we require it. This has great benefits for memory utilization especially for embedded systems as we are only required to store a matrix of standard integers in primary memory at any given time rather than a matrix of much larger variable-precision integers. In the case of lattice dimension 1000, it can be seen that the memory usage would only be approximately 4Mb at any given time. Obviously in such a case, decryption speeds would be I/O bound in the case of loading each matrix in from a hard drive or flash-based storage.

2.4   Improvement of GGH Using CRT

We were able to optimize GGH via CRT by placing R^{-1} into small finite fields and performing the multiplication inside Babai's Round-Off step over these fields. This not only yielded significantly faster decryption speeds, but also provided the platform flexibility discussed above.

Firstly, we define the following scalar function which will be used in our CRT reconstruction.

$$Q(p_i) = \frac{P}{p_i}, P = \prod p_i$$

Secondly, to avoid rational arithmetic, we multiply $R^{-1} \in \mathbb{Q}^{n,n}$ by $\det R$ to obtain $S \in \mathbb{Z}^{n,n}$ ($R^{-1} = S/\det R$). We define a function R' to represent the following:

$$R'(p_i) = R^{-1} \times \det R \times Q(p_i)^{-1} \bmod p_i$$

Since we will be multiplying by $Q(p)^{-1}$ in the reconstruction phase, we are able to save crucial decryption time by precomputing this value in the Key Generation phase.

Algorithm 2. CVP Round-Off using CRT

Input : $v \in \mathbb{Z}^n$ the input vector, $R \in \mathbb{Z}^{n,n}$ a basis of $\mathcal{L} \subseteq \mathbb{Z}^{n,n}$.
Output: $w \in \mathcal{L}$ a close vector of v in the lattice $\mathcal{L}$
begin
    $x \leftarrow 0$
    foreach p_i **do** $x \leftarrow x + Q(p_i) \times (c \times R'(p_i) \bmod p_i) \bmod P$
    $w \leftarrow \lfloor x/det(R) \rceil \times R$
end

Due to the complexity of Micciancio's Nearest-Plane CVP decryption method, we were unable to modify this to work in a CRT environment.

3   New Scheme

We design this new scheme specifically to provide faster operations, in a particular faster implementation, while still maintaining a similar structure to existing cryptosystems. Specifically, this involved consideration in the design for decryption using the CRT round-off method discussed earlier.

3.1   Key Generation

Private Basis. To create the private basis, we use GGH's private basis construction, namely $R \leftarrow bI + M$. This was chosen over Micciancio's basis construction for two reasons. Firstly, it allows us to make generalizations about the bound on the size of $||R||_\infty$ which allows for much faster key generation as we do not need to perform a full matrix inversion. Secondly, and perhaps more importantly, it provides a more orthogonal basis with which to perform decryption, which in turn, decreases the size necessary to ensure correct decryption using CVP Round-off. This in turn allows us to decrease the size of the coefficients while keeping the same security parameter, saving storage and transmission space and increasing efficiency.

Public Basis. To create the public basis, we use Micciancio's method of applying a HNF reduction on the private basis, as this provided a greater level of security, simplified key storage and much smaller public keys. Optionally, an LLL reduction can also be applied to this key to reduce the size of the coefficients, however we found this step to be unnecessary.

Precomputation. We are able to significantly speed up the decryption phase by precomputing all the required values of the functions $Q(p)$ and $R'(p)$ in the key generation phase and storing these values in a look-up table. This increases the speed of the decryption step, at the expense of a lower key generation speed.

Algorithm 3. KeyGenerate

Input : $n \in \mathbb{N}$ the security parameter.
Output: $B \in \mathbb{Z}^{n,n}$ the public key, $R \in \mathbb{Z}^{n,n}$ the private key.
begin

$\quad M \leftarrow 0$ **for** $i, j \leftarrow 0$ **to** $n-1$ **do** $M_{i,j} \leftarrow Rand(-1, 1)$
$\quad b \leftarrow \left\lceil 2\sqrt{2n/3} \right\rceil$ **repeat** $b \leftarrow b+1$ **until** $\|(bI + M)^{-1}\|_\infty \leq 1/2$
$\quad R \leftarrow bI + M$
$\quad B \leftarrow HNF(R)$

end

3.2 Encryption

For encryption, we use Micciancio's method of modulo lattice reduction with the public basis. We felt that this provided excellent speed and provided strong notions of security. We modified Micciancio's construction by limiting the encryption vector domain to $\{-1, 0, 1\}$ and by using $\|R^{-1}\|_\infty < 1/2$, we can ensure (See [5]) that there will not be any decryption error.

3.3 Decryption

Decryption is of a similar form to the improved GGH decryption method using CRT except with a minor change to reflect the encoded message being in the error vector rather than the lattice point. i.e. given the lattice vector w, we calculate the plaintext $p = c - w$.

4 Implementation and Performance Analysis

Since we are primarily concerned with the implementation aspects of the cryptosystems discussed, we have coded both existing cryptosystems as well as our new scheme in C++ using Victor Shoup's NTL [11] compiled against GNU MultiPrecision Library (GMP) [4]. We feel that these libraries are the most appropriate choice for implementation as they are created with runtime speed a major factor in the design while maintaining numerical stability and correctness. With this in mind, the authors feel that this choice of implementation forms a good basis for comparison. These values were obtained on a 2.1Ghz Intel Core 2 Duo platform with 4Gb RAM.

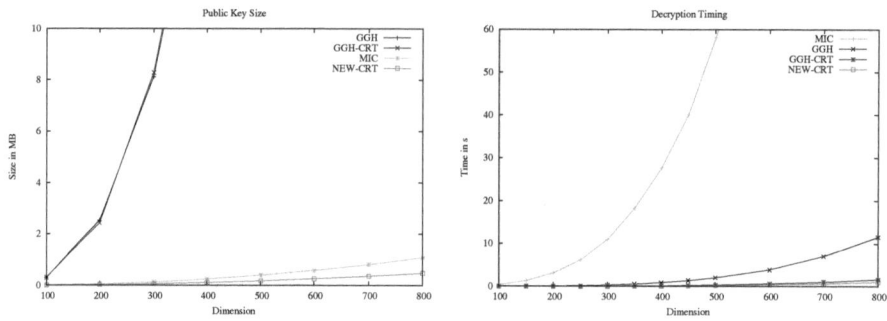

Fig. 1. Performance Results

Table 1. Speeds and keysizes

Dimension	400				800			
	Enc.	Dec.	Pub. Key	Priv. Key	Enc.	Dec.	Pub. Key	Priv. Key
GGH	0.04s	0.86s	18.7 MB	43.0 MB	0.23s	11.57s	153.0 MB	374.4 MB
Micciancio	0.01s	27.69s	241.4 kB	95.4 MB	0.02s	259.94s	1.1 MB	861.6 MB
GGH (CRT)	0.04s	0.18s	19.3 MB	72.3 MB	0.23s	1.63s	150.9 MB	606.4 MB
New Scheme	0.01s	0.11s	109.8 kB	97.0 MB	0.02s	1.05s	476.3 kB	848.0 MB

References

1. Babai, L.: On Lovász' lattice reduction and the nearest lattice point problem. Combinatorica 6, 1–13 (1986)
2. Cohen, H.: A course in computational algebraic number theory. Graduate Texts in Mathematics, vol. 138. Springer, Heidelberg (1993)
3. Van Emde Boas, P.: Another NP-complete Problem and the Complexity of Computing Short Vectors in Lattices. Tech. rep. 81-04, University of Amsterdam (1981)
4. GNU Multiple Precision Arithmetic Library, http://gmplib.org
5. Goldreich, O., Goldwasser, S., Halevi, S.: Public-Key Cryptosystems from Lattice Reduction Problems. Tech. rep., Massachusetts Institute of Technology (1996)
6. Kannan, R., Bachem, A.: Polynomial Algorithms for Computing the Smith and Hermite Normal Forms of an Integer Matrix. J. of Computing 8, 499–507 (1979)
7. Lenstra, A.K., Lenstra, H.W., Lovász, L.: Factoring polynomials with rational coefficients. Mathematische Annalen 261, 513–534 (1982)
8. McEliece, R.J.: A Public-Key Cryptosystem Based On Algebraic Coding Theory. Deep Space Network Progress Report 44, 114–116 (1978)
9. Micciancio, D.: Improving Lattice Based Cryptosystems Using the Hermite Normal Form. In: Silverman, J.H. (ed.) CaLC 2001. LNCS, vol. 2146, pp. 126–145. Springer, Heidelberg (2001)
10. Nguyên, P.Q.: Cryptanalysis of the Goldreich-Goldwasser-Halevi Cryptosystem from Crypto 1997. In: Wiener, M. (ed.) CRYPTO 1999. LNCS, vol. 1666, pp. 288–304. Springer, Heidelberg (1999)
11. Shoup, V.: NTL: A Library for doing Number Theory, http://www.shoup.net/ntl

The Case for Quantum Key Distribution

Douglas Stebila[1], Michele Mosca[2,3], and Norbert Lütkenhaus[2]

[1] Information Security Institute, Queensland University of Technology,
Brisbane, Australia
douglas@stebila.ca
[2] Institute for Quantum Computing, University of Waterloo,
Waterloo, Ontario, Canada
mmosca@iqc.ca, nlutkenhaus@iqc.ca
[3] Perimeter Institute for Theoretical Physics, Waterloo, Ontario, Canada

Abstract. Quantum key distribution (QKD) promises secure key agreement by using quantum mechanical systems. We argue that QKD will be an important part of future cryptographic infrastructures. It can provide long-term confidentiality for encrypted information without reliance on computational assumptions. Although QKD still requires authentication to prevent man-in-the-middle attacks, it can make use of either information-theoretically secure symmetric key authentication or computationally secure public key authentication: even when using public key authentication, we argue that QKD still offers stronger security than classical key agreement.

Keywords: Quantum key distribution, quantum cryptography.

1 Introduction

Since its discovery, the field of quantum cryptography — and in particular, quantum key distribution (QKD) — has garnered widespread technical and popular interest. The promise of "unconditional security" has brought public interest, but the often unbridled optimism expressed for this field has also spawned criticism and analysis [1,2,4,5].

QKD is a new tool in the cryptographer's toolbox: it allows for secure key agreement over an untrusted channel where the output key is entirely independent from any input value, a task that is impossible using classical[1] cryptography. QKD does not eliminate the need for other cryptographic primitives, such as authentication, but it can be used to build systems with new security properties. As experimental research continues, we expect the costs and challenges of using QKD to decrease to the point where QKD systems can be deployed affordably and their behaviour can be certified.

---

[1] All computation must be viewed as taking place in a physical system described by particular laws of nature. By *classical cryptography*, we mean cryptography taking place in a computational and communication system modelled with classical physics (i.e., non-quantum-mechanical and non-relativistic physics).

A. Sergienko, S. Pascazio, and P. Villoresi (Eds.): QuantumCom 2009, LNICST 36, pp. 283–296, 2010.
© Institute for Computer Sciences, Social-Informatics and Telecommunications Engineering 2010

Through the rest of this paper, we restrict our discussion on quantum cryptography to quantum key distribution (QKD). Many other quantum cryptographic primitives exist — quantum private channels, quantum public key encryption, quantum coin tossing, blind quantum computation, quantum money — but almost all require a medium- to large- scale quantum computer for implementation. QKD, on the other hand, has already been implemented by many different groups, has seen attempts at commercialization, and thus its potential role in upcoming security infrastructures merits serious examination.

There are three phases (which are sometimes intertwined) to establishing secure communications:

1. *Key agreement:* Two parties agree upon a secure, shared private key.
2. *Authentication:* Allows a party to be certain that a message comes from a particular party. In order for key agreement to avoid man-in-the-middle attacks, authentication of some form must be used.
3. *Key usage:* Once a secure key is established, it can be used for encryption (using a one-time pad or some other cipher), further authentication, or other cryptographic purposes.

QKD is just one part of this overall information security infrastructure: two parties can agree upon a private key, the security of which depends on no computational assumptions, and which is entirely independent of any input to the protocol.

If we live in a world where we can reasonably expect public key cryptography to be secure in the short- to medium-term, then the combination of public key cryptography for authentication and QKD for key agreement can lead to very strong long-term security with all the convenience and benefits we have come to expect from distributed authentication in a public key infrastructure.

If we live in a world where public key cryptography can no longer be employed safely, we must revert to doing classical key establishment over a private channel, such as a trusted courier, or use QKD. QKD would still require a private channel to establish authentication keys. Instead of just establishing short authentication keys, a private channel could in principle be used to exchange an amount of key comparable to what QKD could produce over a long period of time. However, in this setting QKD can have an advantage because the amount of private communication required is much less and because the session keys output by QKD are independent from the keys transmitted across the private channel, leaving a short time window in which compromised keying material can affect the security of future sessions. How much of an advantage this is in practice will depend on the nature of the private channel in question and the trust assumptions.

If we live in a world where there exist public key agreement schemes that are believed to be secure indefinitely, then there is a reduced case for QKD, but it is still of interest for a variety of reasons. QKD creates random, independent session keys, which can reduce the damage caused by ephemeral key leakage. Other forms of quantum cryptography may also be of interest, especially for the secure communication of quantum information if quantum computing becomes widespread.

Experimental research on quantum key distribution continues to improve the usability, rate, and distance of QKD systems, and the ability to provide and certify their physical security. As public key cryptography systems are retooled with new algorithms and standards over the coming years, there is an opportunity to incorporate QKD as a new tool offering fundamentally new security features.

Related work. This work is motivated as a response to other opinions about the role of QKD, especially the thoughtful note "Why quantum cryptography?" by Paterson, Piper, and Schack [2]. Our discussion on encryption and authentication addresses many of the same points as [2] with an optimistic view of the prospect of post-quantum public key cryptography; we provide additional information on the assumptions for the security of QKD, the current state of QKD implementations, and how the structure of QKD networks will evolve as technology progresses. A response by the SECOQC project [6] addresses related concerns as well, with special attention paid to the networks of QKD links.

Outline. In the rest of this paper, we argue that QKD has a valuable role to play in future security infrastructures. In Section 2, we give an overview of how QKD works, and give an example where its high security is needed in Section 3. We describe the conditions for the security of QKD in Section 4. We then discuss the other parts of the communication infrastructure: encryption in Section 5 and authentication in Section 6. In Section 7, we discuss some limitations to QKD as it stands and how they may be overcome, with special consideration to networks of QKD devices in Section 8. We offer a concluding statement in Section 9.

2   A Brief Introduction to QKD

In this section we provide a very brief overview of quantum key distribution. More detailed explanations are available from a variety of sources [7,6,8].

In QKD, two parties, Alice and Bob, obtain some quantum states and measure them. They communicate (all communication from this point onwards is classical) to determine which of their measurement results could lead to secret key bits; some are discarded in a process called sifting because the measurement settings were incompatible. They perform error correction and then estimate a security parameter which describes how much information an eavesdropper might have about their key data. If this amount is above a certain threshold, then they abort as they cannot guarantee any secrecy whatsoever. If it is below the threshold, then they can apply privacy amplification to squeeze out any remaining information the eavesdropper might have, and arrive at a shared secret key. Some of this classical communication must be authenticated to avoid man-in-the-middle attacks. Some portions of the protocol can fail with negligible probability.

A flow chart describing the stages of quantum key distribution is given in Figure 1.

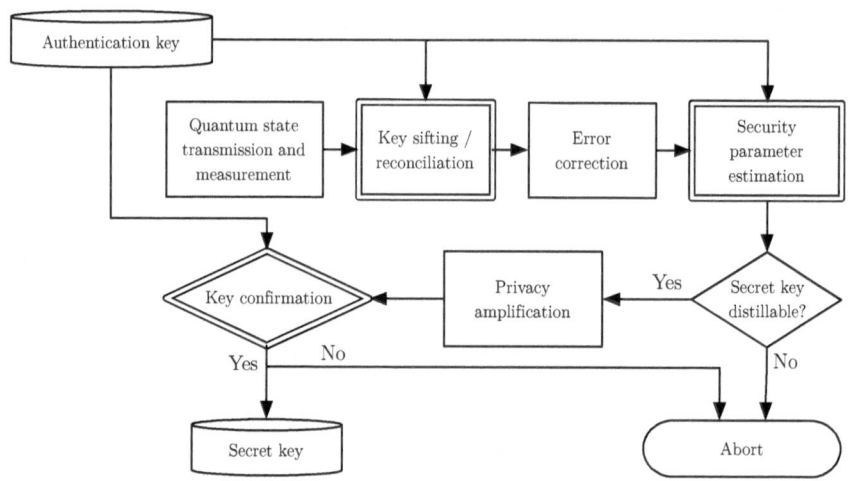

Fig. 1. Flow chart of the stages of a quantum key distribution protocol. Stages with double lines require classical authentication.

Once a secret key has been established by QKD, it can be used for a variety of purposes. The most common approach is to use it as the secret key in a one-time pad to achieve unconditionally secure encryption. The key can also be used for classical authentication in subsequent rounds of QKD.

We can expect that as QKD research continues, QKD devices will become more robust, easier to configure, less expensive, and smaller, perhaps sufficiently miniaturized to fit on a single circuit board.

3   Who Needs Quantum Key Distribution?

It is widely understood that "security is a chain; it's as strong as the weakest link" [1], and cryptography, even public key cryptography, is indeed one of the strongest links in the chain. We cannot trust that a particular computationally secure cryptographic scheme and parameter size will remain secure indefinitely, and many expert recommendations are unwilling to provide guidance for much more than 30 years in the future. While much of the information being encrypted today does not need 30 years of security, some does.

Moreover, it is important to plan well in advance for changes in security technology. Suppose, for example, that a particular application using RSA or elliptic curve cryptography (ECC) needs information to be secure for x years, and it takes y years to retool the infrastructure to a new cryptosystem. If large-scale quantum computers capable of breaking RSA or ECC are built within z years, with $z < x + y$, then we are already too late: we need to start planning to use new cryptosystems long before old ones are broken.

Government, military, and intelligence agencies need long-term security. For example, the UK government did not declassify the 1945 report on its efforts in breaking the Tunny cipher during World War II until 2000 [9], and the US government's current classification regime keeps documents classified for up to 25 years [10, §1.5(b)].

Businesses trying to protect long-term strategic trade secrets may also wish for long-term confidentiality. Situations with long-term deployments but well-specified communication requirements could also benefit from QKD: it is inconvenient and expensive to have to upgrade the 1.5 million automated teller machines (ATMs) worldwide whenever the latest cryptographic protocol is broken or deemed obsolete, but QKD could provide standards less likely to change due to cryptanalysis.

One particular industry likely to require long-term, future-proof security is health care. Health care systems are slowly but irreversibly becoming more electronic, and health care records need privacy for 100 years or more. Securing the storage of these records in data centers is essential, of course, and quantum key distribution does not aim to solve this difficult problem. Equally important, however, is the secure communication of health care records, which can be protected by the information-theoretic security offered by quantum key distribution.

Quantum key distribution is also not the only way to establish information theoretically secure keys. The physical transfer of long, randomly generated keys is also an information theoretically secure key distribution scheme. With hard drive prices approaching US $0.10 per gigabyte, one should not underestimate "the bandwidth of a truck filled with hard drives" (although increases in fuel prices may counteract the cost efficiency of such a communication system). This approach is not appropriate for all scenarios. In some cases, it may be impossible to rekey a system in this manner (e.g., satellites and space probes). It requires assurances that the physical keys were transported securely. It also requires secure storage of large amounts of key until use. QKD requires only a small amount of key, the authentication key, to be securely stored until use. Importantly, QKD can generate fresh encryption keys on demand that need only be stored for the short time period between key generation and message encryption/decryption, rather than needing large secure key storage since the distribution of the systems.

Moreover, research into experimental quantum information is still at such an early stage that one cannot predict the final form of the products that could be developed from this technology, and these systems may come to exceed the expectations and dreams of today's researchers and engineers.

4   The Security of QKD

Quantum key distribution is often described by its proponents as "unconditionally secure" to emphasize its difference with computationally secure classical cryptographic protocols. While there are still conditions that need to be satisfied for quantum key distribution to be secure, the phrase "unconditionally secure" is justified because, not only are the conditions reduced, they are in some sense minimal necessary conditions. Any secure key agreement protocol must make a few minimal assumptions, for security cannot come from nothing:

we must be able to identify and authenticate the communicating parties, we must be able to have some private location to perform local operations, and all parties must operate within the laws of physics.

The following statement describes the security of quantum key distribution, and there are many formal mathematical arguments for the security of QKD (e.g., [11,12,13]).

Theorem 1 (Security statement for quantum key distribution). *If*

A1) quantum mechanics is correct, and
A2) authentication is secure, and
A3) our devices are reasonably secure,

then with high probability the key established by quantum key distribution is a random secret key independent (up to a negligible difference) of input values.

Assumption 1: Quantum mechanics is correct. This assumption requires that any eavesdropper be bounded by the laws of quantum mechanics, although within this realm there are no further restrictions beyond the eavesdropper's inability to access the devices. In particular, we allow the eavesdropper to have arbitrarily large quantum computing technology, far more powerful than the current state of the art. Quantum mechanics has been tested experimentally for nearly a century, to very high precision. But even if quantum mechanics is superseded by a new physical theory, it is not necessarily true that quantum key distribution would be insecure: for example, secure key distribution can be achieved in a manner similar to QKD solely based on the assumption that no faster-than-light communication is possible [14].

Assumption 2: Authentication is secure. This assumption is one of the main concerns of those evaluating quantum key distribution. In order to be protected against man-in-the-middle attacks, much of the classical communication in QKD must be authenticated. Authentication can be achieved with unconditional security using short shared keys, or with computational security using public key cryptography. We discuss the issue of authentication in greater detail in Section 6.

Assumption 3: Our devices are secure. Constructing a QKD implementation that is verifiably secure is a substantial engineering challenge that researchers are still working on. Although the first prototype QKD system leaked key information over a side channel (it made different noises depending on the photon polarization, and thus the "prototype was unconditionally secure against any eavesdropper who happened to be deaf" [15]), experimental cryptanalysis leads to better theoretical and practical security. More sophisticated side-channel attacks continue to be proposed against particular implementations of existing systems (e.g., [16]), but so too are better theoretical methods being proposed, such as the decoy state method [17]. Device-independent security proofs [18,19] aim to minimize the security assumptions on physical devices. It seems reasonable to expect that further theoretical and engineering advances will eventually bring us devices which have strong arguments and few assumptions for their security.

5   Key Usage: Encryption

The most commonly discussed usage for the key generated by quantum key distribution is encryption. There are two ways [2] this key can be used for encryption.

In an *unconditionally secure system*, the private key from QKD is used as the key in a one-time pad. Since the key is information theoretically secure, so too is the encryption of the message: no computer, quantum or classical, will ever be able to decipher the encrypted message. There are challenges to this system, however. First, the one-time pad keys must be carefully stored and managed, as the double-use of one-time keys can seriously compromise security. Second, as we discuss in Section 7, physical QKD systems cannot yet achieve sufficiently high key generation rates to be able to encrypt large messages with one-time pads in real time.

To deal with this second challenge of low QKD key rates, *hybrid systems* have been proposed, where the key from QKD is expanded with a classical stream cipher or block cipher such as the Advanced Encryption Standard (AES) to encrypt long messages. In this setting, the security of the encrypted messages is no longer information theoretic: it depends on the computational assumption that the cipher used is hard to break. While this is not ideal, it may not be too risky either. Historically, cryptographers have been very good at designing block ciphers with few weaknesses: for example, the Data Encryption Standard (DES), designed in the 1970s, is no longer considered secure due to its short key length, but DES has stood up well to over 30 years of cryptanalytic attacks. Under a known plaintext attack, the security of DES is reduced from 2^{56} to about 2^{41}, but, when rekeying is sufficiently frequent, the effect of known plaintext attacks is limited [6, §3.2]. Moreover, quantum computers do not seem to have too much impact on ciphers: while Grover's search algorithm implies that the key length needs to be doubled, the exponentially faster attacks promised by Shor's algorithm and others do not apply to most ciphers.

Even when used in hybrid systems, QKD offers a substantial advantage over classical key agreement: the key from QKD is independent of any inputs to the key agreement protocol. Thus, QKD reduces the number of points of attack: once a key has been established, the only way to attack such a system is to cryptanalyze the encryption. By contrast, a system using classical key agreement could be attacked by trying to take the inputs to the classical key agreement protocol and determining the generated private key (e.g., by solving the Diffie-Hellman problem). However, when using QKD to generate short keys, care must be taken due to finite length effects [20].

Hybrid QKD systems offer enhanced security compared to ciphers used without QKD: the QKD subsystem provides fresh, independent keying material frequently, which can rekey the classical block or stream cipher; with frequent rekeying, we reduce the risk of attacks against the underlying cipher that make use of many plaintexts or ciphertexts encrypted under the same key.

6   Authentication

Quantum key distribution does not remove the need for authentication: indeed, authentication is *essential* to the security of QKD, for otherwise it is easy to perform a man-in-the-middle attack. There are two main ways to achieve authentication: public key authentication and symmetric key authentication. *Symmetric key authentication* can provide unconditionally secure authentication, but at the cost of needing to have pre-established pairs of symmetric keys. *Public key authentication,* on the other hand, is simpler to deploy, and provides extraordinarily convenient distributed trust when combined with certificate authorities (CAs) in a public key infrastructure (PKI). Public key authentication cannot itself be achieved with information theoretic security. We argue, however, that the security situation is more subtle than this: the use of public key authentication can still lead to systems that have very strong long-term security.

A third method for authentication is to use trusted third parties which actively mediate authentication between two unauthenticated parties, but there has been little interest in adopting these in practice. Certificate authorities, which are used in public key authentication, are similar to trusted third party authentication but do not actively mediate the authentication: they distribute signed public keys in advance but then do not participate in the actual key authentication protocol. The difference in trust between trusted third parties and certificate authorities for authentication in QKD is smaller than in the purely classical case since the key from QKD is independent of the inputs.

6.1   Symmetric Key Authentication

Parties who already share a short private key can use an unconditionally secure message authentication code to authenticate their messages. The first such approach was described by Wegman and Carter [21] and has been refined for use in QKD (for example, [22]). It is for this reason that quantum key distribution is sometimes called *quantum key expansion*: it can take a short shared key and expand it to an information-theoretically secure long shared key.

Interestingly, the universal composability of quantum key distribution implies that we can use some of the key generated by QKD to authenticate the messages in the next round of QKD with a negligible decrease in security. Thus we can continue QKD (more or less) indefinitely having started only with a relatively short (on the order of a few kilobytes) authentication key.

6.2   Public Key Authentication

While symmetric key authentication promises unconditionally secure authentication, it is difficult to deploy because each pair of communicating parties must share a private key. Public key infrastructures allow for distributed trust and have been essential to the success of electronic commerce. While many advocates of quantum cryptography dismiss the role of computationally secure public key authentication in QKD, we argue that public key authentication will be vital in a quantum key distribution infrastructure and can still provide meaningful security statements.

Public key authentication schemes, being computationally secure, tend to be broken, and invariably sooner than we expect. In 1977, Rivest speculated [23] that it could take 40 quadrillion years to solve the RSA-129 problem (factoring a 129-decimal-digit RSA modulus), but it was broken only 17 years later [24]. While the popular press still occasionally uses expressions such as "more than a quadrillion years" [25] to describe the security of number-theoretic schemes, technical recommendations [26,27] are more nuanced and tend not to speculate too far beyond 2030. Notably, these recommendations tend to "assume [...] (large) quantum computers do not become a reality in the near future" [27, p. 25].

Large scale quantum computers are widely believed to be some time off, but there appears to be no reason at present to doubt their eventual efficacy. Quantum computers, however, are not the only threat against public key authentication. Computers do become faster and new algorithms do help speed cryptanalysis. However, we are not so pessimistic to think that all public key authentication is doomed forever. In fact, we believe that public key authentication will continue to play a vital role in communication security indefinitely, even in the presence of quantum computing.

Although today's popular public key schemes — RSA, finite field discrete logarithm, and elliptic curve — would be broken by a large scale quantum computer, other "post-quantum" schemes do not immediately fall to quantum algorithms, and other schemes are sure to be developed (cf. [28]). It seems to us, then, that public key schemes in the future are likely to go through a lifecycle in which a new primitive is proposed, it appears secure against current attack techniques, reasonable parameter sizes are proposed, adopted, and then computing technology and cryptanalysis advances chip away at the security until a newer scheme provides better tradeoffs. It is not too hard to imagine a 20-year window in which a public key scheme, along with a particular set of parameter sizes, is considered viable.

It is in this scenario, where a particular public key authentication scheme is only deemed to be secure for a 20-year period, that quantum key distribution can thrive. A public key authentication infrastructure provides the large scale usability that we have come to expect from PKIs, and when combined with quantum key distribution can offer strong security promises. In quantum key distribution, the authentication — in the form of public key authentication — only needs to be secure up to and including the initial connection. Once the QKD protocol has output some secret key, a portion of this secret can subsequently be used for symmetric key authentication. In fact, even if the original authentication keys are revealed after the first QKD exchange, the key from QKD remains information theoretically secure. In other words, we have the following statement:

If authentication is unbroken during the first round of QKD, *even if it is only computationally secure*, then subsequent rounds of QKD will be information-theoretically secure.

By contrast, classical public key exchange schemes do not have this feature. Although one can employ a protocol in which a new key is transmitted encrypted under the old key, an eavesdropper who logs all communications and subsequently breaks the first key can read all future communications. With QKD, new session keys are completely independent of all prior keys and messages.

7   Limitations

Two undeniable limitations of present quantum key distribution schemes are distance and key rate. Because of the fragile nature of the quantum mechanical state that is transmitted during quantum key distribution, the longer the distance that the photons have to travel, the more photons that are lost to decoherence and noise and hence the lower the rate of secret key formation. Distance and key rate are a tradeoff, but progress is being made on improving the overall tradeoff.

Distance. The longest QKD experiments to date have acheived secure key generation over a 184.6km fiber optic link [29] and over a free-space link spanning a distance of 144km at a rate of 12.8 bits/second[30] . This free-space distance is considered sufficient to communicate between any two points on the surface of the Earth via orbiting satellites, the feasibility of which is the subject of a proposed experiment [31].

Quantum repeaters [32] would also overcome the distance limitation, allowing shared quantum states to be established between distant parties. While these systems are not yet operational, they are easier to implement than full-scale quantum computers; theoretical and experimental work progresses on their development.

Key rate. While long distance experiments achieve very low key rates on the order of a few bits per second, shorter distance experiments have demonstrated very high key rates. Experimental groups have achieved key rates of over 4 MB per second over 1km of fibre [33] and 1 Mb per second at 20km [34]. These key rates are an impressive accomplishment are coming closer to the rates needed to secure real communication channels.

When a QKD key is used for encryption, current key rates may not be sufficient for a one-time pad and hybrid schemes need to be used, in which the QKD key is used as the private key in a symmetric encryption algorithm such as the block cipher AES. However, as we have argued in Section 5, even hybrid QKD systems offer enhanced security compared to classical key agreement since the keys generated by QKD are independent of any inputs to the key agreement procedure and since many symmetric encryption algorithms are resistant to known attacks by quantum computers. Key rate can always be negatively impacted by an adversary disturbing the quantum channel, but such an adversary can not impact the security of the key agreement.

8   QKD Networks

As QKD technology progresses, the structure of deployed QKD systems will progress in four stages to reduce distance limitations and increase commercial applicability:

1. *Point-to-point links:* Two QKD devices are directly connected over a relatively short distance.
2. *Networks with optical switches:* Multiple QKD devices are arranged in a network with optical switches to allow different pairs of interaction. Optical switches, however, do not increase communication distance. The switches need not be trusted. One example of such a network is the DARPA quantum network [35].
3. *Networks with trusted relays:* Multiple QKD devices are arranged in a network. Intermediate nodes in the network can act as classical relays which relay information between distant nodes. The relay nodes need to be trusted, although trust can be reduced by having the sender use a secret sharing scheme [36]. This type of QKD network would be suitable for scenarios where the operator of the network is also the user of the network, for example, a bank creating a network among its many branches, each of which is individually trusted. One example of such a network is the SECOQC quantum network [6,37].
4. *Fully quantum repeater network:* Multiple QKD devices are arranged in a network with quantum repeaters [32]. Although individual links are still distance-limited, the quantum repeater nodes allow entanglement to be linked across longer distances, so QKD can be performed between distant parties. The quantum repeaters need not be trusted, and this type of QKD network corresponds to the service provider scenario.

9   Conclusion

Quantum key distribution makes use of the eavesdropper-detection power offered by quantum mechanics to establish a shared key that is verifiably secure and independent of any other data, provided the communicating parties share an authentic channel. The security of the system depends on no computational assumptions and thus has the potential to offer security against present or future attackers with unbounded classical or quantum computational power.

There are many scenarios, such as government, military, and health care, in which information needs to remain secure for 25, 50, or even 100 years. Using QKD reduces the assumptions about the cryptographic system and produces a shared secret key that, by the properties of quantum mechanics, is independent of any other data, including the input.

It is important to consider how QKD fits into the larger cryptographic infrastructure. When used with public key authentication, QKD provides strong security with the convenience of distributed authentication using public key infrastructures; the public key authentication scheme need only be secure up until

QKD occurs, but the key from QKD will remain secure indefinitely. If public key authentication is not possible, shared secret authentication can still be used to give enhanced security compared to classical key expansion.

The present limitations of QKD — distance and key rate — will be further mitigated as experimental research in QKD continues, and quantum repeaters promise fully quantum long distance networks.

We believe that, as the technology continues to improve, QKD will be an increasingly valuable tool in the cryptographer's toolbox for building secure communication systems.

Acknowledgements

The authors gratefully acknowledge helpful discussions with Romain Allaume, Daniel J. Bernstein, Hoi-Kwong Lo, Alfred Menezes, and Kenny Paterson. Research performed while D.S. was at the University of Waterloo. D.S. was supported in part by an NSERC Canada Graduate Scholarship. M.M. is supported by a Canada Research Chair. The authors acknowledge funding from the Ontario Centres of Excellence (OCE), Canada's NSERC, QuantumWorks, MITACS, CIFAR, Ontario-MRI, and Sun Microsystems Laboratories.

References

1. Schneier, B.: Crypto-Gram: Quantum cryptography (December 2003), http://www.schneier.com/crypto-gram-0312.html#6
2. Paterson, K.G., Piper, F., Schack, R.: Why quantum cryptography? Published as [3] (June 2004), http://arxiv.org/abs/quant-ph/0406147
3. Paterson, K.G., Piper, F., Schack, R.: Quantum cryptography: A practical information security perspective. In: Zukowski, M., Kilin, S., Kowalik, J. (eds.) Proc. NATO Advanced Research Workshop on Quantum Communication and Security. NATO Science for Peace and Security Series, Sub-Series D: Information and Communication Security, vol. 11. IOS Press, Amsterdam (2007); See [2]
4. Schneier, B.: Schneier on Security: Switzerland protects its vote with quantum cryptography (October 2007), http://www.schneier.com/blog/archives/2007/10/switzerland_pro.html
5. Schneier, B.: Quantum cryptography: As awesome as it is pointless. Wired (October 2008)
6. Alléaume, R., Bouda, J., Branciard, C., Debuisschert, T., Dianati, M., Gisin, N., Godfrey, M., Grangier, P., Länger, T., Leverrier, A., Lütkenhaus, N., Painchault, P., Peev, M., Poppe, A., Pornin, T., Rarity, J., Renner, R., Ribordy, G., Riguidel, M., Salvail, L., Shields, A., Weinfurter, H., Zeilinger, A.: SECOQC white paper on quantum key distribution and cryptography (January 2007), http://www.arxiv.org/abs/quant-ph/0701168
7. Nielsen, M.A., Chuang, I.L.: Quantum Computation and Quantum Information. Cambridge University Press, Cambridge (2000)
8. Scarani, V., Bechmann-Pasquinucci, H., Cerf, N.J., Dusek, M., Lütkenhaus, N., Peev, M.: The security of practical quantum key distribution. To appear in Reviews of Modern Physics (2008)

9. Good, J., Michie, D., Timms, G.: General report on tunny. Technical report, Government Code and Cypher School (1945); Declassified September 28, 2000, by Pulic Records Office, UK, documents HW 25/4 and HW 25/5

10. Bush, G.W.: Executive Order 13292. Further amendment to Executive Order 12958, as amended, Classified National Security Information (March 2003)

11. Mayers, D.: Unconditionally secure quantum bit commitment is impossible. Physical Review Letters 78(17), 3414–3417 (1997)

12. Lo, H.K., Chau, H.F.: Unconditional security of quantum key distribution over arbitrarily long distances. Science 283(5410), 2050–2056 (1999)

13. Gottesman, D., Lo, H.K., Lütkenhaus, N., Preskill, J.: Security of quantum key distribution with imperfect devices. Quantum Information and Computation 4(5), 325–360 (2004)

14. Barrett, J., Hardy, L., Kent, A.: No signaling and quantum key distribution. Physical Review Letters 95(1), 010503 (2005)

15. Brassard, G.: Brief history of quantum cryptography: A personal perspective. In: IEEE Information Theory Workshop on Theory and Practice in Information-Theoretic Security 2005, pp. 19–23. IEEE, Los Alamitos (2005)

16. Zhao, Y., Fung, C.H.F., Qi, B., Chen, C., Lo, H.K.: Quantum hacking: Experimental demonstration of time-shift attack against practical quantum key distribution systems. Physical Review A 78(4), 042333 (2008)

17. Hwang, W.Y.: Quantum key distribution with high loss: Toward global secure communication. Physical Review Letters 91(5), 057901 (2003)

18. Mayers, D., Yao, A.C.: Quantum cryptography with imperfect apparatus. In: Proc. 38th Annual IEEE Symposium on Foundations of Computer Science (FOCS) 1997, pp. 503–509. IEEE Press, Los Alamitos (1997)

19. Pironio, S., Acin, A., Brunner, N., Gisin, N., Massar, S., Scarani, V.: Device-independent quantum key distribution secure against collective attacks. New Journal of Physics 11(4), 045021 (2009)

20. Cai, R.Y.Q., Scarani, V.: Finite-key analysis for practical implementations of quantum key distribution. New Journal of Physics 11, 045024 (2009)

21. Wegman, M.N., Carter, J.L.: New hash functions and their use in authentication and set equality. Journal of Computer and System Sciences 22(3), 265–279 (1981)

22. Peev, M., Nölle, M., Maurhardt, O., Lorünser, T., Suda, M., Poppe, A., Ursin, R., Fedrizzi, A., Zeilinger, A.: A novel protocol-authentication algorithm ruling out a man-in-the-middle attack in quantum cryptography. International Journal of Quantum Information 3(1), 225–231 (2005)

23. Gardner, M.: Mathematical games: A new kind of cipher that would take millions of years to break. Scientific American, 120–124 (August 1977)

24. Atkins, D., Graff, M., Lenstra, A.K., Leyland, P.C.: The magic words are squeamish ossifrage (extended abstract). In: Safavi-Naini, R., Pieprzyk, J.P. (eds.) ASIACRYPT 1994. LNCS, vol. 917, pp. 265–277. Springer, Heidelberg (1995)

25. Lysyanskaya, A.: Cryptography: How to keep your secrets safe. Scientific American, pp. 89–94 (September 2008)

26. NIST: Recommendations for key management – Part 1: General (revised) (March 2007),
    http://csrc.nist.gov/publications/nistpubs/800-57/
    sp800-57-Part1-revised2_Mar08-2007.pdf

27. Babbage, S., Catalano, D., Cid, C., Dunkelman, O., Gehrmann, C., Granboulan, L., Lange, T., Lenstra, A., Nguyen, P., Paar, C., Pelzl, J., Pornin, T., Preneel, B., Rechberger, C., Rijmen, V., Robshaw, M., Rupp, A., Smart, N., Ward, M.: ECRYPT yearly report on algorithms and keysizes (2007-2008) (July 2008)

28. Bernstein, D.J., Buchmann, J., Dahmen, E. (eds.): Post Quantum Cryptography. Springer, Heidelberg (2009)
29. Hiskett, P.A., Rosenberg, D., Peterson, C.G., Hughes, R.J., Nam, S., Lita, A.E., Miller, A.J., Nordholt, J.E.: Long-distance quantum key distribution in optical fibre. New Journal of Physics 8(9), 193 (2006)
30. Schmitt-Manderbach, T., Weier, H., Furst, M., Ursin, R., Tiefenbacher, F., Scheidl, T., Perdigues, J., Sodnik, Z., Kurtsiefer, C., Rarity, J.G., Zeilinger, A., Weinfurter, H.: Experimental demonstration of free-space decoy-state quantum key distribution over 144 km. Physical Review Letters 98(1), 010504 (2007)
31. Perdigues Armengol, J.M., Furch, B., de Matos, C.J., Minster, O., Cacciapuoti, L., Pfennigbauer, M., Aspelmeyer, M., Jennewein, T., Ursin, R., Schmitt-Manderbach, T., Baister, G., Rarity, J., Leeb, W., Barbieri, C., Weinfurter, H., Zeilinger, A.: Quantum communications at ESA: Towards a space experiment on the ISS. Acta Astronautica 63(1-4), 165–178 (2008)
32. Briegel, H.J., Dür, W., Cirac, J.I., Zoller, P.: Quantum repeaters: The role of imperfect local operations in quantum communication. Physical Review Letters 81(26), 5932–5935 (1998)
33. National Institute of Standards and Technology: Quantum information networks (2006), http://www.antd.nist.gov/qin/
34. Dixon, A.R., Yuan, Z.L., Dynes, J.F., Sharpe, A.W., Shields, A.J.: Gigahertz decoy quantum key distribution with 1 Mbit/s secure key rate. Optics Express 16(23), 18790–18979 (2008)
35. Elliott, C., Colvin, A., Pearson, D., Pikalo, O., Schlafer, J., Yeh, H.: Current status of the DARPA quantum network (2005), http://arxiv.org/abs/quant-ph/0503058
36. Beals, T.R., Sanders, B.C.: Distributed relay protocol for probabilistic information-theoretic security in a randomly-compromised network. In: Safavi-Naini, R. (ed.) ICITS 2008. LNCS, vol. 5155, pp. 29–39. Springer, Heidelberg (2008)
37. Salvail, L., Peev, M., Diamanti, E., Alleaume, R., Lütkenhaus, N., Laenger, T.: Security of trusted repeater quantum key distribution networks. To appear in Journal of Computer Security (April 2009), http://arxiv.org/abs/0904.4072

On QKD Industrialization

J. Dávila, D. Lancho, J. Martinez, and V. Martin

Facultad de Informática, Univ. Politécnica de Madrid
Campus de Montegancedo, Boadilla del Monte
Madrid 28660, Spain
Vicente@fi.upm.es

Abstract. During the 25 years of existence of the first protocol for Quantum Key Distribution, much has been said and expected of what came to be termed as Quantum Cryptography. After all this time, much progress has been done but also the reality check and analysis that naturally comes with maturity is underway. A new panorama is emerging, and the way in which the challenges imposed by market requirements are tackled will determine the fate of Quantum Cryptography. The present paper attempts to frame a reasonable view on the issues of the security and market requirements that QKD should achieve to become a marketable technology.

Keywords: Quantum Key Distribution, Security assurance and standardization, Market requirements.

1 Introduction

Setting aside the historical paper of Wiesner about quantum money, the birth of Quantum Cryptography could be associated with the BB84 protocol, actually in 1983. Although Quantum Cryptography is a broader field, it was the Quantum Key Distribution (QKD) schemes, initiated by this protocol, which shaped the field as we know it today. At present, it is only QKD to which a reasonable degree of technological maturity and market relevance can be ascribed. QKD protocols serve the purpose of growing a preshared secret among two parties. The preshared secret serves to guarantee the integrity of the protocol in the first transaction, while the quantum properties of nature are used to guarantee, with any threshold we would like to pose, the privacy of the generated key. In the QKD implementation proposals to date, part of this new generated key is used to check the integrity of the next protocol round, a practice that should be carefully reconsidered from a practical perspective. From a security standpoint, integrity control and key generation are two basically different processes, hence they should be kept separated [2]. Concepts of separation and controlled information flow are a well acknowledged practice [3] that has made its way in modern high security architectures as exemplified by MILS (Multiple Independent Levels of Security/Safety [4]) and that could have prevented or, at least, alleviated the possible impact of recently discovered weaknesses in QKD [5]. Simple integrity

A. Sergienko, S. Pascazio, and P. Villoresi (Eds.): QuantumCom 2009, LNICST 36, pp. 297–302, 2010.
© Institute for Computer Sciences, Social-Informatics and Telecommunications Engineering 2010

control techniques well regarded in practice, like seeding a pseudo random number generator with an initially shared secret, are robust and demand only a small secret to run for a long time. A simple XOR among the strings obtained by this method and the quantum key used for the same purpose would provide the best of both worlds.

Key management is the provisions made in a cryptography system design that are related to generation, exchange, storage, safeguarding, use, vetting, and replacement of keys. It includes cryptographic protocol design, key servers, user procedures, and other relevant protocols. Key management is essential for any security infrastructure and QKD can be certainly a very powerful primitive to strength many operations relevant to the security market. However, while QKD protocols can be proven theoretically secure under simple assumptions, these are not, and cannot be, backed by implementation under any known industrial process. The widely spread view that QKD could achieve perfect secrecy in real applications is clearly flawed, certainly from the point of view of the conventional cryptography community, as opposed to the quantum cryptography one, where many were pushing absolute security as the spear head of the new field. The history of cryptography is full of examples of good ideas that have claimed to offer higher levels of security, and then reality has put them in the curiosities corner [6]. These excessively triumphant views, together with the fact that shared secrets are a relatively small part of the whole security market and can be achieved by other means [7], led also to the early dismissal of QKD by many security practitioners. At most, it was relegated to an immature technology status that could be relevant to some niche markets in the future.

As QKD technology advanced and made its way out of the laboratory and began to be marketed and tested in competition with more traditional technologies, issues about its actual security level, market relevance, reliability, cost/benefit, etc. started to arise.

The first fact to realize in this commercial environment is that in general, and in one component in particular, absolute security is not really an interesting goal to pursue in itself. Security is a general property of the system that is build up over many components and strengthening one of them does not necessary makes the full system more secure. Application always dictates the security level requirements but usability, reliability, interoperability and cost are many times as relevant as security needs. Some are started to be addressed by the QKD community using rules akin to those applied to conventional systems. Certification is a case in point. To build trust on the final user, similar methods to those that have already proven its validity must be used. Intensive and detailed independent evaluation, strict quality control, good acceptance by the insurance companies and adequate information campaigns help to market a security product, but these do not cover all the bases. To base QKD devices certification on well known standards like FIPS 140, Common Criteria [8,9], etc. as is being done in the current work at the Quantum Industry Specification Group of the European Telecommunications Standards Institute [10,11] is a reasonable and necessary move, maybe the only one possible. These certifications are routinely applied to all kinds of electronic devices and their application to the corresponding part in QKD devices should be straightforward. Its use concerning the optical subsystem is an unexplored field in need of being addressed.

QKD modules will include microcontrollers, electronic memories, buses and many other elements common to the general microelectronics market. QKD modules cannot be safer than the software that runs inside them and controls all their functionalities. Everything in the software that can be reprogrammed, updated or maintained in any way inside a QKD module has to be specially protected because its integrity must be guaranteed all along the module service. Using general purpose hardware and software components (microprocessors, memories, operating systems, drivers,...) has many advantages, in particular those related to the final cost and maintenance, but can also introduce security breaches in the system.

QKD modules require specific purpose software that implements the protocols, controls the optoelectronic hardware and is responsible for the administrative and operational interfaces. If we accept the software security as an upper security bound in a QKD system, as it is with all embedded systems, then the software has to be secure by design and has to be evaluated, inspected and certified at a high level of security if we want to see the QKD technology in the high security market shelves.

Definitively, security and risk always go together; consequently, security has to be as multi-valuated as the risk is in real scenarios. In many infrastructures, different security levels are defined, and the products used inside them have to match the security level specified. Because of this, different certification levels could be adapted to different applications. QKD technology can provide different security levels at different costs using different technologies and settings. This flexibility must be made available in real QKD systems to exploit its commercial horizon because it could match many different scenarios. In principle, Common Criteria methodology profiles will help to discover and add some flexibility to QKD so as to meet the different demands of those various potential markets.

However, certification is not the security holy grail, and one must bear in mind that, for example, Common Criteria higher levels do not necessarily equate with higher security, but claims have been more thoroughly evaluated. For instance, Windows XP operating system is EAL4+ certified [12,13], despite the continuous patches needed due to the almost daily discovery of security failures. The sets of claims for QKD must be carefully crafted to be meaningful for the intended market. Certification, however, fulfills an important role for QKD since it translates QKD jargon and claims to the language used by its potential customers.

Usability and interoperability are also requirements that could prove essential for the QKD success. QKD must offer a set of characteristics compelling enough to be the technique of choice. For an extremely secure application with only a point to point link, usability and interoperability could be of secondary interest compared to the increase in security; however, for a company seeking to introduce QKD in an already deployed platform, these two could prove as essential as the perceived security increase.

QKD devices generate keys to be used outside the QKD device itself and, because of that, interoperability with other security systems is absolutely necessary. For example, in a high security environment, the QKD link would generate the same key at both ends on the quantum link but, probably, the key will be fed into an Electronic Key Management System (EKMS) or fill device [1] that will distribute it for its final use. In such case, QKD equipment has to be fully compatible with all key management systems it pretends to connect to and operate with.

Interoperability would be also of primary interest in the case of a network provider selling QKD services to its clients. For a customer, interoperability is a must. It allows for various QKD providers, meaning more market competition and thus lower price. It also means not to be locked with just one manufacturer. In fact, advanced security models advocate for increasing modularity to allow a better security scrutiny but this also means lower maintenance risks and costs if you have different providers for each module. QKD systems have to be interoperable with all the systems they will work with, and they also require full interoperability, standardization and security certification of all its internal modules and components (optic fibers, laser, diodes, phase shifters, delay lines, etc.).

Reliability comes hand in hand with low maintenance. One of the advantages of QKD is the possibility of low maintenance costs if the system is reliable. When using standard devices, a master key is needed to operate the system. At a given moment in time, there is no more entropy in the system than that originally in the master key. Hence, the need to balance the security level with the frequency of master key update. The procedures to change the master key in high security systems are rather involved and an acknowledged weakness. In a reliable QKD system, this weakness is confined to just the first installation. After a correct install, the system can work unattended as long as the device does not fail or the channel is interrupted. The system can even raise an alarm in case of attack, an example of an advantage of a QKD system over a conventional one. Potential low maintenance in QKD systems illustrates also how different the new markets for QKD can be from those expected at first sight. In fact, the use most commonly cited: as an extremely safe device producing keys to be used in a Vernam-Mauborgne cypher, would probably be one of the least used. Cyphering large amounts of data through a high speed link with a symmetric block cypher like AES would be much more likely. Even for a high speed channel, changing the key a few times per hour would suffice to keep a much higher security level than the attained nowadays. Hence, a low key generation rate but in a much more reliable and interoperable system would be the preferred choice, as opposed to the ever higher key generation rate philosophy pursued above all in current developments. High key rate would be useful in a scenario in which just one QKD link is used to feed keys to many data channels. On the contrary, low key rate systems able to withstand high optical losses would be much more suited for

---

[1] A fill device is an electronic module used to load cryptographic keys into electronic encryption machines. Fill devices are usually hand held and battery operated.

network integration in standard networks, the preferred scenario for a network operator.

Cost comes also with reliability, as it is also a product of physical integration in robust and compact devices. When considering the maximum cost of any security device, the first thing to mind is that investments in a security system should never cost more than the assets that is trying to protect. Then, the reliability and maintenance of the system have to be included.

When we compare the relative cost of a conventional device to a QKD system, several things must be taken into account. In conventional electronic security devices or systems, main budget expenses go to pay high quality design, manufacture, inspection, quality control and secure delivery. QKD systems will add to those charges the specific expenses related to its optical and optoelectronic subsystems. This is an additional cost because QKD modules also include electronic subsystems that are equivalent to those used in actual conventional security devices. At this respect, QKD systems have in their optical and quantum subsystems an additional handicap compared with the conventional devices in use for key generation and key distribution.

In general, we can conclude that other issues are more or equally important than extremely high security levels, and new QKD developments should take this into account.

In order to see a QKD industry pleasantly installed in the telecommunications market, many issues are to be addressed. Some of them are already being worked out, whereas others will take more time. Once all will be solved, there is certainly a range of applications in which QKD will fit nicely. Whether its use will be widespread or not, it will depend as much on technological advances with wisely chosen development targets as on a correct market approach.

Acknowledgment

This work was supported by CDTI, Ministry of Trade and Industry of Spain under project Segur@, CENIT-2007 2004 and UPM 178/Q06 1005-127.

References

1. Bennett, C.H., Brassard, G.: Quantum Cryptography: Public Key Distribution and Coin Tossing. In: Proceedings of IEEE International Conference on Computers Systems and Signal Processing, Bangalore, India, December 1984, pp. 175–179 (1984)
2. Barker, E., Barker, W., Burr, W., Polk, W., Smid, M.: Recommendation for Key Management Part 1: General. NIST Special Publication 800-57 (March 2007), http://csrc.nist.gov/groups/ST/toolkit/documents/ SP800-57Part1_3-8-07.pdf
3. Rushby, J.: Design and Verification of Secure Systems. In: Proc. 8th ACM Symposium on Operating System Principles, pp. 12–21 (1981)
4. Alves-Foss, J., Harrison, W.S., Oman, P., Taylor, C.: The MILS (Multiple Independent Levels of Security/Safety) Architecture for High Assurance Embedded Systems. International Journal of Embedded Systems (2007) (in press)

5. Cederloff, J., Larsson, J.: Security Aspects of the Authentication Used in Quantum Cryptography. IEEE Transactions on Information Theory 54, 1735 (2008)
6. Knudsen, L., Rijmen, V.: Two Rights Sometimes Make a Wrong. In: Proceedings of SAC 1997, Fourth Annual Workshop on Selected Areas in Cryptography, School of Computer Science, Carleton University, pp. 213–223 (1997)
7. Menezes, A.J., van Oorschot, P.C., Vanstone, S.A.: Handbook of Applied Cryptography. CRC Press, Boca Raton (2001)
8. FIPS PUB 140-2, Security Requirements For Cryptographic Modules. Federal Information Processing Standards Publication (2001)
9. ISO/IEC 15408, Common Criteria for Information Technology Security Evaluation and annex Common Methodology for Information Technology Security Evaluation, the technical basis for the international agreement known as Common Criteria Recognition Agreement
10. European Telecommunications Standards Institute, Quantum Industry Specification Group. *portal.etsi.org*
11. Länger, T., Lenhart, G.: Standardization of quantum key distribution and the ETSI standardization initiative ISG-QKD. New J. Phys. 11, 055051 (2009)
12. Woodie, A.: Windows Server 2003 Earns EAL 4 Certification from U.S. Government. The Windows Observer, January 11 (2006),
http://www.itjungle.com/two/two011106-story05.html
13. http://www.commoncriteriaportal.org/files/epfiles/
20080303_st_vid10184-vr.pdf

CTES Factorization Algorithm

Vincenzo Tamma[1,2], Heyi Zhang[1], Xuehua He[1], Augusto Garuccio[2],
and Yanhua Shih[1]

[1] Department of Physics, University of Maryland, Baltimore County, Baltimore,
Maryland 21250, USA
tammav1@umbc.edu

[2] Dipartimento Interateneo di Fisica, Università degli Studi di Bari, 70100 Bari, Italy

Abstract. We introduce a new factorization algorithm, based on the
analogue determination of the periodicity of a single generalized continuous truncated exponential sum (CTES) interferogram. We demonstrate
that this algorithm allows, in principle, to factorize arbitrary numbers
exploiting a remarking rescaling property of the recorded CTES interference pattern. Such an interferogram can be realized taking advantage of
multi-path optical interference, using a polychromatic light source and
a spectrometer. The resulting interference pattern, when observed as a
function of wavelength, contains the information about all factors of any
arbitrary number N. This information is encoded in the location of the
maxima of the interferogram.

Keywords: Factorization algorithm, interference, continuous truncated
exponential sums, Gauss sums, cryptography.

1   Introduction

To find the factors of a large integer number N is a rather difficult problem in
computation theory. Indeed, the security of codes relies on this fact. The most
celebrated algorithm for factorization is Shor's algorithm, which takes advantage
of quantum systems [1]. In the present paper we present a new factorization
algorithm, in which both number theory and a physical process allow to solve
the problem of factorization. Such an algorithm has a similar working principle
with respect to Shor's case: factorization by exploiting the periodicity of a known
function, which, in our case, is a generalized truncated continuous exponential
sum (CTES), as a function of a continuous variable.

2   Main Challenge in Factorization

In order to develop an effective factorization algorithm, it is important to understand what is the main challenge in factorization that this algorithm needs
to overcome. We want to show that such a challenge consists in the computation
of the ratio

$$f(\xi) \doteq \frac{1}{\xi}, \tag{1}$$

A. Sergienko, S. Pascazio, and P. Villoresi (Eds.): QuantumCom 2009, LNICST 36, pp. 303–310, 2010.
© Institute for Computer Sciences, Social-Informatics and Telecommunications Engineering 2010

as a function of the continuous parameter ξ, with $0 < \xi < 1/\sqrt{N_{min}}$, where N_{min} is the smallest number to be factored. In fact, once we know such a function, we have information about all the possible functions N/ξ associated with all the possible numbers $N > N_{min}$ that we want to factorize. We need simply to look at f as a function of the new variable, obtained by the scaling relation:

$$\xi_N \doteq N\xi, \tag{2}$$

with $0 < \xi_N < N/\sqrt{N_{min}} > \sqrt{N}$. In this way we obtain:

$$f(\xi_N) = \frac{N}{\xi_N}. \tag{3}$$

For each possible N, the factors are given by the values $\xi_N = l$, where l is a trial factor, such that

$$f(l) = \frac{N}{l} = k, \tag{4}$$

with k positive integer. In terms of computation, the introduction of ξ_N involves the multiplication of each possible value of ξ by the constant value N, for each number N we want to factorize. This means that no division operations are required once we know the function $f(\xi)$ in Eq. (1).

We have demonstrated that the factorization process turns out to be very fast in terms of computations if we know the function $f(\xi)$ in Eq. (1). Unfortunately, determining if Eq. (4) is satisfied is not an easy task. In fact, there are trial factors l for which $f(l)$ is very close to an integer. In the next section we will show how the constructive/destructive periodical interference associated with a generalized continuous truncated exponential sum (CTES) allows an easy distinction between factors and non factors. Such an approach is substantially different respect to the usual factorization approach using truncated exponential sums [2,3,4,10,11,7,8], which experimental realizations [12,13,18,14,15,17] present a precalculation of the ratio between N and l[19].

3   New Factorization Algorithm by Exploiting the Periodicity of a CTES

We want to show that the function $f(\xi)$, in Eq. (1), and so all the possible functions $f(\xi_N)$, can be extracted by determining the periodicity of the modulo squared of the generalized continuous truncated exponential sum (CTES) $\mathcal{C}^{(M,j)}(\xi)$, defined as:

$$|\mathcal{C}^{(M,j)}(\xi)|^2 \doteq |\frac{1}{M} \sum_{m=1}^{M} \exp\left[\phi_{m,j}(\xi)\right]|^2, \tag{5}$$

with the phase terms $\phi_{m,j}(\xi)$ given by:

$$\phi_{m,j}(\xi) = 2\pi i(m-1)^j f(\xi). \tag{6}$$

In particular, for each possible number N to factorize, the CTES in Eq. (5) can be rescaled in the following way:

$$|\mathcal{C}^{(M,j)}(\xi_N)|^2 = |\frac{1}{M} \sum_{m=1}^{M} \exp\left[2\pi i(m-1)^j \frac{N}{\xi_N}\right]|^2.$$ (7)

The factors of an arbitrary number N are the integer values $\xi_N = l$, which correspond to dominant maxima of such a rescaled sum.

In Fig. 1, it is represented the modulo squared of the rescaled CTES in Eq. (7), with $M = 3$, $j = 2$, as a function of the variable $\xi_N \in [330.74, 337.21]$, for the factorization of $N = 111547$. We can see that the two factors $l = 331, 337$ (represented by stars) give complete constructive interference. On the other hand, for the other trial factors (represented by triangles), there is partially destructive interference. Moreover, there are absolute maxima (represented by points) which do not correspond to integer trial factors.

We have also represented, in Fig. 2, the modulo squared of the rescaled CTES for the same value of N and M and the same range of values of ξ_N, in the case of $j = 3$. It turns out, as expected, that, as the order j of the exponential sum increases, the peaks associated with the absolute maxima become sharper. On the other hand, increasing the order j, also the values of the second order maxima in the interference pattern increase. In order to suppress such maxima it is necessary to increase the number of terms M in the sum.

In the wavelength range in Fig. 1 and Fig. 2, the more a non factor is near to a factor, the less is the correspondent value of intensity. We analyze now

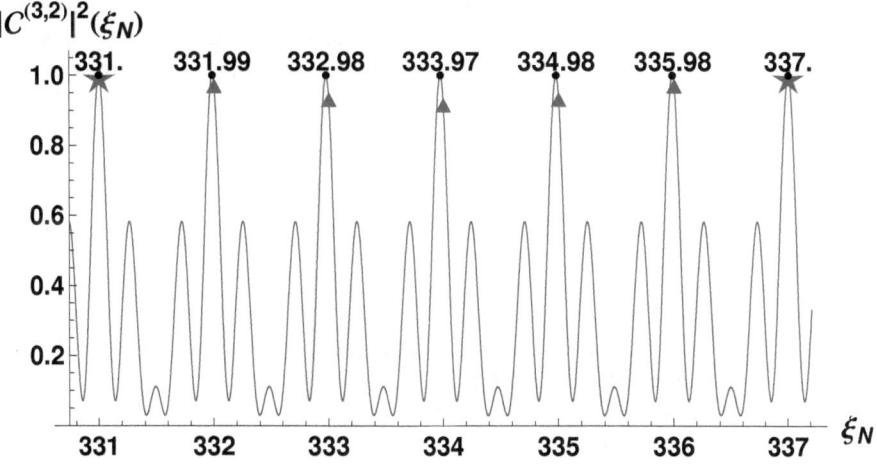

Fig. 1. Modulo squared of the rescaled CTES, in Eq. (7), for $N = 111547$, with $M = 3$ and $j = 2$, as a function of the variable $\xi_N \in [330.74, 337.21]$. We can see that the two factors $l = 331, 337$, represented by stars, give complete constructive interference, despite the other trial factors, represented by triangles, which present partially destructive interference.

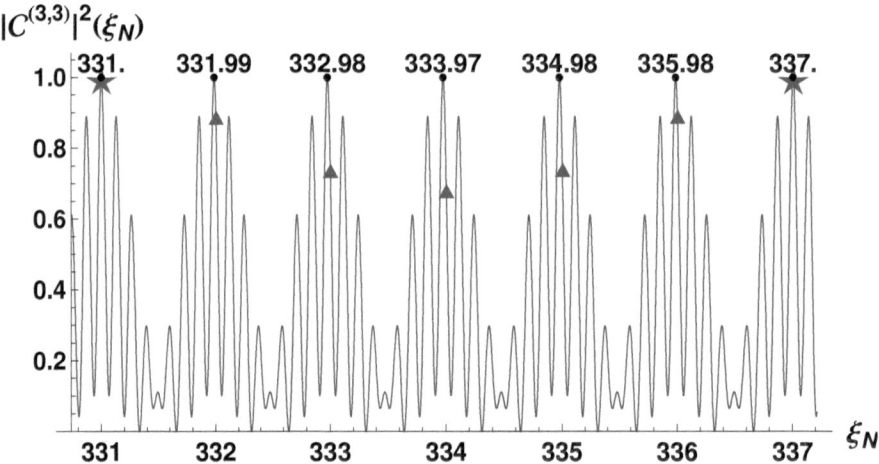

Fig. 2. Modulo squared of the rescaled CTES in Eq. (7), for $N = 111547$, with $M = 3$ and $j = 3$, as a function of the variable $\xi_N \in [330.74, 337.21]$. As expected, the peaks associated with the absolute maxima, in the case $j = 3$, are sharper than the respective peaks, in the case $j = 2$, represented in Fig. 1. On the other hand, increasing the order j, increase the value of the maxima of second order in the interference pattern.

the case of integer wavelengths far from the factors. In Fig. 3, it is shown, for example, a simulation of the interference pattern in the range $[230.9, 237.1]$. We can observe, as expected, that there a is large probability of finding trial factors with associated relatively limited value of intensity. This allows us to speed-up the general selection of the factors among all the possible trial factors, simply disregarding all the value of ξ_N below a suitable threshold value. In this way, the number of trial factors to be considered in order to determine the factors is sensibly reduced.

We gain some insight into the behavior of the function $\mathcal{C}^{(M,j)(\xi)}$, in Eq. (5), as a function of the continuous variable ξ when we represent $f(\xi)$ as

$$f(\xi) = k(\xi) + \frac{1}{2}\tau(\xi), \tag{8}$$

where both the integer k and the continuous parameter τ, which extends from -1 to $+1$, depends on the value of ξ.

When we substitute this representation of $f(\xi)$, in the CTES expression in Eq. (5), we find

$$|\mathcal{C}^{(M,j)}(\xi)|^2 = |s^{(j)}(\tau(\frac{\lambda}{u_N}, N))|^2, \tag{9}$$

where

$$s^{(j)}(\tau) = \frac{1}{M}\sum_{m=1}^{M} \exp\left[\pi i(m-1)^j \tau\right]. \tag{10}$$

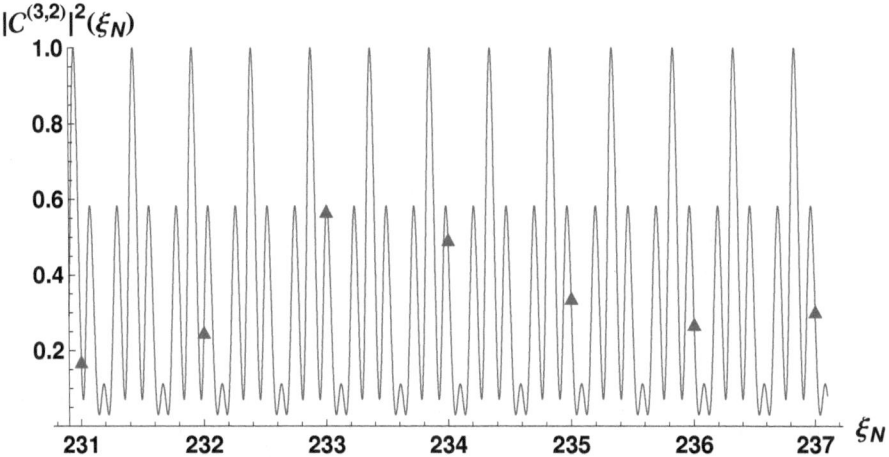

$|C^{(3,2)}|^2(\xi_N)$

Fig. 3. Modulo squared of the rescaled CTES, in Eq. (7), for $N = 111547$, with $M = 3$ and $j = 2$, as a function of the variable $\xi_N \in [230.9, 237.1]$. We can clearly see that all the integer values of wavelengths in such a range have a relatively limited value of intensity, so that they can be easily disregarded as possible factors.

Hence, the CTES in Eq. (5) is a sequence of the same symmetric function $s^{(j)}$, whose variable τ is determined by Eq. (8). In the case of $j = 0$ this function is closely related to the curlicue function which has a dominant maximum at $\tau = 0$ and oscillations on the sides. The function $s^{(j)}$ repeats periodically for each value ξ such that $f(\xi)$ is equal to an integer k. Consequently the period of repetition of $s^{(j)}$, in the interference pattern, determines the dominant maxima. When a maximum corresponds to an integer value q of $\xi_N = \frac{\xi}{N}$ then q is a factor of N.

In conclusion, the CTES, in Eq.(5),which is independent from the number N to be factored, allows us to recognize all the values ξ, corresponding to an integer value of $f(\xi)$, as dominant maxima in the interference pattern. The factors of an arbitrary number N are given only by the values ξ such that ξ_N, in Eq. 2, is an integer.

4   Analogue Realization of a CTES with a Multi-path Interferometer and a Spectrometer

We have shown that the implementation of a CTES would allow to factorize, in principle, arbitrary numbers. Unfortunately the calculation of such a sum would require an exponential number of divisions associated with the computation of the function $f(\xi)$. On a digital computer, for which division is a rather costly process, such a computation turns out to be very slow. Therefore it would be interesting to reproduce the sum in Eq. (5) with an analogue technique, to solve the problem quickly.

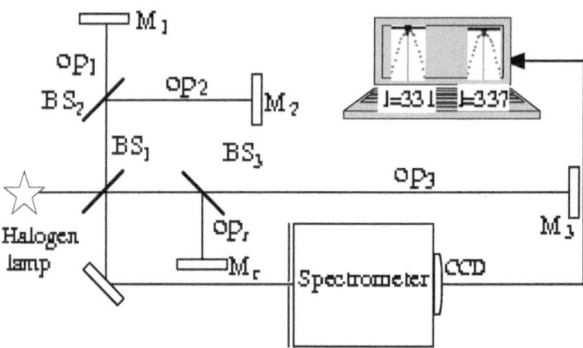

Fig. 4. Experimental setup: generalized symmetric $M+1$-path Michelson interferometer for the realization of truncated exponential sum with truncation parameter $M = 3$. Such a setup consists of a polychromatic source (halogen lamp), M balanced beam splitters BS_1, BS_2 and BS_3, the mirrors M_r and M_m, with $m = 1, 2, 3$, and a spectrometer connected to a CCD camera. The $M = 3$ interfering paths can be varied respect to the reference path op_r, by moving longitudinally the mirrors M_m, with $m = 1, 2, 3$, respect to the reference mirror M_r, so that the relative difference is given by $op_m \equiv m^j u$, with $u = N u_N$ where N is the number to factorize.

We introduce an interesting analogue procedure which allows to reproduce the interfering phases terms in Eq. (6). Such a procedure is based on the wave nature of light. In fact the light emitted by a source is characterized by electromagnetic phases of the form $\phi(\lambda) = 2\pi x/\lambda$, which allow to encode the exponential phases $(m-1)^j$ in the optical paths x and the continuous variable ξ, associated with all the possible trial factors, in the wavelengths λ. Moreover, a polychromatic source of light contains a broad range of wavelengths and thereby allows us to test trial factors simultaneously.

In particular, the algorithm described in the previous sections can be implemented, using an $M+1$-path symmetric Michelson interferometer in free space, shown in Fig. 4, for the case $M = 3$[1]. The system includes M balanced beam splitters and $M+1$ mirrors. The M interfering paths, whose values op_m, with $m = 1, 2, .., M$, are measured relative to a reference path op_r. We can encode the phase terms $(m-1)^j$ in Eq. (6), with $m = 1, 2, 3$, in the relative optical paths:

$$op_m^{u,j} \equiv (m-1)^j u, \tag{11}$$

for $m = 1, 2, 3$, with j integer larger than 1, and u suitable unit of length, leading to the interfering phase terms

$$\phi_m(\lambda/u) \equiv 2\pi(m-1)^j/(\lambda/u). \tag{12}$$

For $m = 1$ the optical path is equal to the reference path.

---

[1] The actual experimental results, obtained implementing this procedure, will be presented in an incoming paper [5].

We can use a polychromatic source so that the output interference pattern, measured by a spectrometer connected to a CCD, at the output port of the interferometer, is a continuous function of the wavelengths λ associated with the bandwidth of the source.

Such an intensity pattern $I(\lambda)$ is given by the superposition of the $M = 3$ interfering terms $\exp[i\,\phi_m(\lambda)]$. When we normalize the output intensity respect to the source intensity, we obtain

$$I(\xi) \equiv |\mathcal{C}^{(M,j)}(\xi)|^2, \tag{13}$$

i.e. the modulo squared of the CTGS $\mathcal{C}^{(M,j)}(\xi)$, in Eq. (5), as a function of the dimensionless real parameter

$$\xi \equiv \lambda/u. \tag{14}$$

As stated before, the periodicity of the recorded interferogram in Eq. (13) as a result of a destructive/constructive interference effect, allows us to extract all the information in the function $f(\xi)$, in Eq. (1), necessary to factorize an arbitrary number N, by exploiting the scaling law in Eq. (2).

In fact, we can rescale the obtained intensity pattern for the factorization of an arbitrary number N:

$$I(\xi_N) \equiv |\mathcal{C}^{(M,j)}(\xi_N)|^2, \tag{15}$$

with $|\mathcal{C}^{(M,j)}(\xi_N)|^2$ given by Eq. (7), and ξ_N rescaled variable in Eq. (2).

The presented analogue procedure allows to test all trial factors simultaneously by using a polychromatic source interferometer. The resulting interference pattern, when observed as a function of wavelength, contains the information about all factors of any arbitrary number N. We only have to extract the information from this pattern. The information is encoded in the location of the maxima of the interferogram. When a maximum is at an integer value l of the variable ξ_N in Eq. (2), we have found a factor of $N = p \cdot q$.

5   Conclusion

In the present paper we have introduced a new factorization algorithm, based on the analogue determination of the periodicity of a generalized CTES exploiting a physical interference process. We have demonstrated that the key of such algorithm stands on the destructive/constructive interference associated with the recorded CTES interferogram, which allows to extract all the information about the factors of arbitrary numbers, contained in the experimentally computed ratio $f(\lambda) \doteq u/\lambda \equiv f(\xi) \doteq 1/\xi$, by exploiting the scaling relation in Eq. 2 which defines the variable ξ_N we use for representing f. Such a scaling property of the interferogram allows to rescale the periodicity of the same recorded pattern, in order to factorize, in principle, arbitrary numbers N, looking at the maxima at integer values of the corresponding variable ξ_N.

References

1. Shor, P.: Proceedings of the 35th Annual Symposium on Foundations of Computer Science, Santa Fe, NM, November 20-22, pp. 124–134. IEEE Computer Society Press, Los Alamitos (1994)
2. Merkel, W., Wölk, S., Schleich, W.P., Averbukh, I.Sh., Girard, B.: Factorization of numbers with Gauss sums and laser pulses: I. Mathematical background (to be published)
3. Merkel, W., Wölk, S., Schleich, W.P., Averbukh, I.Sh., Girard, B., Paulus, G.G.: Factorization of numbers with Gauss sums and laser pulses: II. Suggestions of implementation (to be published)
4. Merkel, W., Averbukh, I.Sh., Girard, B., Paulus, G.G., Schleich, W.P.: Fortschr. Phys. 54, 856–865 (2006)
5. Tamma, V., Zhang, H., He, X., Garuccio, A., Schleich, W.P., Shih, Y.: Factoring numbers with a single interferogram. Submitted to Nature Photonics
6. Wölk, S., Feiler, C., Schleich, W.P.: J. Mod. Opt. (2009)
7. Clauser, J.F., Dowling, J.P.: Phys. Rev. A 53, 4587–4590 (1996)
8. Summhammer, J.: Phys. Rev. A 56, 4324–4326 (1997)
9. Rangelov, A.A.: J. Phys. B: At. Mol. Opt. Phys. 42, 021002 (2009)
10. Stefanak, M., Merkel, W., Schleich, W.P., Haase, D., Maier, H.: New J. Phys. 9(370), 1–18 (2007)
11. Stefanak, M., Merkel, W., Schleich, W.P., Haase, D., Maier, H.: J. Phys. A: Math. Theor. 41, 304024 (2008)
12. Mehring, M., Müller, K., Averbukh, I.Sh., Merkel, W., Schleich, W.P.: Phys. Rev. Lett. 98, 120502 (2007)
13. Mahesh, T.S., Rajendran, N., Peng, X., Suter, D.: Phys. Rev. A 75, 062303 (2007)
14. Gilowsky, M., Wendrich, T., Muller, T., Jentsch, Ch., Ertmer, W., Rasel, E.M., Schleich, W.P.: Phys. Rev. Lett. 100, 030201 (2008)
15. Bigourd, D., Chatel, B., Schleich, W.P., Girard, B.: Phys. Rev. Lett. 100, 030202 (2008)
16. Sadgrove, M., Kumar, S., Nakagawa, K.: Phys. Rev. Lett. 101, 180502 (2008)
17. Weber, S., Chatel, B., Girard, B.: EPL 83, 34008 (2008)
18. Peng, X., Suter, D.: EPL 84, 40006 (2008)
19. Jones, J.A.: Phys. Lett. A 372, 5758–5759 (2008)

Author Index